Emerging Technology and Management for Ruminants

International Stockmen's School Seminars

Emerging Technology and Management for Ruminants

edited by Frank H. Baker and Mason E. Miller

The *Proceedings* of the 1985 International Stockmen's School Seminars, *Emerging Technology and Management for Ruminants,* includes approximately fifty technical papers given at this year's Stockmen's School, sponsored by Winrock International. The authors are outstanding animal scientists, agribusiness leaders, and livestock producers who are experts in animal technology, animal management, and general fields relevant to animal agriculture.

The goal of these *Proceedings* is to present advanced technology in problem-oriented form readily accessible to livestock producers, operators of family farms, managers of agribusiness, and scholars and students of animal agriculture. This year's volume on technology and management for ruminants has three major sections: a section on pasture, range, and forages; a larger section on advanced technologies of reproduction and genetics; and a third section on production and management of small ruminants. The species covered include cattle, sheep, and goats.

Frank H. Baker, director of the International Stockmen's School at Winrock International, is program officer of the National Program. An animal production and nutrition specialist, Dr. Baker served as dean of the School of Agriculture at Oklahoma State University, president of the American Society of Animal Science, president of the Council of Agricultural Science and Technology, and executive secretary of the National Beef Improvement Federation.

Mason E. Miller is communications officer at Winrock International. A communications specialist, Dr. Miller served as communication scientist with the U.S. Department of Agriculture; taught, conducted research, and developed agricultural communications training programs at Washington State University and Michigan State University; and produced informational and educational materials using a wide variety of media and methods for many different audiences, including livestock producers.

A Winrock International Project

Serving People Through Animal Agriculture

This book is composed of papers presented at the
International Stockmen's School
January 6–9, 1985, San Antonio, Texas
sponsored by Winrock International

A worldwide need exists to more productively exploit animal agriculture in the efficient use of natural and human resources. It is in filling this need and carrying out the public service aspirations of the late Winthrop Rockefeller, Governor of Arkansas, that Winrock International bases its mission to advance agriculture for the benefit of people. Winrock's focus is to help generate income, supply employment, and provide food through the use of animals.

1985 Livestock Seminars, International Stockmen's School

Emerging Technology and Management for Ruminants

edited by Frank H. Baker
and Mason E. Miller

CRC Press
Taylor & Francis Group
Boca Raton London New York

CRC Press is an imprint of the
Taylor & Francis Group, an **informa** business

First published 1985 by Westview Press

Published 2018 by CRC Press
Taylor & Francis Group
6000 Broken Sound Parkway NW, Suite 300
Boca Raton, FL 33487-2742

CRC Press is an imprint of the Taylor & Francis Group, an informa business

Visit the Taylor & Francis Web site at
http://www.taylorandfrancis.com

and the CRC Press Web site at
http://www.crcpress.com

Library of Congress Catalog Card Number: 84-62209

ISBN 13: 978-0-367-00589-4(hbk)

This book is dedicated to Forrest Bassford, veteran agricultural journalist, a truly great leader of the industry, a premier livestock journalist and philosophical leader. Mr. Bassford has been named Honor Guest for the 1985 International Stockmen's School Seminars.

Bassford, whose career has spanned 46 years, is currently executive director of the Livestock Publications Council headquartered in Encinitas, California. For 30 years, he was vice-president and part owner of Nelson R. Crow Publications, Inc., publishers of such magazines as the *Western Livestock Journal* and *Western Dairy Journal*. Bassford, who was raised on farms in Oklahoma, Texas, Idaho, and Wyoming, says his lifetime aim was to contribute to the "betterment of the livestock industry through expansion and improvement of magazines, newspapers, and newsletters devoted to that industry."

In recent years, Bassford has received awards from 15 livestock- and farm-related associations and from his alma mater, Colorado State University. In addition to the hundreds of articles he has written over the years, he is author of a booklet, "Wyoming Hereford Ranch, 1883-1983, Century of Endurance."

CONTENTS

Part 4. FORAGE, RANGE, AND PASTURE

Part 5. PRODUCTION AND MANAGEMENT OF SHEEP, GOATS, AND CERTAIN OTHER SMALL ANIMALS

PREFACE

This book, *Emerging Technology and Management for Ruminants*, includes presentations made at the International Stockmen's School Seminars, January 6-9, 1985. The faculty members of the School who authored this volume are scholars, stockmen, and agribusiness leaders with national and international reputations. The papers are a mixture of technology and practice that presents new concepts from the latest research results of experiments in all parts of the world. Relevant information and concepts from many related disciplines are included.

The School was held annually from 1963 to 1981 under Agriservices Foundation sponsorship; before that it was held for 20 years at Washington State University. Dr. M. E. Ensminger, the School's founder, is now Chairman Emeritus. Transfer of the School to sponsorship by Winrock International with Dr. Frank H. Baker as Director occurred late in 1981. The 1983 School was the first under Winrock International's sponsorship after a one-year hiatus to transfer sponsorship from one organization to the other.

The five basic aims of the School are to:

1. Address needs identified by commercial livestock producers and industries of the United States and other countries.
2. Serve as an educational bridge between the livestock industry and its technical base in the universities.
3. Mobilize and interact with the livestock industry's best minds and most experienced workers.
4. Incorporate new livestock industry audiences into the technology transfer process on a continuing basis.
5. Improve the teaching of animal science technology.

Wide dissemination of the technology to livestock producers throughout the world is an important purpose of the publications and the School. Improvement of animal production and management is vital to the ultimate solution of hunger problems of many nations. The subject matter, the style of presentation, and the opinions expressed in the papers are those of the authors and do not necessarily reflect the opinions of Winrock International.

ACKNOWLEDGMENTS

Winrock International expresses special appreciation to the individual authors, staff members, and all others who contributed to the preparation of the book. Each of the papers (lectures) was prepared by the individual authors. The following editorial, secretarial, and word processing staff of Winrock International assisted in reading and editing the papers for delivery to the publishers:

Editorial Assistance

Jim Bemis, Editor
R. Katherine Jones, Editor
Essie Raun, Assistant editor
Venetta Vaughn, Illustration editor
Melonee Baker, Proofer
Beverly Miller, Proofer

Secretarial Assistance and Word Processing

Patty Allison, General coordinator
Ann Swartzel, Secretary
Tammy Henderson, Secretary
Shirley Zimmerman, Word processing
Darlene Galloway, Word processing
Tammie Chism, Word processing

Part 1

GENETICS AND BREEDING OF CATTLE

1

MUSCLE + BONE + FAT = BEEF

Rex M. Butterfield

Beef, the final product of all cattle production, is achieved through the growth process. Without growth we have nothing! The first requirement for beef production, therefore, is that the animal must grow and thereby increase in live weight. Both producers and scientists have been very interested in the achievement of maximum live weight growth and, although much research effort in the U.S. has been on the yields of salable beef of various types, less study has been made of the growth of muscle, bone, and fat. In the last 20 yr, there has been a tendency in other countries to base more and more research effort on the growth of the body tissues rather than on the final product. Studies of the underlying biological processes that cause animals to grow and develop in the way that they do are understood only when each body tissue is studied separately. However, studies of composite tissues, such as meat, reveal little in the way of underlying principles because the changes in one tissue may completely mask the changes in another.

On the other hand, much of the work in the U.S. is more relevant to the problems of the day than the basic work in other countries. It is always stimulating to discuss the difference between U.S. and Australian approaches to the fascinating task of understanding more about the growth of meat animals and its application to meat production. The discussion here will be confined to the simple facts of how, and why, tissues grow the way they do.

We are inclined to forget that cattle do not contemplate the prospect of ending up as beef when they are growing, but rather are aiming their whole growth and existence at the prospect of a happy old age after having reproduced themselves as often as possible to ensure the survival of the species. So there are no rules of nature laid down by beef production; we must look to the process of growth through the challenges of birth, survival, and reproduction to find the reasons why animals grow as they do.

Let us first consider the growth process system by system and then later put the three carcass tissues (muscle, bone, and fat) together to try to make some sense in terms of beef production.

MUSCLE GROWTH

The calf at birth already has gone through an interesting phase of muscle growth in that he arrives equipped with a set of over 200 muscles that enable him to survive; and, at this stage, survival is his only interest. First, his musculature must be of such shape and size that he can be born. Interestingly, when he is born, he has about the same proportion of live weight to muscle as he will have throughout life; but the calf muscle has quite a different structure from that of the mature animal.

To the newborn calf, the mature animal is a distant and uninteresting goal. He must first survive. And that means he must be able to chase his mother and latch onto that milk bar with its four elusive outlets that go past at an alarming rate as they swing in the breeze. So -- the problem that confronts the muscle system of the newborn calf is to be able to propel that calf along in a manner, no matter how shaky, that will allow him to home-in on those life-preserving teats -- and once he scores a hit he must be able to suckle the milk so freely available.

What sort of muscle system does he need? First, he must be able to stand and to chase his mother. We find that the muscles of the lower parts of the limbs are extremely well developed at birth so that the calf can walk. To put it another way, they are closer to their final size and strength than most of the other muscles in the body, and, in simple terms, we regard them as "early developing."

Next, the calf must be able to suckle, and so the muscles of the head, and particularly the jaws, must also be much more developed than the rest of the musculature (figure 1).

Figure 1. A newborn calf needs well-developed muscles in his legs and jaws.

Let us now presume that our calf has survived and is entering the next phase of life, which is to grow as fast as possible and enjoy itself. In other words, it wishes to become as big as possible and be as functionally efficient as possible while grazing, chasing other calves, and, in general, enjoying the period of adolescence. In this phase, we see that the relatively poorly developed muscles at birth now have the opportunity to grow and reach their full potential. These are, for example, muscles that make up the belly wall. They had little work to do in the newborn calf because it had little weight in its gut. Now there is a mass of grass and other things pouring into the gut that must be carried around for a day or so; therefore, the muscles of the belly wall begin to grow very fast to enable them to cart around the contents of the abdomen.

There are many other muscles that must mend their sleepy prebirth ways by growing rapidly to meet the functional requirements of our bouncing adolescent. The muscles in the upper part of the back leg, for example, just could not be well developed at birth because they never would have passed through the pelvis of the cow. But now these muscles are needed, if our calf is to compete in the adolescent games of healthy, frisky calves. And so the back leg muscles, as well as the large muscles of the back, grow at a terrific rate for a time following birth to provide the driving force behind the gambolling of our calf.

Even though it is generally a risky business to make decisions on how muscle weight is distributed over the bodies of animals merely by looking at them, it is nevertheless clear in comparing the outline of a day-old calf with that of a one-month-old calf (figure 2) that dramatic changes have taken place. Like so many features of the calf's musculature, these changes are most dramatically displayed in muscular European breeds.

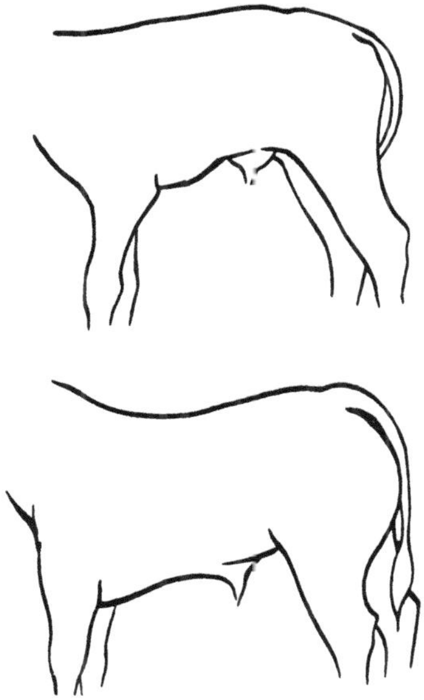

Figure 2. The outlines of a newborn Piedmont calf (above) and a one-month-old Piedmont calf (below).

Once the musculature changes have taken place, we have a fairly stable condition that remains throughout life in castrated males but which will be altered in both females and intact males as time goes by. The changes in the females depend upon the well-known phenomenon of "falling pregnant," a time when the demands on the muscles of the abdominal wall become extremely large and these muscles must speed up their growth.

Bulls show the most dramatic changes in muscle growth as maturity approaches. The bull, at least under range conditions, must fight for the right to mate with the available female population, and he can do this only if he changes the layout of his muscle strength. He needs massive muscles in his neck to enable him to use his horns (or poll) to eliminate his rivals for the favors of the cows. The result is that the neck muscles of bulls, and particularly one specific muscle, grow at an incredible rate as sexual maturity is approached. This growth can be very clearly seen in the outline of the mature bull, which shows that the crest on the bull's neck is much more developed than is that of the cow, steer, or young bull (figure 3), and this fact is reflected in the proportion of total muscle in the forequarter (figure 4).

The changes in the musculature take place in a very logical manner as cattle are born, survive, and reproduce. Fortunately, the other two major carcass systems (bone and fat) are less influenced by functional changes.

BONE

Bone is also a tissue that is very strongly influenced by functional demands, but many of the interesting changes occur before birth.

In summing up the changes in bone, it is obvious that the bones of the limbs must be very well developed at birth for the calf to perform the functions we described for muscle.

We call all of the bones of the limbs, except one, "early developing;" that is, they have grown much closer to their final weight than has the total skeleton at birth. The only exception is the scapula or shoulder blade, which tends to lag in growth. Clearly, the more essential a bone is to the simple functions of early life, the better developed it is at birth. Thus, the bones lower down in the limbs tend to be a little further developed than are the bones higher up in the limbs.

Of course, the bones in the head, which are essential for the early harvest of food, must also be developed early, but the remainder of the skeleton can develop to its adult proportion later during the growth phase.

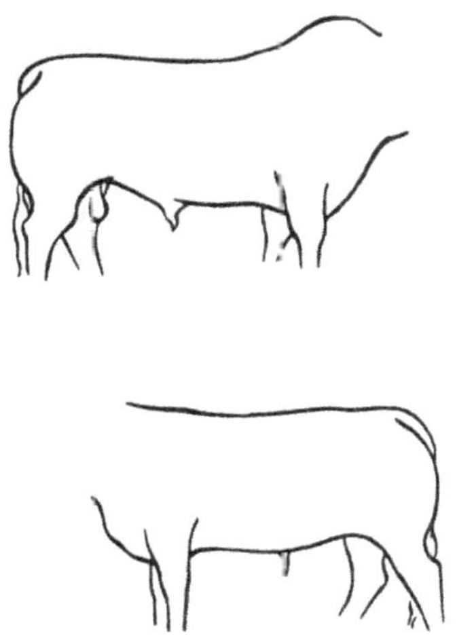

Figure 3. The outlines of a bull (above) and a steer (below).

Figure 4. Increased muscle weight in the front end of male cattle is a bonus -- it displaces fat.

FAT

Fat is not as functionally active as are muscle and bone. It is, nevertheless, vital to the survival of the animal as an energy reserve rather than in any way related to mechanical function. If the animal does not find immediate sustenance at birth, its store of fat can be very rapidly exhausted. However, if the newborn has, as is the usual condition, a rapidly available fast-food outlet accompanying him wherever he goes, then fat becomes of minor functional importance in the growth process during the suckling period. It is only when weaned and left to its own devices that the animal has a real need for a mobile inbuilt energy store in the form of fat. Thus, in the growth process following birth, fat growth is very restrained unless the animal obtains huge amounts of milk that are surplus to its maintenance and growth requirements.

Usually we find that animals put on little fat until after the major periods of fast growth have been achieved in all other body tissues and organs. Hence, a postnatal period of slow fat growth is followed by a fattening phase when the survival of the animal has been assured by a developed system of vital organs, muscle, and bone.

Not all animals or breeds will have the same fattening pattern within their bodies, and some will store fat in different places. In general, we find that the animals that have a high potential for milk production, or a high potential to produce large numbers of young, tend to lay down more of their fat internally. Those breeds or animals that produce only single progeny and(or) produce low quantities of milk, put more of their fat in the carcass. Because most cows tend to produce single calves, the level of milk production is important in cattle.

When comparing two breeds of different size, scientists often discuss whether differences in the amount of fat found in the various depots around the body are in fact real genetic differences or merely a reflection of the stage of maturity of the animals of the two breeds. The answer is yet to be determined for cattle. We have, however, just completed studies of sheep in Australia that indicate that all the various fat depots within the body move towards their mature weights at almost the same rate. This suggests that differences in the proportion of total fat in each of the fat depots will be similar whether the animals are compared at the same weight or the same stage of maturity. The answer likely will be the same for cattle and, if so, it suggests that the rate of growth of all the various fat depots in the body are similar.

So much then for the way the tissues grow in isolation. Let us now look at the effect this growth has on the carcass composition (figure 5).

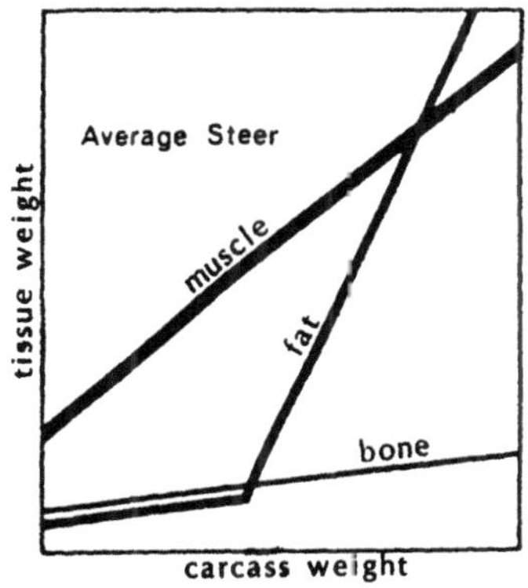

Figure 5. Growth of muscle, bone, and fat in an average steer.

Both total muscle and total bone tend to grow in straight lines in relationship to the growth of the whole carcass. Muscle grows so much faster than bone that the ratio of muscle to bone continues to increase throughout life. Thus, as the total amount of muscle and bone becomes heavier in an animal, the muscle:bone ratio usually increases. In other Stockmen's School papers, I discuss the implications of such changes to both meat quality and reproductive efficiency. But for the moment we will establish the various changes that take place.

While muscle and bone are making their major dash following birth, fat is just ambling along, but once the environmental needs of locomotion are met in steers and in heifers, the fat depots take over in absorbing the huge intake of food that the animal has learned to harvest. Thus, the calf enters the fattening phase, during which the major harvest of quality beef is made.

To market "quality" beef, the cattleman needs to understand the changes that take place in the fattening phase.

Let us look at the different types of cattle available that govern the most basic of all decisions for the producer: What breed?

<u>Early maturing cattle</u>. These traditionally have been the greater proportion of the British breeds, and, although the link is not necessarily unbreakable, we normally associate early-maturing cattle with small cattle. These cattle have dominated the non-European beef cattle world primarily because of their ability to fatten at lighter weights and thereby produce beef of the composition that has been sought by most consumers over the last century (figure 6).

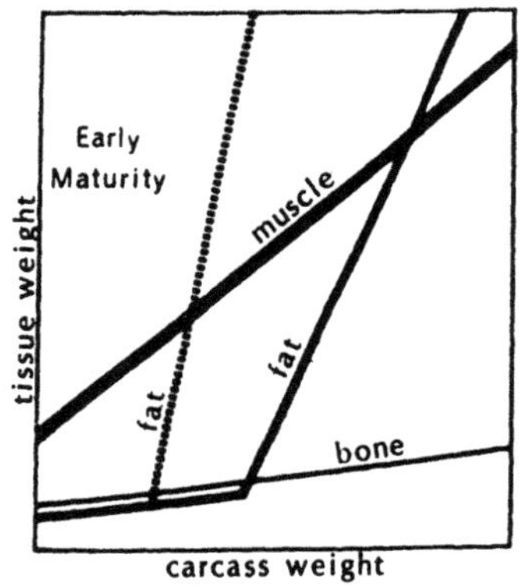

Figure 6. Early maturity. The only difference is that the animal fattens at lighter weights (dotted line).

Late maturing cattle. The first late-maturing types of cattle to become significant in beef production in those countries in which British cattle reigned supreme were the large dairy breeds, particularly the Friesian, which even today produce a large proportion of the world beef because of the huge numbers of surplus male calves coming from the dairy industry. However, these cattle do not have the tinge of respectability needed to make them acceptable as producers of beef in the traditional beef-producing countries. In addition, because of their unnecessarily high milk production (for calf-rearing), which triggers all kinds of reproductive and survival problems, they are totally unsuitable for extensive beef production (figure 7). And so the even later-maturing Continental breeds have swept out around the world to help meet the demands for low-fat carcasses under both grass and feedlot conditions. As shown in figure 8, not only are they later maturing but also they usually have heavier bone and higher muscle:bone ratio.

The problems that have been created, and the problems that have been solved, by this change in traditional types of cattle in North America and other countries are the basis of my other papers at the Stockmen's School.

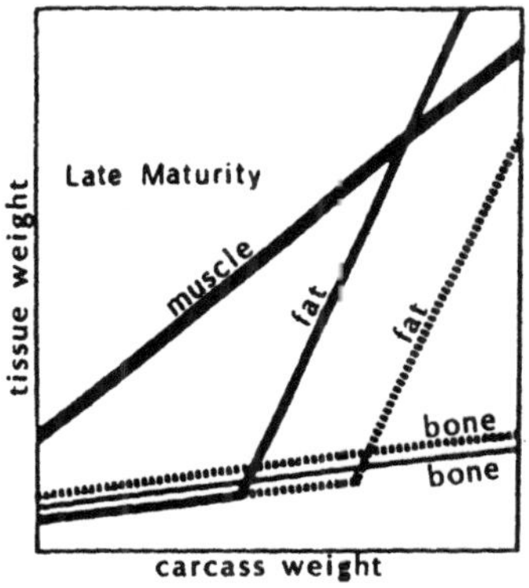

Figure 7. Late maturity. The major difference is that the animal fattens at greater weight. It usually needs slightly heavier bone.

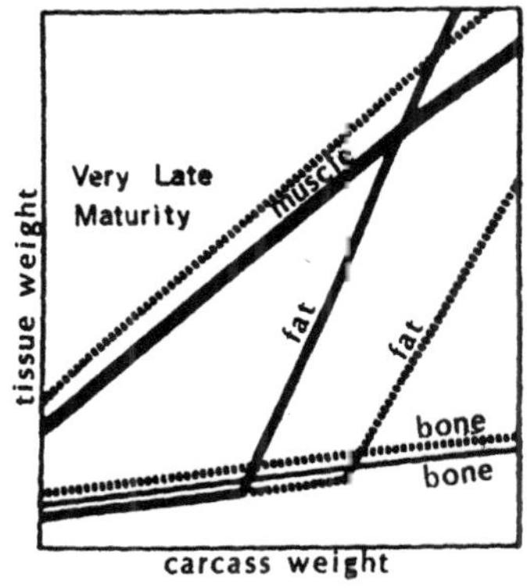

Figure 8. Very late maturing breeds, such as many European breeds, usually fatten when very heavy and have a high muscle to bone ratio.

2
INFORMATION-BASED BEEF BREEDING

R. L. Willham

A gold cloud of dust rolled up from the prairie as another herd of longhorns snaked its way across Indian territory toward the railhead in Kansas. What an awesome sight. The romance and nostalgia still awake in us a proud heritage. Now a bleary-eyed computer programmer is seated by an output-strewn desk. The last bug in the program came out from under the chip and soon the beef breeder will be making selections based on all available information banked in the computer. The excitement of our heritage and new frontiers is still present. If only we can make an orderly transition into our mysterious destiny!

Each one of us involved in the beef industry has the opportunity to guide and capitalize on the dynamics if we study what is most likely to transpire. The purpose of this paper is to outline some of the possible genetic technology and to suggest how this new technology may be assimilated into our beef industry. In particular, the role of the breeding stock producer is considered.

The livestock industries we know today are the product of the industrial age. Livestock have served industrial society quite well; we know the limits. The problem today is that the world is in transition from an industrial society to an information society. The computer is the prime mover in the information society and it has permeated the breeding stock industry already. In the early 1960s, the computer began to write pedigrees. The speed of transition is what is so frightening. Mundane job solutions are fine, but computers can now accomplish what we perceived as our job of weighted decisionmaking! It sometimes looks like the computer can make us bystanders in agriculture's transition to the information age.

To understand the magnitude of our discomfort, consider cattle breeding in the 18th century. Country gentlemen of England such as Robert Bakewell refined the art of cattle and stock breeding. They took a local stock that showed promise of filling the needs of the expanding urban markets and set about to make improvement in the stock. In wishing to make dramatic change, they made wide outcrosses. Then the selected stock from the crosses were inbred to produce a degree of genetic uniformity. The resulting stock were more prepotent, having the ability to transmit desirable properties uniformly to their offspring, which usually were crosses. Inbreeding depression also occurred, especially in the reproductive complex, and it produced refinement. But refinement was desirable since the draft ox was being selected for earlier maturity.

In time, over rural England, there developed groups of stock that were welded into what we now call breeds. As Lush (1945) put it, "Here we already have a useful and profitable breed. We should protect its purity and our own interests as possessors of this valuable breeding stock and the interests of the purchasers who want genuine animals of this breed." At first, herdbooks were begun to record the ancestry of the stock and soon breed societies were formed to keep the herdbooks. Mating decisions consumed hours of self-debate. The points of all the ancestors were studied. It was a highly developed art form. An aura of mystery surrounded the successful breeder. Inbreeding in small herds was a matter of chance as to which of the genes became fixed; therefore, the successful breeder was the lucky one!

Breed societies believed that pedigree allied with the use of eye judgment for securing adherence to formalized breed type was the basis of successful breeding. Over time, the livestock shows became social gatherings where breeders came to present their show string of fitted animals. Judges of these shows were the highly respected cattlemen of the age. As Darlow (1958) aptly put it, they were master teachers. The judges ordered the cattle from best to worst. When breeders' livestock were ranked for all their peers to see, breeders paid attention. The large purple rosettes of the winning stock meant good sales of breeding stock for the breeders.

As the process was codified and the detail enlarged upon, breeders in the United States far outstripped the humble beginnings in rural England and Scotland. The International Livestock Exposition in Chicago was an incredible event for generations of farm kids. Not only were the shows extravaganzas, but through them the breed associations in the United States grew powerful. At one stage, the associations ran the beef industry. However, after some 150 yr of selecting for thick calves of early maturity, a dwarf gene was snagged and multiplied. This simple recessive gene taught cattle breeders about genetics the hard way. Animal geneticists at the land-grant agriculture colleges were called on for help. They did help but simultaneously they also presented the purebred beef industry with the concept that animals could be given tests that would reflect their ability to perform as parents for the commercial producer.

Performance testing was born amidst the chaos induced by dwarfism, but it had been well researched long before on the range near Miles City, Montana. A few purebred breeders and commercial producers embraced the idea. For these "nuts," or early innovators, the price was excommunication from the possibility of ever belonging to the mighty main stream of the industry of that era.

Meanwhile other species had seemingly been transformed by the application of new breeding technology. From geneticist to retailer of eggs and broilers, the poultry industry became a transformation from the farm flock. Swine were fast becoming crossbred and the dairy industry had utilized sire selection and artificial insemination to improve milk production 1% to 2% per year.

Slowly but surely through the 1960s the performance evaluation of beef grew -- partly because judges of the era realized that the rapidly expanding feeding industry of the Southwest did not think that small was beautiful. Lean tissue growth by feeding cheap concentrates was the issue. The judges saw quickly that the obvious difference between performance cattle and the conventional ones was size. And the tall and lean animal burst into vogue. Showmen saw the shift and purchased performance cattle for their show strings and won. This popularized the performance herds of the innovators and eventually many herds not only became elite but their germ plasm repopulated their respective breeds.

Not to be outdone by the state performance associations or those giant white cattle of French name, several breed associations embraced performance recording as a priority in their total program. Sire evaluation was developed by breed associations through the efforts of the Beef Improvement Federation. With the importation of exotic breeds via Canada, the traditional British breeds became leaders in utilization of performance records. The field data sire evaluations were educational experiences for all the participating breeders. The evaluations statistically ranked the sires on the prediction of the future performance of their progeny based on the available records of progeny. This provided fair comparisons among all the sires of a breed. The reports were a bit like the major shows -- the sires were ranked but the breeder had to promote the product of his breeding program. No longer was there suspense, a slap on the rump, and the roar of the spectators in the amphitheater as the champion was named. Performance became a part of many purebred herds.

Consider the purebred breeder of today with a creative breeding program. Calving is seasonal to ensure large contemporary groups. Calves are weighed at birth and at weaning. Then the males are put on a gain test while the heifers are grown out so that both sexes get yearling weights. Live fat measures are sometimes taken and even hip height measured. Once each test is completed (weaning and yearling) the records are sent into the association and the analysis is returned. Weight ratios, growth breeding values, and maternal breeding values are computed for animals available for selection in the herd. Armed with all that his association knows about his cattle, the breeder goes out among his cattle and selects replacements, prospective herd sires, and his sale offering. Not as much time as in the past is spent working out each individual mating. Then the breeder fills his sale catalog with performance data. Even so, some bull buyers still make selections during a walk through the pens. A lot of the art and mystique of the cattle breeder is gone. The association computer applies the mystery to the simple records to regress them and account for relatives' performances.

Still there is no clear cut direction to cattle breeding. Some cattle really have a high early growth rate of lean tissue but just are not tall enough to be in vogue. And there is the matter of reproductive ability. Big testicles appear to be the new measure. Then how does the breeder merchandise what was measured?

In the future, it will soon be possible to rank the herds of a breed on the measurements of performance that are on record at the breed association. Then much of the mystery of beef breeding will lie in the sophisticated analysis procedure done on the computer. Some breeders will simply husband the cattle, record the performance tests, and await the outcome from the computer. More time can be spent mating the right cow to the right bull since the computer now untangles the contribution from the sire and the dam simultaneously.

Sitting in the confines of his office stacked high with computer output, what is the breeder's reward for knowing he owns one of the truly elite herds of the breed? Where is the electricity in the air as his bull is given the nod in front of all his peers?

What does the breeder do after years of work and hours at the scales when he realizes from the genetic-trend graph that his herd is below average? What will his motivation be to record the next animal? Record keeping is so revealing. The breeder thinks about those university types who led him down the garden path twice. First it was the path to dwarfism and now it is the computer path.

The design of beef-breeding programs in the last decades of the 20th century is our next consideration. Until now that which could be done by breeders to produce genetic change in animal populations was circumscribed by the known individuals having a genome that could be introduced into the population by sexual reproduction. This circle has been expanded considerably by the new genetics. In the near future, it will be possible to introduce desirable genes and gene complexes into existing genomes to transform the genetic ability of an individual and then a population. These gene complexes may or may not be from the same species. It remains to be seen whether these introductions by man, into a relatively fine-tuned genome, create unsuspected interactions in function. Nevertheless the potential does exist. When desirable individual animals are finally created, their propagation should follow existing population dynamics. Exploiting a good sire is not difficult today using artificial insemination.

Cloning and embryo splitting give new potentials for animals with identical genomes. All breeders must realize that the segregation of the genes in gamete formation and subsequent recombination to form new individuals maintains the genetic variation so essential in sexually reproduced species.

Where does the traditional beef breeder fit into such exotic breeding programs? This may depend on whether there is a multiplication phase needed to supply germ plasm to the beef producer. If proprietary stock is possible and necessary for the conduct of the program, money in the venture may be the prerequisite to capitalize on the exercise. Seed corn and poultry breeding have gone this direction with much less sophisticated genetics to exploit.

Applying the new genetics to beef improvement may be long in coming. The economic opportunity may be judged too small by corporations having the necessary capital. Even so, much new technology is still available for use in breeding beef cattle.

Breed associations are seeking ways to remain relevant in the transition to the information age. In the early 1960s, computers produced three generations of pedigrees on the registration certificate. Then performance records were banked and analyzed for return to the breeders for their use in selection. During the 1970s, estimated breeding values for growth and material performance were produced. The values for each animal used the performance of close relative groups. Then national sire evaluation, first in designed progeny tests and later using all available field records, became a part of the association function. Artificial insemination made national sire evaluation possible since the popular sires tied the bulls together so that fair comparisons across herd lines could be made.

Thus, breeders of today have the opportunity to fairly compare a large number of available sires and, therefore, greatly enlarge the potential of sire selection in their herd. That is, the effectiveness of sire selection has been enhanced by giving all breeders equal potential for success instead of just the breeders with large numbers of animals to control. Soon the cow will come under similar evaluation and astute breeders will be able to improve the reproductive potential of their superior females through embryo transfer. Next will come the comparison of young sires by using their own performance and that of their sire and dam. By identifying breed-improving sires at a young age and using artificial insemination, the rate of genetic change per unit of time will be increased. Further, the interchange of superior sires through AI provides better comparisons of individuals from different herds. It is a bit like a snowball collecting size as it goes down hill.

However, these opportunities for breeders to acquire more comparative data on all herds of a breed comes at a price. And that price is the loss of the mystery of how to be a successful breeder. Clearly, to be on the top in a sire report is not as rewarding as being chosen in front of all your peers at a national show. Nor is it particularly rewarding to have everybody know where your herd ranks for a trait, since half the herds will be below breed average. Yet differences are and always have been the issue in cattle breeding.

There are creative ways for traditional breed associations to function so that their breeders are benefited both monetarily and socially. Just how this will be done in the face of new technology application will require honest effort and a clear understanding of the ramifications of each new venture.

The human animal is such that new technology will be developed and it will be utilized, many times, prior to its real need. Existing organizations that creatively adopt the new and remain relevant in service will grow. Those that do not will fade into the proud past. The question is not whether new technology should be embraced, it is simply a question of how it can be used the most creatively. Used it will be!

The creative breeder will use the successful methods of the present and seize the opportunity to make newer technology work for him. Profit comes first -- and so the use of commonplace technology is a matter of keeping up. The choice of direction is always the first step. Then having the resolve to adhere to a well-conceived direction over time is the worst stumbling block. Even by using current genetic methods, cattle breeding is a long-time venture involving decades, not just years. The creative breeder senses the needs of the future and sets a simple direction to produce stock that meet those needs.

In the second step, the breeder learns to predict the breeding performance of his stock for important traits. The association has recorded some traits of importance so that the breeder can get fair predictions for the majority of sires, dams, and yearling stock of that breed for the near future. The predictions are most relevant when the breeder has used enough outside sires to tie his herd to the breed base. For the traits not on record at the association, the breeder can only compare them among his herd. These unrecorded traits could be among the most important to the breeder's direction.

After all the direction and prediction choices have been made, the remaining step is to select the parents. Selection involves using the predictions to move the herd in the desired direction. Selection, when broadened to include aspects of the new genetics, will remain the only force available to the breeder to make directional change in his biological population. There is no other!

Traditionally, breeders have incorporated into their initial population or herd enough genetic variation so that genetic change can be made. Until recently, this incorporation was defined by what could be crossed in and the resulting offspring of both sexes still be fertile. Divergent genetic groups, such as types, breeds, or lines within a breed, have been used to incorporate genetic variance. Incorporation of exotic genes or gene complexes fits into this category when trait change becomes a reality.

Breeders have selected intensely within their population or herd using the best predictors of parental performance available to them. Being able to join gametes from the same sex of parent now offers some real potential. This will happen because sires can be evaluated with much larger progeny numbers and in a much shorter time than dams. Thus, if sperm from two well-predicted sires could be combined into an offspring, the possible increase in genetic gain could be larger than the current procedures. Whether individual breeders could do this remains to be seen, but bull studs surely can. Breeders with such sires could benefit. Breeders that wish to contribute to their breed's relevance need to consider all avenues available to them to increase their rate of genetic change per unit of time.

Breeders often have mated close relatives or inbred their stock because the master breeders did so. Which gene pairs become fixed in inbreeding is a matter of chance, and so the change is nondirectional but the dispersion amount is predictable. Inbreeding must be accompanied with intense selection just to stay even -- relative to the direction of traits desired by the breeder. Inbreeding has been used by breeders to make stock more prepotent so when used as parents the performance of the offspring is more predictable. Inbreeding has also been used in the process of linebreeding, which is selection of the individuals in the pedigree. Linebreeding is done to increase the relationship of animals to a particular ancestor. To do this the other ancestors are methodically removed as the linebreeding proceeds. At present, mating full brother with full sister is the most intense inbreeding that can be done in stock. However, two sperm from the same desirable sire could be joined to produce an offspring and the result would be selfing. Inbred line formation in animals could have the same advantages as corn breeders have had. Were propagation techniques developed, cross-line commercial production could become a reality. It is unlikely that one breeder could develop sufficient lines but cooperatively several might. If, on the other hand, the new genetics field yields ways to produce commercial zygotes, circumventing segregation and recombination in sexual reproduction, there is no need to predict parental performance or to inbreed for prepotency.

Admittedly, our predictions for the future of beef breeders are fuzzy, but the opportunity appears to be virtually unlimited. How individual breeders can promote their product is likewise only conjecture.

As we contemplate our romantic heritage and question the present, the fact is we really revere the past innovators and those who took our industry on to the next step. It is not those who were left around the near empty amphitheater of Chicago in 1975 that mattered; it is the group of men who gathered together to start the international livestock exposition in 1900 that we remember. Who will the creative breeders be who will utilize every scrap of new technology and replace the computer's function with new ideas? The breeds blessed with such breeders will still be relevant long into the future.

REFERENCES

Darlow, E. A. 1958. Fifty years of livestock judging. J. Anim. Sci. 17:1058.

Lush, J. L. 1945. Animal Breeding Plans. Iowa State College Press, Ames.

3
MATERNAL SELECTION

Roy G. Beeby

The history of breeds of cattle in the U.S. reveals that the commercial industry has intermittently turned to the breed that could solve periodic economic problems. In the beginning, the Longhorn was used because it could survive in West Texas and walk to a market in Kansas. The first English breed used by U.S. cattlemen was the Shorthorn, which had fleshing ability. Then the Hereford breed was used extensively because of its wide adaptability. Angus cattle were introduced for their maternal traits and carcass qualities. Brahman cattle became popular in southern areas because of their hybrid vigor contribution and their unique resistance to heat and insects. In recent years, a number of European breeds have been introduced for increased size, growth rate, and leanness. The popularity of the European imports, because of their size and gain, caused many of the British breeds to move rapidly toward size, performance, and terminal-sire production. This lecture will deal with some of the effects this move toward performance has had on the maternal traits of U.S. beef cattle.

Can one beef cow herd produce terminal-sire herd bulls and functionally efficient, maternal females? Negative genetic relationships among traits make this an extremely hard approach for beef improvement. Intense breeding for size can create problems in maternal trait selection. A more realistic procedure is a breeding program designed to produce a specialized maternal herd. The primary objective of the Red Angus program at Prairie City Farms in Marshall, Oklahoma, is maternal female and maternal sire production. The motivation for this maternal program is the belief that it would be beneficial if all feeder calves had a maternal dam and a terminal-sire. To accomplish this would create a tremendous demand for maternal cows that are more effectively produced from maternal bulls. Both maternal cows and maternal bulls come in a wide variety of sizes. This is good because the beef producing ecology of this world is very diverse.

The maternal-herd concept of Prairie City Farms has been influenced by a long family tradition of sound economics. The family operation, older than the state, began 18 yr before statehood, on April 23, 1889, when my grandparents made the run into Oklahoma and staked their claim. Their first dwelling was a dugout, partly underground, partly walled up with sod with an earthen floor and a roof of overlapping sod. Range riders from near-by ranches would stop by the homestead, but some of the most interesting visitors were bands of Cheyenne Indians from the west and Ponca Indians from the east. Even before statehood, our family had started in the cattle business. Since the homestead was only a mile from the "Cherokee Strip," open range was rented from the Indians to graze cattle in the summer.

The year of the "Run," in 1889, marked the beginning of one of the worst drought cycles Oklahoma has ever known, and a disastrous nationwide financial depression. To the settlers, it brought actual hunger. For more than 3 yr, they worked their land without raising anything but kafir corn and one very light crop of $.40 wheat. For those who stayed, cattle and native grass brought them through. Reviewing hard times in the cattle business 100 yr ago should help us evaluate our priorities for survival today.

The unique character of the early cattle breeding program at Prairie City Farms was its harmony with nature. The natural resources were fully utilized and little outside supplementation was required. The cattle that developed were adapted to their environment. Any cow that was fat in the fall was sold for cash flow. Cows that gave plenty of milk and calved regularly never got fat. Cows that did not raise a calf or were poor milkers got fat and went to town. Although this practice was not designed as selection criterion for regularity of calving and milking ability, it served that purpose.

There were other beliefs, true or false, that were strongly held by the pioneers at Prairie City Farms. One was that the registered bulls purchased for the herd often introduced more problems than benefits. The second was that cattle shows provided breeders with a forum to exaggerate extremes in type and were often counterproductive to beef production. The third was that "blue sky" promotion of breeding stock had little relationship to the real world of cattle breeding. There is some truth in these traditional beliefs. We still try to avoid any problems or activity that compromises reproduction.

The commercial cattle industry through the years has used the breeds that promised to solve current economic problems. There have always been genuine seed stock producers striving to breed animals of value for the commercial industry. Also, promoters of popularity trends have always been available to carry new directions to extremes.

Problems that surface in the beef industry stimulate changes in direction. This could be referred to as the change cycle in beef cattle. Economics force a change until promotion of the change forces economics to force a change.

The change cycle in seed stock production repeats itself periodically. Today in seed stock production, we could be at the beginning of another change cycle. Breeds or families within breeds that have the ability to solve economic problems in the industry will be used. In the future, cattle that fill some special need in the beef industry will have their day.

At the time we founded our Red Angus herd almost 30 yr ago, the industry used the terminology "beef cattle production testing." This phraseology soon gave way to the term "performance testing." This change in name may describe a transition in philosophy from total herd production testing to the performance evaluation of sires. Performance testing began to mean postweaning feed tests that increased adult size, growth, and gain in beef sires. Standards of perfection in the show ring also emphasized the theme that "bigger-is-better." The annual selection for maximum weaning weights was also biased for unlimited milk and growth. Female production in the performance movement has been almost a by-product of sire selection motivated by priorities for unlimited size.

ANTAGONISMS BETWEEN SIZE AND MATERNAL SELECTION

The industry in the past 25 yr has been very successful at increasing adult size. The maternal traits have paid a price for this progress. Occasionally, size should be referred to as "frame size" or "weight size" to be more accurate in the use of the term. A tall animal that is the same age and weight as a short-legged animal is often considered to have more size. In addition, it has been my observation that weight size is not as negative as frame size in maternal selection.

Maternal selection is more than milk production reflected in weaning weights. All of the things that make up motherhood in a cow and calf program are included: regular reproduction, gestation length, trouble-free calving, mothering ability, survival of progeny, foraging ability, maximum intake of forage, adaptation to adversity, and weaning a desirable calf without heavy supplementation. These are all a part of the total influence of a maternal beef cow.

It is no secret that antagonistic genetic relationships among traits exist in beef production. Increased size in bulls can increase calving difficulties. Increased size and milk production in cows does increase maintenance requirements, which require the producers' greatest expense. Increased frame size can delay sexual maturity and may limit regular reproduction.

It should not be surprising that a change in one trait can cause changes in other traits. Sir Isaac Newton's third law of motion states: "For every action, there is an equal and opposite reaction." This law of physics may not be completely applicable to animal breeding. However, it can remind us that a change in a selection criterion should be evaluated for the total effect of all traits of economic importance. We may be on the threshold of a real need for specialization in trait selection in herds and breeds in beef cattle. It will be hard for individual animals to make significant genetic contributions in all of the economic traits of efficient beef production. There is a real need in the industry today for specialized maternal breeds and maternal families to support the terminal-sire herds.

Prairie City Farms has participated in all phases of the performance movement with 28 yr of weaning weight data, birth weight, dam weight, yearling weight, central bull tests, home bull tests, carcass evaluation, national multiple-sire progeny tests, performance pedigrees, and sire and dam progeny summaries. This herd has had the high-gaining sire on test station multibreed 140-day feed tests. Progeny from test winners have been the high-gaining pen of bulls on subsequent tests. Herd-average weaning weights have increased dramatically through the years. This is mentioned to show the Prairie City Farms has used the conventional performance programs available to the industry.

MATERNAL SELECTION PROGRAM

The maternal selection program at Prairie City Farms has been in progress for over a quarter of a century. The objective of the "maternal" approach has been to produce the optimum cow whose nutritional requirement does not interfere with regular reproduction in unpampered range conditions. This is accomplished by using as many sons as possible out of the regular-calving, top-producing cows of the herd. A direct route to genetic balance is provided by identifying those cows that can produce the maximum amount of calf on a regular basis in a natural environment and intensify their desirability in the herd through their sons.

Selection for fault-free cattle is the first priority in building a maternal herd, i.e., functional and structural soundness that includes trouble-free udders, feet, legs, eyes, reproductive systems, and disposition. These traits may not be the most important but you do not want to propagate them into your program. Correct the faults in your herd by selection. You should never have to trim a hoof or treat an udder. Breed your herd until it is structurally and functionally correct. Your progress in developing a sound herd will be apparent when you do not make an excuse for any animal you own.

An April 1976 Prairie City Farms advertisement described an ideal maternal cow as follows:

- Be fast maturing and sound
- Be able to conceive on exposure
- Be able to calve trouble-free
- Raise an outstanding calf every year
- Wean calves over 50% of her body weight

There may be a better description of an ideal maternal cow, but there is little doubt that a cow that fits this description is maternal. A new procedure is now being tested at Prairie City Farms for the selection of maternal cows that have optimum productivity in their environment. (See "Maternal Selection: Producing Cows with Optimum Productivity for Their Environment" in this book.)

An observation was made in our May 1976 Red Angus advertisement concerning the selection of a maternal herd sire: "When you select a herd sire for maternal traits, you do not have to make great sacrifices in weaning weights, postweaning rate of gain, or mature weight. But you cannot select on these traits alone and obtain a bull that will produce maternal females." Bulls produced in our program have convinced me that there is truth in this statement. A maternal bull is one that produces maternal females. Later in the lecture some techniques will be discussed for the selection of maternal bull prospects.

A September 1980 Prairie City Farms advertisement described what was felt to be the frontier for cattle breeders -- maternal reliability. The virgin territory in cattle breeding today is maternal reliability. It cannot be measured in the showring, on the scales, or in the cooler. Selection by current procedures, feed test gains, or adjusted yearling weight will not improve it. Few sire progeny tests include it. Yet maternal reliability adds more net profit to commercial cow herds than any other trait. Maternal reliability combines regularity of calving, short calving interval, trouble-free calving, and the ability to raise a calf to weaning. The cattle industry today does not have enough interest or emphasis aimed at improving the regularity of production in beef cows. This is why maternal reliability is a frontier in cattle breeding.

Another effect of maternal reliability is longevity. A cow decreases her chances of being culled if she is structurally and functionally sound. To have optimum producers, breed cows with the following ideals:

- High-producing cows that will not quit
- Cows that can stand the stress of producing more milk on grass and forage and still breed back in time to calve during the same month the next year
- Cows that produce heavier-than-average calves for the herd every year and still remain sound without developing udder or feet problems
- Cows that do not develop health problems or die due to the stress that milk production and regular calving places on them
- Cows that continue to perform on a regular basis and lead the herd on weaning weights during the drought years when feed is short but still breed back
- Cows that don't foul up when they are entitled to -- cows that just won't quit!
- Cows that are bred with these ideals in mind are optimum regular producers, not quitters.

Table 1 describes some selection techniques used by Prairie City Farms for producing and maintaining a maternal herd.

TABLE 1. SIX ESSENTIALS FOR PRODUCING AND MAINTAINING A MATERNAL HERD

1. Each year, eliminate any cow that fails to calve trouble free or wean a calf.
2. Select replacement heifer calves from cows that breed early in the calving season and calve unassisted.
3. Keep the bred heifers that breed early in the breeding season and calve unassisted.
4. Select herd sires with acceptable birth weights from the highest-producing females in no. 1 and no. 3 above.
5. Breed the bulls in no. 4 above to the females in no. 2 and no. 3 and continue to practice no. 1.
6. Do not engage your cattle in any activity or program that interferes or compromises rules no. 1 through no. 5.

This formula for producing and maintaining a maternal herd is certainly not for everyone and is easier to establish in a functionally sound herd. It is not compatible with a show program. Current sire-evaluation programs are not presently being utilized successfully to produce maternal bulls. The performance of a sire's daughters should receive primary consideration in maternal sire selection.

In the past, maternal bulls could only be discovered through the function and production of their daughters. Successful maternal bull selection can be enhanced when prospects are picked from the higher performing, regularly producing females in the herd. The dams that have the ability to function regularly and efficiently within the herd fit the environment for the herd.

Cows that are not regular in their production within the herd may have one of the following adaptation problems: cow size too large; too much milk production; lack of appetite; poor foraging ability; timid disposition. Any one or combination of these traits, or others that cause a cow to be irregular in her production, should not be transmitted to the herd through her son.

Females produced out of sires whose dams are functionally efficient within herd should be more compatible to the program than those sired out of outcross bulls from different environments.

The beef industry should have programs specifically designed to produce specialized strains of seed stock. The industry needs to give a higher priority to the production of maternal strains of beef cows. (Refer to "Maternal Selection: Producing Cows with Optimum Productivity for Their Environment.")

4

MATERNAL SELECTION: PRODUCING COWS WITH OPTIMUM PRODUCTIVITY FOR THEIR ENVIRONMENT

Roy G. Beeby

CRITERIA FOR SELECTION

Maternal selection for optimum productivity in diverse environments is a formidable challenge for beef-cattle seed stock producers today. Before outlining this challenge, perhaps I should first define the way I am using the words "optimum productivity." "Optimum" is used as the best results obtainable under specific conditions and "productivity" as: creation of something of economic value. Thus optimum productivity in a cow can be defined as consistent efficient production of a product of economic value.

Cow productivity combines all of the things that make up motherhood in a cow and calf program: regular reproduction, gestation length, trouble-free calving, mothering ability, survival of progeny, foraging ability, maximum intake of forage, adaptation to adversity, weaning a desirable calf without heavy supplementation. A shorter definition for cow productivity is optimum production with maximum reproduction in her natural environment. The selection criteria should not be guided by anything that interferes with reproduction. The role that cattle will play in commercial production should determine the choice of priorities, and the selection objectives should be in harmony with the environment in which the cattle will be raised.

PAST TRENDS

Since the 1950s, the "performance movement" has directed much effort to improving growth and carcass characteristics, and because these traits are intermediate and high in heritability, they have responded to selection very well. In addition, breeders have derived some "glamour" and promotional benefits from these traits through feed tests, carcass contests, and the show ring. Ideally animals would be evaluated on the basis of physiological maturity rather than chronological age as occurs in the previous events.

In contrast, heritability is relatively low for reproduction. The selection emphasis for increased frame size has been negative for total reproduction efficiency. Moreover, maternal selection has not been glamorous. You may feel as lonesome as the proverbial Maytag repairman if you have had a maternal selection program for the past 25 yr.

The economic pressures to increase outputs, including weight at weaning and yearling weights, have been very compelling in the beef industry. Performance breeders have produced successive generations of cattle with each generation heavier at maturity than its parents. This has increased the maintenance cost of the herd. However, once optimum levels of size and milk production are achieved, additional selection for size can be counterproductive to the reproductive traits and can only be justified in the production of terminal sires.

FUTURE NEEDS

Crossbreeding systems in the commercial industry correctly use terminal sires on maternal cow herds. In the future, the need for terminal sires could be met with a relatively small part of the seed stock population, while there is a much greater need for maternal cow production. Many of the traditional maternal female seed stock sources that were available at the beginning of the performance movement have been incorporated into terminal sire upgrading and crossbreeding programs.

Currently, economic conditions suggest future changes in beef production in the U.S. Such changes may include more beef produced on grass and roughage. The most valuable cows may be the ones that can beat their competition in conceiving and raising calves on forage for a profit. Cows that are superb grazers should play an important role in food production in the future.

Although all cows will eat forage, some cows eat more forage per unit of cow weight and many utilize it more efficiently than others. A production test for a cow could be described in the following terms:

- Produces a calf every 12 mo
- Gives enough milk to raise the calf
- Provides enough inherent growth to the calf for a profit
- Performs the above on grass and forage with a minimum amount of supplementation

Weight is a good description of size. Frame size is a popular substitute today. A long-legged 1,200 lb cow has been considered larger than a short-legged 1,200 lb cow. Breeders often refer to frame size rather than weight size. The other dimensions in animal size are length and thickness; both contribute to an animal's weight. Although an animal's height can be measured in inches, its "largeness" must be determined on the scales. Weight records the accumulation of all dimensions of size. (I have observed that weight size is not as negative a factor as is frame size in maternal selection.)

Many producers have made signficant improvements in weaning weights, thinking there was no limit to progress, when suddenly they could advance no further or the cows showed irregular reproduction. Such findings could be a sign that the genetic ability of their cattle has outgrown their available energy. The environment can place a ceiling on milk production, and(or) cow size, and(or) reproduction. When this happens, one option is to improve the environment, but this is seldom profitable for the operation or desirable for the commercial customer. The alternative is to select females that have the ability to produce calves regularly and effectively in their environment. Energy conversion by a cow is independent of her size.

Whereas cow efficiency is independent of cow size, reproduction is influenced by her size. Cows of different sizes can possess the inherent traits that make them efficient producers. Some of these traits deal with a cow's ability to utilize forage effectively such as: ravenous appetite, effective foraging ability, large forage capacity, and adaptation to adverse conditions.

Factors that increase cow size but are negative to optimum reproduction are: slower sexual maturity, irregular calving, and excessive supplementary feeding. These causes of larger cow size increase maintenance requirements and may have a negative effect on the lifetime number of calves weaned.

Seed stock producers may be required to show some restraint in keeping performance trait improvement at a level that is compatible with the operations of customers. Competition among seed stock producers for maximums in performance trait development can be counterproductive to commercial customers. Increased birth weights resulting in calving problems and increased size, and milk production resulting in increased maintenance and decreased regularity of calving, is "progress" that most commercial customers cannot afford.

My experience through the years suggests that the selection for heavier weaning weights and postweaning gain has made it more difficult to maintain appropriate birth weights. In our herd size selection, we should not pick herd-sire prospects with maximum gain if they have an unacceptable birth weight. A seed stock producer must not rationalize that increased birth weights might be justified, even if his herd shows no associated difficulties. The calving difficulties experienced by his commercial customers must govern his tolerance for increased birth weights.

The economic importance of early puberty in heifers is well documented. Long-term breeding studies have compared the lifetime production in heifers that calved first at 2 yr of age with those that calved at 3 yr of age. The data indicate that those heifers bred as yearlings to calve first at 2 yr of age will produce about one more calf in their lifetime as compared to heifers bred to calve first at 3 yr of age. Research studies also have shown that heifers that calve early in their first calving season continue to calve early and wean heavier calves throughout their lifetime.

When outcross sires are from areas that are dissimilar to those where they will be used, there may be negative efforts for selection for environmental adaptation. Artificial insemination provides an extensive choice of sires. Sire selection within-herd from outstanding, regularly producing females is a more positive approach to producing females adapted to the operation.

Beef cattle shows may also have a negative influence on selection for a specific environment; show animals are raised in a near-perfect environment and many areas where they will be used are not so ideal.

Registered seed stock producers should not lose touch with the problems of the commercial beef industry. A lesson can be learned from the poultry industry. Registered chicken breeders once supplied seed stock for commercial chicken producers. Today, 100% of the breeding stock for broiler and egg production is produced by the commercial industry. Registered chicken breeders are no longer in the real world of food production. Many of the associations for poultry breeding still exist and their breeders still produce, exhibit, and sell breeding stock to each other, but they are in a world apart from the commercial broiler and egg producers.

Promotion of registered breeding stock will be a necessary part of the cattle business until such promotion leaves the real world of profitable beef production. Commercial beef herds can progress without registered cattle; it is up to registered breeders to see that the commercial beef herd can continue progressing with registered seed stock.

SOME SPECIFIC GOALS

How successful has the beef industry been in developing a beef-improvement program for all environments and breeding objectives? Performance testing took the correct direction 30 yr ago; today there are indications that some modifications should be made. Some signs of beef improvement limitations in beef cow herds are:

- Birthweight increases, resulting in increased calving problems
- Cows failing to breed back, resulting in irregular calving
- Cows harder to maintain while nursing, requiring increased supplementation
- More stress, resulting in increased health problems
- Shorter average longevity in the herd
- Heifers breeding later, resulting in fewer calves weaned during lifetime
- Bull calves reaching sexual maturity later
- Mature bulls too large for heifers in natural service

The goal of beef improvement has been to help seed stock producers develop maximum potential for growth -- which results in an increase in size without an upper limit. Maternal selection for functional efficiency requires an upper limit on growth dictated by ecology, management, and breeding objectives.

The limitations in conventional beef improvement programs to provide an alternative program for maternal selection has led us to seek and test alternatives in procedures. We are testing a software program that offers alternative selection choices to Beef Improvement Federation (BIF) procedures. It is called MATE (maternal analyses = total efficiency), and its goal is maximum reproduction and fast early growth to optimum composition within existing environment and management. The basic principle is that what a cow does in the interaction of her inheritance and her environment is what she is.

Whereas current beef improvement procedures use the same age-of-dam adjustments in all herds and environments, MATE measures and uses the actual difference in contemporary age groups within the herd. MATE also removes the positive-progeny-parent-size relationship that interferes with optimum cow-size selection. It aids in stabilizing the cow size for optimum production with maximum reproduction in different ecological zones. Maternal breeding values of sires, their probable rate-of-maturity, and their adult size can be estimated from the records of cows as shown in their pedigrees long before the records of their progeny are available. In a MATE data bank, many more outstanding sires would receive recognition. One of the greatest potentials of the system is the maternal analysis that writes specifications for different ecologies while identifying proper replacements.

MATE generates meaningful selection criteria and management assistance from a minimum number of weights and measurements. An early-age calf weight and cow weight are the only measurements needed in addition to regular BIF procedures. The conventional and alternative procedures should be put to the same test to evaluate differences. The MATE system allows you to evalute both procedures, concurrently, in a side-by-side comparison of findings.

SUMMARY

The MATE program offers an alternative to age-of-dam adjustments that does not disturb the mean and produces distributions that are more nearly normal and that have a nonsignificant relationship to weight-of-dam. MATE, an alternative maternal selection program, helps provide:

- Individual herd analyses and program designs to fit the environment and herd selection choices
- Stabilized cow sizes for different ecologies and herd objectives

- Maximum herd reproduction and early calf growth to optimum composition
- Selection aid for replacement females and herd sires that excell in reproduction and whose calves grow rapidly to physiological maturity
- Early estimates of maternal breeding values of sires, their probable rate-of-maturity, and their adult size from the records of cows before they are progeny tested
- Supplemental information that can be used with and compared to BIF guidelines for total herd analysis

REFERENCES

Gosey, J. A. 1984. Beef cattle selection opportunities for net merit. In: F. H. Baker and M. Miller (Ed.) Beef Cattle Science Handbook Vol. 20. p 370. Westview Press, Boulder, CO.

Gosey, J. A. 1984. Fitting cow size and efficiency to feed supply. In: F. H. Baker and M. Miller (Ed.) Beef Cattle Science Handbook Vol. 20. p 476. Westview Press, Boulder, CO.

Totusek, R. 1981. What is most important in the beef peformance package? In: F. H. Baker and M. Miller (Ed.) Beef Cattle Science Handbook Vol. 20. p 5. Westview Press, Boulder, CO.

Totusek, R. 1984. Pounds versus profit, size and milk in beef cattle. In: F. H. Baker and M. Miller (Ed.) Beef Cattle Science Handbook Vol. 20. p 456. Westview Press, Boulder, CO.

Wilkes, D. 1984. Herd replacement rate -- factors to consider. NCA Pub.

5

SELECTION OF REGISTERED BRAHMANS

Dieter Plasse

PURPOSE

Of the registered *Bos indicus* cattle in Venezuela more than one-half are Brahmans. Thus, genetic progress of the commercial beef cattle population depends heavily on the amount of genetic progress possible in the bull-producing herds of this breed.

To determine the value of sequential selection under tropical conditions, a genetic-trend analysis of growth traits was conducted in two Brahman herds in which a selection program had been designed and supervised within a research program. Appropriate health programs are followed in both herds.

DESCRIPTION OF HERDS, LOCATION, AND MANAGEMENT

Herd A is a registered Brahman herd owned by the Facultad de Ciencias Veterinarias Universidad Central de Venezuela (UCV) and located at the La Cumaca Experiment Station. The herd originated from 145 cows imported from the USA during 1960 and 1961 and was subjected to a rigid selection program from 1964 to 1976. During this time, it increased from 182 to 228 brood cows and was genetically closed in 1967.

The experiment station is located in the northcentral part of Venezuela in a tropical climate with mean annual rainfall of 1,700 mm. Cows in Herd A graze guinea grass *(Panicum maximum)* throughout the year, mineral supplementation and molasses only during the dry season. The quality of pasture improved during the first years of the project then decreased dramatically because of inadequate management practices and overstocking. During the first 5 yr, the breeding season was gradually reduced to 3 mo so that calves were born at the end of the dry season and beginning of the rainy season. They were weaned at 7 mo of age and maintained on pasture with mineral supplementation and 2 kg/day of concentrate until the next rainy season. Calves were weighed again at 18 mo.

Herd B is a privately-owned registered Brahman herd located in the state of Portuguesa in a tropical climate with mean annual rainfall of 1,652 mm. It was established in 1959 with 100 heifers imported from the U.S., and in recent years has varied from 290 to 350 brood cows. The herd has been subjected to a rigid selection program since 1966 and forms part of a combined research and technical assistance project.

Cattle in Herd B graze improved pasture throughout the year. All animals receive minerals throughout the year and hay and molasses during the dry season. During the first 4 yr, the breeding season was reduced to 4 mo. Calves were born at the end of the dry season and beginning of the rainy season. They were weaned at 8 mo of age, then received 1 kg/day of concentrate until the next rainy season while being maintained on pasture. The calves were weighed again at 18 mo of age.

GENETIC PROGRAM

Breeding

Breeding in both herds has been mainly at random with males and females entering the breeding herds at 2 yr of age. Since 1973, half of the heifers of Herd A have been artificially inseminated, and in Herd B, heifers and nonlactating cows have been artificially inseminated since 1975. Whereas Herd A was closed genetically in 1967 and has used only bulls bred in the herd, Herd B has used its own bulls, as well as proven bulls from outside the herd.

Male Selection

Strict programs of production testing on pasture (Plasse, 1982) have been maintained in both herds and data have been fully computerized and analyzed each year. Replacement bulls were selected mainly on 18-mo weight and the result of their semen test. They also were required to have high-level weaning weights, as well as a dam with high reproductive efficiency and good maternal ability. Every year, three to five young selected bulls in each herd were bred randomly to 20 to 23 cows. Since 1973, least-squares analyses have been performed yearly, and herd sires have been culled according to the progeny results for weaning and 18-mo weight, as well as according to cow reproductive efficiency.

Female Selection

About 20% of the heifers were culled annually mainly because of low weights and problems in their reproductive organs. All other heifers were put into the herd to replace low-producing cows. These culled cows were replaced primarily on the basis of reproductive efficiency and maternal ability according to the following criteria:

- Positive reaction to test for brucellosis or other reproductive diseases
- Open after the first breeding season
- Open during two consecutive years
- Open in alternating years
- Low maternal ability
- Open after the last breeding season as well as some previous year

Mean age of cows weaning calves in both herds decreased from a mean value of 6.5 to 5.5 yr during the study period.

PHENOTYPIC ENVIRONMENTAL AND GENETIC TREND

Reproductive Efficiency

Reproductive efficiency, measured by pregnancy percentage, increased in Herd A from 56% to 83% during the first 10 yr. In Herd B, pregnancy percentage ranged between 66% and 81%. Only phenotypic trends are known for the two herds.

Growth Rate

Table 1 shows the results of statistical analyses of phenotypic, environmental, and genetic trends in birth weight, 205-day weaning weight, and 18-mo weight (Plasse et al., 1979; Plasse et al., 1983). Data used for the analyses correspond to the years from 1965 to 1976 in Herd A, and from 1968 to 1979 in Herd B. For each weight and herd, the number of observations (no.) and the adjusted mean value of the trait ($\bar{x}$) are shown along with the phenotypic (Δ_P), environmental (Δ_E), and genetic (Δ_G) trends. These trends are estimated in two ways: 1) the mean annual change shown in kilograms, and 2) as a percentage of the average value ($\bar{x}$). The sum of Δ_G and Δ_E is the phenotypic trend Δ_P. In analyzing the values of Δ_P, mean annual change of birth weight was found to be .4 in Herd A and .2 in Herd B. This is a favorable change because birth weight (which has a negative effect on viability and growth) is usually low in tropical beef cattle populations (Plasse, 1978). Weaning values were: Herd A, 1 kg; Herd B, 2.3 kg. Such values at 18-mo were: Herd A, 2.5 kg; Herd B, 4.4 kg. The environmental trend of all weights was slightly negative in Herd A due to bad pasture conditions, mainly during the last years. In Herd B, the environmental trend was positive but close to zero.

The annual genetic trend was similar in both herds. In Herd A, this trend was .6 kg for birth weight, 1.7 kg for weaning weight, and 3.5 kg for 18-mo weight, while in Herd B these values were .2 kg for birth weight, 2.7 kg for weaning weight, and 3.7 kg for 18-mo weight. The genetic trend for 18-mo weight was 1.2% (Herd A) and 1.3% (Herd B) of the adjusted mean. These findings agree well with the results of 11 studies of *Bos taurus* cattle in the U.S. and Canada summarized by Plasse (1979). Most of these research results had mean annual genetic trends slightly below 1% for post-weaning weight and slightly lower for weaning weight. Phenotypic changes were variable, but only in a few cases ranged beyond 1%. This compares with the findings reported for 484 Brahman calves in Florida where the trends for weaning weights were: phenotypic, 1.6%; environmental, .8%; and genetic, 2.5% (Franke, 1974).

TABLE 1. MEAN VALUES AND PHENOTYPIC, ENVIRONMENTAL, AND GENETIC TRENDS FOR BIRTH WEIGHT, WEANING, AND 18-MONTH WEIGHT IN BRAHMAN HERDS A AND B

			Δ_P		Δ_A		Δ_G	
Trait	no.	$\bar{x}$ (kg)	kg	%	kg	%	kg	%
Herd A								
Birth weight	1405	27.1	.4[b]	1.3	-.3[a]	-1.0	.6[c]	2.3
Weaning weight	1371	160.0	1.0[b]	.6	-.8	-.5	1.7[c]	1.1
18-mo weight	1177	292.0	2.5[c]	.9	-1.0	-.3	3.5[c]	1.2
Herd B								
Birth weight	2862	27.3	.2[b]	.8	.1	.2	.2	.6
Weaning weight	2588	155.0	2.3[b]	1.5	-.5	-.3	2.7[b]	1.8
18-mo weight	2487	280.0	4.4[a]	1.6	.6	.2	3.7[b]	1.3

Source: D. Plasse, J. Beltran, O. Verde, N. Marquez, L. Arriojas, T. Shultz, N. Braschi, and A. Benavides (1979); D. Plasse, R. Hoogesteijn, J. Bastardo, C. Verde, and P. Bastidas (1983).
[a] P<.05
[b] P<.01
[c] P<.001

The relatively high genetic trend in birth weight in Herd A can be attributed mainly to the high genetic correlation that normally exists between this trait and 18-mo weight. Although an increase in birth weight is still desirable in these populations, future selection programs must closely monitor the correlated genetic response of this trait because extremely high birth weights are not desirable.

The annual genetic increases of 3.5 kg (Herd A) and 3.7 kg (Herd B) in 18-mo weight suggested genetic gains of 35 kg and 37 kg during the first 10 yr of the program. Herd A realized only 25 kg in phenotypic change because of the negative environmental trend; whereas Herd B produced a phenotypic gain of 44 kg due to a small positive environmental trend. Thus, results from Herd A demonstrate that genetic improvement must be accompanied by adequate environmental conditions if benefits are to translate into phenotypic increase.

Analysis of the raw data from Herd A over the first 10 yr shows an increase in calf weight per cow of 133% at weaning and of 153% at 18 mo. The increase can be attributed to greater reproductive efficiency and growth rates, and to a decrease in mortality. In Herd B, similar measures of productivity over 6 yr reflected an increase of 41% at weaning and of 57% at 18 mo.

CONCLUSIONS

The rigid selection programs have improved genetic and phenotypic growth rates and reproductive efficiency. However, improvement programs must take into account pasture production as well as genetic aspects: an increase in the number of animals weaned, combined with higher growth rate, can result in a higher stocking rate that requires an increased forage supply.

Sequential selection using individual and progeny performance data has proven adequate for genetic improvement of Brahman cattle under tropical conditions.

ACKNOWLEDGEMENT

Dr. Lucia Vacarro's collaboration in the revision of this manuscript is deeply appreciated.

REFERENCES

Franke, D. E. 1974. Tendencia genetica de peso al destete en ganado Brahman. ALPA, Mem. 9:52 (Abstr.).

Plasse, D. 1978. Aspectos de crecimiento del *Bos indicus* en el tropico americano. (Primera parte). World Rev. Anim. Prod. XIV, 4:29.

Plasse, D. 1979. Aspectos de crecimiento del *Bos indicus* en el tropico americano. (Segunda parte). World Rev. Anim. Prod. XV, 1:21.

Plasse, D. 1982. Performance recording of beef cattle in Latin America. World Anim. Rev. 41:11.

Plasse, D., J. Beltran, O. Verde, N. Marquez, L. Arriojas, T. Shultz, N. Braschi, y A. Benavides. 1979. Tendencias fenotipicas, geneticas y ambientales de tres pesos en ganado Brahman. ALPA Mem. 14:149 (Abstr.).

Plasse, D., R. Hoogesteijn, J. Bastardo, O. Verde, y P. Bastidas. 1983. Tendencias fenotipicas, geneticas y ambientales del crecimiento en un rebano Brahman registrado. IX Reunion ALPA Mem. G-21 (Abstr.).

6

SIRE SELECTION BASED ON BREED ASSOCIATION TOOLS

Jim Gibb

Never before has the demand been so great for predictable genetics. Feedlot operators and packers are sending this message to commercial cow-calf producers with increasing emphasis. They want cattle with predictability that will consistently meet their narrowing specifications. As a result, cow-calf producers are demanding seed stock with predictability, not only to satisfy the needs of the feedlot industry, but also to meet their own needs for cattle with reliable productivity. They cannot afford the risk of buying bulls that will not work.

Fortunately, beef cattle producers now have at their disposal the most informative, accurate selection tools ever available. These new tools, such as estimated breeding values and sire summaries supply objective information to help both seed stock and commercial producers make better-informed decisions.

TERMINOLOGY

Breeding value is the value of an individual as a parent. It is the true genetic makeup of an animal for each trait that can be transmitted to its offspring. Breeding value is what seed stock breeders sell and commercial producers buy. Since it is currently not possible to identify all the genes that affect a trait, the best method of estimating breeding values is to measure the outward expression of each of the economically important traits. The two most common estimates of true breeding value are EBV (estimated breeding value) and EPD (expected progeny difference).

Estimated breeding value (EBV) combines pedigree, individual, and progeny information into one concise value. Estimated breeding values may be categorized into two areas: growth (direct) and maternal. The main differences between the two categories are the sources of data used in their calculation and the information they provide. Growth EBVs are calculated using 1) the individual's own record, 2) the record of his paternal half-sibs, 3) the records of his maternal half-sibs, and 4) the records of his progeny.

Maternal EBVs are calculated using the weaning weights of progeny of 1) own daughters, 2) daughters of the sire, 3) daughters of the paternal grandsire, and 4) daughters of the maternal grandsire. Each source of information is weighted to provide the best estimate of an individual's true breeding value.

Birth, weaning, and yearling weight EBVs fall into the growth category, while maternal EBV obviously fits into the maternal category. **Birth weight EBV provides an excellent estimate of calving ease, whereas weaning and yearling weight EBVs estimate preweaning growth and growth through 12 mo of age, respectively. The best overall estimate of growth is yearling weight EBV.**

Maternal EBV is an estimate of an animal's genetic makeup for maternal influence on weaning weight. A sire affects progeny weaning weights through growth genetics, while a cow contributes growth as well as other performance factors such as mothering ability and milk. Therefore, maternal EBV is not a direct estimate of milk production; instead it represents a combination of factors associated with the cow's influence on her calf's weaning weight. It is true, however, that daughters of bulls with higher maternal EBVs will likely give more milk. Since beef producers are paid for pounds of beef, not pounds of milk, maternal EBV is indeed economically more significant than any individual estimate of milk production and can be used to improve weaning weights as affected by the cow herd.

Birth weight EBVs are inverted in that higher EBVs correspond to lighter birth weights, while higher weaning and yearling EBVs correspond to more weight. Higher maternal EBVs equate to higher genetic value for maternal influence on weaning weight.

Table 1 shows how you can use EBVs to compare the expected differences in performance of progeny from two bulls produced in the same herd.

TABLE 1. EXPECTED DIFFERENCES IN PROGENY PERFORMANCE

	Birth weight		Yearling weight		Maternal performance	
Bull	EBV	Progeny weight (lb)	EBV	Progeny weight (lb)	EBV	Daughters' progeny weaning weight (lb)
A	100	80	100	1,000	100	500
B	90	84	108	1,040	106	515

Note that the progeny weight differences represent only one-half the differences in the EBVs. This is because a sire supplies one-half of a progeny's breeding value, the other one-half comes from the dam. If each bull were randomly mated to a group of cows, progeny differences would represent just one-half of the differences in the true breeding value in the sires. In the example given, the difference between the two sires in yearling weight EBV is 8%. Since only one-half (4%) of that difference is transmitted to progeny, the actual difference in yearling weight is 40 lb.

EBVs have been developed to improve the accuracy of in-herd selection, not across-herd selection. EBVs on animals from different herds are not directly comparable for two reasons. First, EBVs are not adjusted for the level of competition. This means that a young bull could have inflated EBVs if his sire is used in herds in comparison with below-average bulls. While this is not typical, it should be considered. In general, progeny of sires used in several herds have more representative EBVs than those that are sons of bulls used in only a few herds.

Second, growth EBVs include the individual's own ratio in the calculation (unless it is an embryo-transplant calf). For example, young Bull A from Herd I with a yearling-weight EBV of 106 would not necessarily be genetically superior to Bull B from Herd III that has a yearling-weight EBV of 104. Herd I could have a lower genetic level for yearling weight than Herd II.

Expected progeny difference values (EPDs) given in sire summaries are adjusted for competition and include only progeny data. Therefore, EPDs are directly comparable and should be used instead of EBVs to compare proven sires listed in sire summaries. For this reason, use of EBVs for comparing older, proven sires will not be discussed here.

Maternal EBVs are more directly comparable across herds than growth EBVs because an individual's own record is not included in the EBV calculation. However, the differences in sire competition among herds still exists, making maternal EBVs less than perfect. Despite their shortcomings, EBVs are extremely useful and are considerably better estimates of genetic value than other factors previously available, such as simple ratios and adjusted weights.

Expected progeny difference (EPD) values are reported in sire summaries in actual units. EPD is an expression of how future progeny sired by the subject bull can be expected to perform compared to the sires in the listing and is calculated from progeny data plus information on close relatives. A bull that has a weaning-weight EPD of +30 lb can be expected to sire calves with 205-day adjusted weights that are 30 lb heavier than calves sired by a breed average bull with a weaning-weight EPD of .0 lb. Birth-weight EPDs are not inverted, as is the case with EBVs printed on performance pedigrees. Higher EPDs correspond to more weight.

Table 2 shows an example of how the EPDs of two bulls may be compared to estimate the difference that can be expected in progeny performance.

TABLE 2. COMPARING EPDs OF TWO SIRES

	Birth, EPD	Weaning, EPD	Yearling, EPD
Sire A	+3.5	+35.0	+60.5
Sire B	-1.0	+20.0	+40.0
Difference	4.5 lb	15.0 lb	20.5 lb

Based on the values given, you would expect the weights of progeny by Sire A to exceed the weights of progeny by Sire B by 4.5 lb at birth, 15.0 lb at 205 days and 20.5 lb at 365 days. This is assuming that both bulls are randomly mated to a group of cows and all progeny receive equal treatment.

Differences between EBVs and EPDs:

EBVs	EPDs
Given in herd reports and performance pedigrees	Given in sire summaries
Reported as ratios (100 = average)	Reported in actual units (.0 = breed average)
True progeny difference of an individual = 1/2 his EBV	Already an expression of expected progeny difference
Directly comparable	Not directly comparable

Accuracy (ACC) may range from .00-1.00 and is a measure of the reliability of the EBV or EPD. As ACC approaches 1.00, the EBV or EPD becomes a more reliable estimate of true breeding value. ACC is primarily influenced by 1) the amount of information, 2) the source of the information, and 3) the heritability of the trait. Table 3 contains some general guidelines for determining reliability based on ACC.

Figure 1 is an example of how EBVs are listed in the APHA weaning reports while figure 2 shows how EBVs and ACC values are presented on Performance Pedigrees. Figure 3 is an example of how EPDs and ACCs are listed in the APHA Sire Summary..

TABLE 3. RELIABILITY LEVELS

Level of reliability	ACC		
	EPD	Growth EBV	Maternal EBV
Low	<.70	<.80	<.85
Medium	.70-.80	.80-.90	.85-.94
High	>.80	>.90	>.94

IDENTIFICATION					BIRTH				WEANING							MATERNAL EBV
CALF'S ID	SEX	BIRTH DATE	DAM'S ID.	SIRE'S ID	C.E.	ADJUSTED WEIGHT	WEIGHT RATIO	EBV	MGT CODE	AGE	ADJUSTED WEIGHT	WEIGHT RATIO	EBV	A	B	
N8 22476888	B	02/25/81	79G 21455704	21455694	1	102	87	95	1	237	515	110	107			103
N14 22476786	B	03/03/81	F002 21168436	21455694	1	90	99	100	1	231	480	102	104			100
N18 22476801	B	03/04/81	K9 21951281	21811396	1	84	106	105	1	230	521	111	100			99
N23 22476959	B	03/05/81	H23 21560236	21490925	1	98	91	94	1	229	471	100	101			102

Figure 1. Example of a listing in APHA Guide Lines Program.

ESTIMATED BREEDING VALUES		ACC.
BIRTH	93	71%
WEANING	106	67%
MATERNAL	106	53%
YEARLING	109	76%

Figure 2. Example of a EBV and ACC listing on the APHA Performance Pedigree.

AR SUPERBULL X22000000 (G. Seal)
2/12/76 S: AR Supersire
B: Acme Ranch Gill, KS
O: Acme Ranch Gill, KS
Green Pasture Farm Spur, PA

BIRTH		WEANING		YEARLING		CARCASS CUTABILITY		MARBLING SCORE		LEAN YIELD		MATERNAL	
EPD	ACC	EPD	ACC	EPD	ACC	EPD	ACC	EPD	ACC	EPD	ACC	EBV	ACC
+1.5	.68	+36.9	.85	+54.1	.75	+.001	.68	−.002	.68	+7.5	.75	105	.92

Figure 3. Example of a listing in sire summary.

SETTING OBJECTIVES

Objective setting is a must if EBVs and EPDs are to be effectively utilized by either seed stock breeders or commercial producers. **Using EBVs and EPDs without a sound management system and clearly defined breeding objectives is like arming a 2-yr-old with a machine gun.**

Factors that seed stock breeders should consider are 1) the role of their breed(s), 2) customers' needs, 3) resource availability, and 4) environment. Commercial producers should evaluate 1) their resource availability, such as labor and feed, 2) marketing requirements and market alternatives, 3) and breeds that best fit their management system for crossbreeding.

After a comprehensive analysis of the above considerations, producers may proceed to establish goals and, ultimately, selection levels for bulls that will allow them to adequately pursue their objectives.

SELECTING AI SIRES USING SIRE SUMMARIES

The procedure for selecting AI sires depends on whether you wish to obtain maximum gain in a single trait or improve two or more traits simultaneously. The more traits you select for, the less improvement you will make in any one trait.

However, when net profit is considered, maximization of a single trait without concern for other traits may be costly. An example would be single-trait selection for yearling weight without regard for the calving problems associated with larger birth weights. The mini summary and selection examples that follow in table 4 illustrate how sire summaries can be used for different goals.

Breeder no. 1 has decided to maximize growth regardless of birth weight or maternal performance and thus has chosen Sire C. Breeder no. 2 wishes to improve growth while maintaining an adequate level of maternal performance. His choice is Sire B. Breeder no. 3 wants to improve growth and maternal performance while minimizing increases in birth weight. Sire A is his choice. Breeder no. 4, who wants to improve maternal performance, maintain acceptable growth, and reduce birth weights, selected Sire D. Four different goals, four different sires. These goals could also apply to commercial producers selecting bulls to meet the needs for a terminal sire, rotational sire, or all-purpose sire. Many combinations of selection criteria are possible, including carcass data that was left out in the above example for the sake of simplicity. Note that minimum ACC values were indicated for birth weight EPD and maternal EBV. This was done to ensure reliability for birth weight and maternal performance. EPDs for weaning weight and yearling weight are not listed unless they have an ACC of at least .70. Obviously, unless your culling levels are very strict, several bulls may meet your criteria. You can then select those that are more affordable.

TABLE 4. MINI SUMMARY

Sire	Birth EPD	Birth ACC	Weaning EPD	Weaning ACC	Yearling EPD	Yearling ACC	Carcass Cutability EPD	Carcass Cutability ACC	Marbling Score EPD	Marbling Score ACC	Lean Yield EPD	Lean Yield ACC	Maternal EBV	Maternal ACC
A	+1.5	.80	+33.5	.86	+52.1	.80	+.214	.63	+.061	.63	+3.2	.70	106	.88
B	+5.6	.75	+39.1	.85	+65.2	.78							103	.85
C	+8.5	.71	+45.2	.82	+80.6	.75	+.352	.65	−.031	.65	+8.5	.72	96	.91
D	−0.5	.73	+24.3	.78	+42.0	.75							110	.90

Some Selection Examples

Breeder No.	Selection Goals	Possible Culling Levels: Birth EPD	Birth ACC	Weaning	Yearling	Maternal EBV	Maternal ACC	Bull Selected
1.	Maximize growth	none	none	maximize	maximize	none	none	C
2.	Improve growth and maintain adequate maternal performance	none	none	+35.0	+60.0	102	.85	B
3.	Improve growth, improve maternal performance, minimize increase in birth weight.	+3.0	.70	+30.0	+50.0	105	.85	A
4.	Improve maternal performance, reduce birth weight, maintain acceptable growth.	0.0	.70	+20.0	+35.0	108	.85	D

STACKING PEDIGREES TO MEET YOUR OBJECTIVES

After a breeding program has been established and sire selection criteria have been identified, the next step in sire utilization is pedigree building to enhance predictability. When stacking pedigrees, it is important to consider the influence that both the sire and dam sides of the pedigree have on the resulting progeny. While the genetic makeup of the current calf crop is a representative one-half of the breeding value of the sires and dams utilized, it is essential to remember the effect that previously used sires have on the breeding value of the current calf crop. This is represented in figure 4.

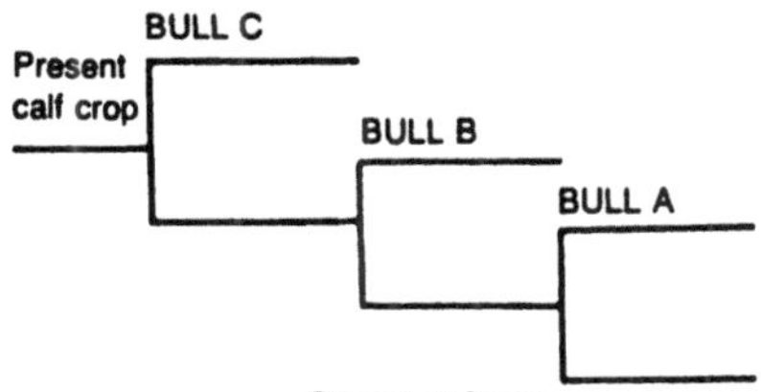

Bull	Percent of Genes Contributed
A	12½
B	25
C	50
Total	87½

Figure 4. Genetic contribution of three bulls to the present calf crop.

The last three successive sires contribute 87.5% of the genetic makeup of the current calf crop. Sires used back in the early 1970s or late 1960s still have a significant genetic influence on the 1983 calf crop since they likely are the sires or grandsires of several cows still in production. The right selection today will pay dividends for years to come, while the wrong decision may be a long-time liability. By defining bulls consistent with your goals and utilizing them generation after generation, it is possible to stack pedigrees and build a cow herd capable of producing predictable breeding value in demand by your customers.

Figure 5 gives an example of a stacked pedigree through three generations. This pedigree was stacked for a balance of traits. Nearly 90% of the genetic makeup of Generation III is a result of definitive sire selection. This is sometimes referred to as "genetic engineering." The reason each sire used is shown to be out of a benchmark dam is to stress the importance of complete functionality associated with a bull's dam and paternal half sisters.

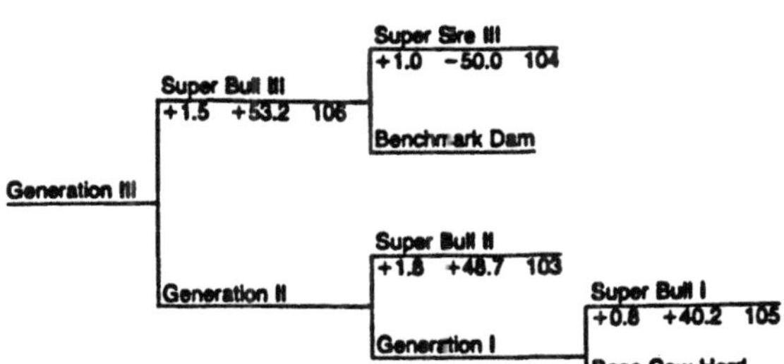

Figure 5. Stacking pedigrees for herd improvement using the sire summary.

The beauty of pedigree building is that it does not require as much time as you might expect. If you were to start in 1984, Generation III could be on the ground by 1989, thus requiring only 5 yr of engineering. Remember, pedigree building contributes to predictability. Predictability is the name of the game when selling or buying breeding value.

The value of sire summary data in pedigree building is demonstrated in table 5, where the results are shown from the Oklahoma BEEF Inc. (OBI) bull test, which ended February 11, 1983. Bulls that were direct sons or grandsons of +40 bulls for yearling weight had significantly more growth than those sired by bulls that were -40 or not listed in the sire summary at all. While these examples are growth oriented, the same results may be realized from a balanced trait selection scheme.

TABLE 5. PERFORMANCE BY CATEGORY[a]

Category[b]	No.	ADG	Adj 365	Index
Sire	6	3.66	1,149	103.4
Paternal grandsire	12	3.70	1,139	103.2
Other	17	3.46	1,056	96.4

[a] OBI bull test that ended February 11, 1983.
[b] Sire = own progeny of +40 bull; paternal grandsire = progeny of son or grandson of +40 bull; other = progeny and(or) grand progeny of bull with less than +40 yearling EPD or not listed in sire summary.

Table 6 shows the results of the OBI test that ended March 24, 1983. In that test three bulls had stacked pedigrees for growth because they were direct sons of +40 bulls, plus their dams were sired by +40 bulls. Again, the same relationship may be observed since the bulls with stacked pedigrees had higher indexes than the other bulls on test.

SELECTING HERD SIRES FOR NATURAL SERVICE

The first priority in selecting young, unproven bulls is to identify breeders whose objectives are similar to yours. For example, if your objectives were to develop a highly functional set of problem-free cows and sell breeding stock with the same characteristics, you would not want to buy a bull from a breeder with a 180-day calving season that never culled a cow for fertility or bad udders. You would likely be more attracted to a herd with a 60-day calving season that had been rigidly culled for functional problems.

TABLE 6. PERFORMANCE BY CATEGORY[a]

Category[b]	No.	ADG	Adj. 365	Index
Stacked	3	3.74	1,231	105.7
Sire	9	3.81	1,133	102.3
Paternal grandsire	23	3.67	1,107	99.5
Other	15	3.60	1,108	98.5

[a] OBI bull test that ended March 24, 1983
[b] Sire = own progeny of +40 bull; paternal grandsire = progeny of son or grandson of +40 bull; other = progeny and(or) grand progeny of bull with less than +40 yearling EPD or not listed in sire summary.

Moreover, whether a commercial producer or purebred breeder, you can use sire-summary information to evaluate the programs of breeders from which you would consider buying bulls for natural service. An evaluation of sire-summary data for bulls in use and(or) their sires will provide insight into a breeder's program. This is demonstrated in table 7. Breeders A and B appear to be emphasizing growth and total performance, respectively, while Breeder C may not be a reliable source of bulls with stacked pedigrees for any particular goals.

TABLE 7. USING SIRE SUMMARIES TO EVALUATE A BREEDER'S PROGRAM

	AI sires in use			
	Birth EPD	Yearling EPD	Maternal EBV	Selection emphasis
Breeder A	+9.0	+75.0	98	Growth
	+4.2	+68.0	104	
	+6.0	+80.0	106	
Breeder B	-0.5	+45.0	104	Total performance
	+1.5	+60.0	108	
	+1.2	+55.0	103	
Breeder C	+5.9	+85.0	96	?
	-1.0	+25.0	105	
	+7.2	+50.0	100	

The point is that your judgment plus sire summaries may be used to locate good breeders. Find out as much as possible about a breeder's program including his management system, his goals, and the EPDs of the bulls he is using. This will go a long way toward eliminating sire-selection mistakes. A bull is no better than his breeder.

After locating qualifying breeders, the next step is to identify bulls that will work. Once you go onto a breeder's place, eliminate from consideration all bulls with structural and(or) reproductive unsoundnesses (there should not be many if you went to the right breeders). Identify the frame size range and scrotal circumference you will accept. Then bulls can be sorted for acceptability using EBVs. The information in table 8 relates to the bulls considered in this example. Accuracy (ACC) is included to ensure an adequate degree of reliability for the EBVs. Note, however, that EBVs for young, unproven bulls will have relatively low ACC values.

Based on the established selection levels for EBV and ACC in table 2, only Bull C qualifies. If the selection objective had been to maximize growth, Bull B would have been the choice. Moreover, Bull D would have been picked to maximize calving ease and Bull A to maximize maternal performance. If you are going through a large number of bulls and end up with too few or too many on your "keep" list, you can adjust the selection levels to give you the number of desired bulls.

TABLE 8. YEARLING BULL SELECTION

	EBVs (accuracy)		
	Birth weight	Yearling weight	Maternal performance
Selection levels	100 (.65)	102 (.65)	104 (.50)
Bull A	95 (.70)	108 (.72)	107 (.51)
Bull B	90 (.68)	110 (.69)	98 (.55)
Bull C	102 (.65)	104 (.65)	106 (.61)
Bull D	108 (.71)	98 (.68)	102 (.59)

COMBINING EBV AND EPD INFORMATION FOR ADDED PREDICTABILITY

Sire summaries may also be put to work to improve the accuracy of herd bull selection for natural service. If more than one bull meets your EBV selection levels and all the other criteria, you may wish to select the bull that has the best stacked pedigree. Figure 6 contains the data for two young bulls. Bull A has a well stacked pedigree while

Bull B does not. You would expect Bull A to be more predictable in his siring ability than Bull B because his genetic value is well documented and more "fixed." You can have more confidence in what he will do as a herd sire.

Young sires that are similar to Bull A and have desirable EBVs and stacked pedigrees can be the basis for building a predictable cow herd without the use of AI.

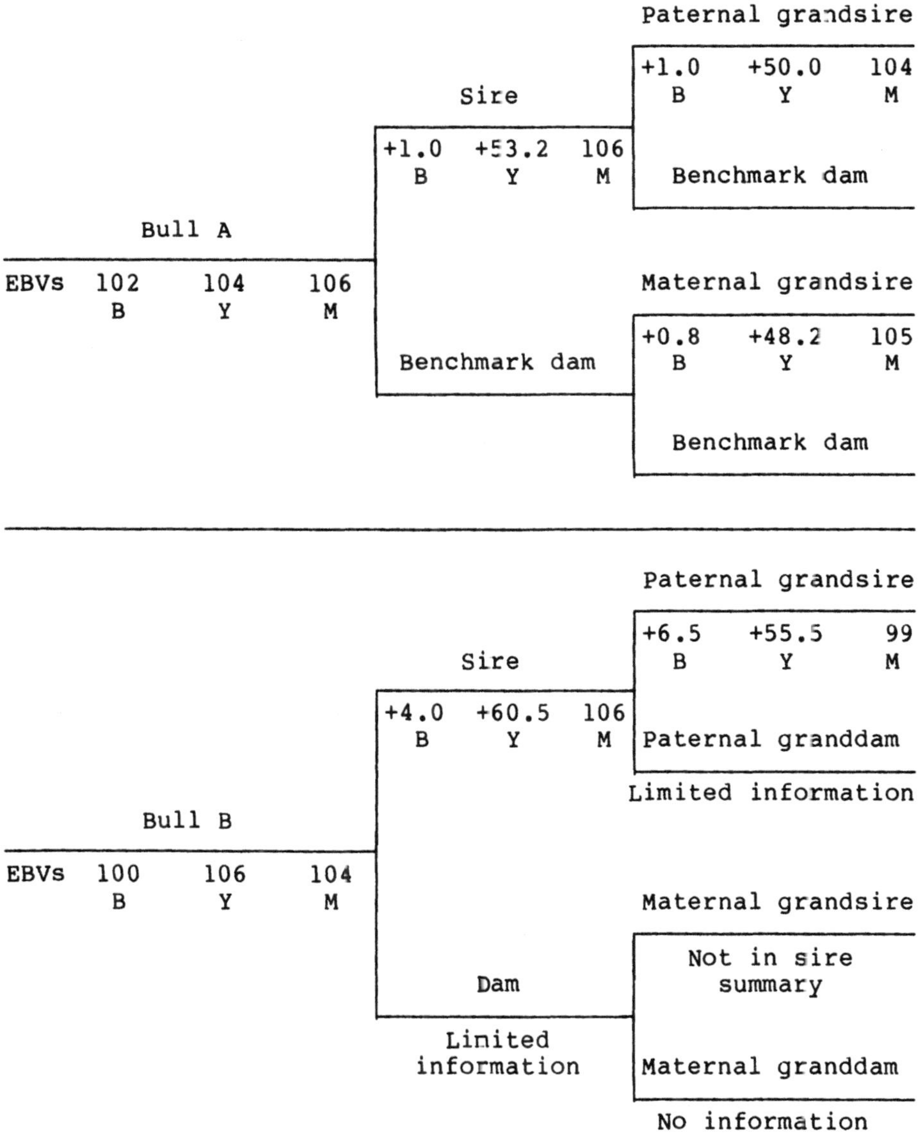

Figure 6. Finding herd bulls with stacked pedigrees.

7

WHAT THE COMMERCIAL RANCHER SHOULD KNOW ABOUT BULLS

H. H. Stonaker

CHANGES IN THE 1980S

Over 50 yr ago, two Hereford bulls from the DeBerard Ranch in Kremmling, Colorado, were purchased for the establishment of the now famous Line One Herefords. Today, a half century later, another dramatic shift seems to be underway in cattle breeding methodology. We are seeing disenchantment with past performance methods and taking another look at the direction we want to go. Fast rate of gain and greater size have been emphasized by bull-testing stations; however, current thinking takes to a more holistic view of cattle production. There is a greater appreciation for the broad variety of ecological and economic niches into which beef cattle production systems can best fit. We know, more intuitively than factually, that our cattle are more productive, especially in feedlot performance, then were those in the 1940s. However, no control herd was established in those early days, and we cannot accurately measure changes.

We feel, however, that not all of those changes have been positive. Systems analyses, measuring input resources relative to the production output of the herd, are indicating that different situations do not necessarily use the same breeds or the same sizes of cattle to optimize production. Type studies conducted in the early 1950s, and the new crossbreeding studies documenting input and output relationships, show that pounds of production relative to cows exposed to bulls may not be the best index of total efficiency of feed use. When 90% of the feed intake of the beef cow is used for her maintenance, and (in the total production system) when 70% of the feed required is used for maintenance, we need to reconsider productivity relative to cow size.

In the 1940s, it was estimated that we were losing about 1% of our calf crop due to the dwarf gene. In the 1980s, it has been estimated that there is a 6.7% loss of calves at calving time, of which 80% probably results from calving difficulty. About half of such losses could have been prevented by making sound management decisions; that is by pairing the right bulls with those yearling heifers. Of course, calving difficulties occurred 50 yr ago, but certainly at a lower incidence than today with the introduction of the large breeds and the drive toward frame size.

Cattlemen and researchers are working on this problem, and I expect that some solutions will soon be found. The commercial cattleman has an enviable position relative to overall evaluation and planning of breeding programs. He usually has stood back with a rather healthy skepticism of new ideas and techniques. This skepticism stems from his constant association with the risks in breeding, production, and economics. Cattlemen have liked crossbreeding, which has provided great increases in productivity. They also have liked buying bulls at central performance test stations.

In 1949, Hesperus, Colorado, had one of the first performance test stations. There are now five or six of these stations in Colorado and the number of bulls tested annually must be in the thousands. In addition, several large breeders holding their own sales have a similar kind of performance information for their own ranches or farms. The "100" Index System is commonly used. Culling levels are established at some sales on the basis of rate of gain or weight per day of age on a "100" index basis. Cattle falling much below this "100" are automatically eliminated from the sales. By and large, the criteria favor the exotic breeds -- those likely to be chosen as terminal crosses without much data on the known maternal characteristics of the bulls on test. As I suggested earlier, modification of this approach seems likely.

WHAT IS NEXT?

A recent summation (Heinze, 1984) of current advice to commercial producers has been prepared by Dr. Brinks at Colorado State University (figure 1, artwork courtesy of Colorado Rancher & Farmer, Denver). In this approach, replacement females come from a backcrossing system using only two breeds, such as the Angus and the Hereford. Large terminal-sire breeds, from which all calves are marketed, provide the sires for the calves from older and poorer cows. Guidelines are then designed for purchasing bulls for these three different groups of cattle within the herd. Dr. Brinks has outlined these guidelines in table 1. Indications are that there would be 20 bulls in this herd of 500 cows and heifers; four would be used on replacement yearlings, six would be used on the major part of the cow herd that would produce the replacement females, and 10 would be used on the other half of the herd that would, in theory, be bred to terminal sires from which all calves would be marketed.

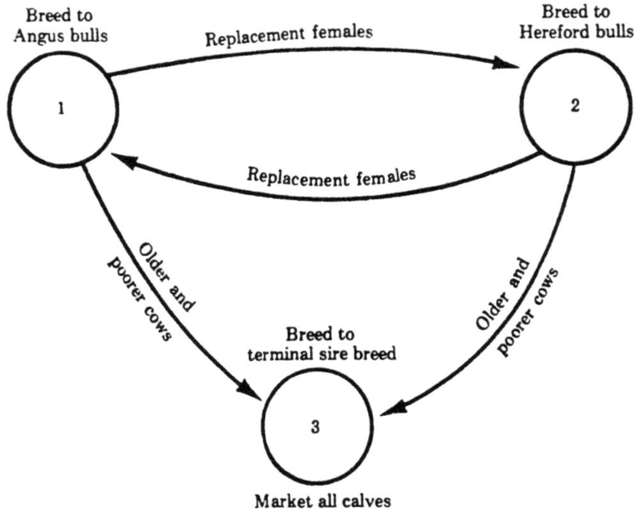

Figure 1. Combination system using two-breed rotational system with terminal sires on one-half of herd.

For simplification, Dr. Brinks has indicated certain levels of performance below which bulls would not be selected. In putting this approach into operation, however, few bulls can be found in central test stations to fit these established culling levels. For example, considering the fairly low heritability of birthweight, one would question whether bulls of large breeds having birthweights below these levels would sire calves sufficiently low in birthweight to minimize calving problems. Central test stations do not offer the desired kind of "bull power" for an entire herd. We will have to look elsewhere to find bulls that will sire calves that can be calved easily and lower the death losses and rebreeding problems in a herd due to dystocia. Unfortunately, there are few experimental findings to help in our search except for radical solutions such as going to the Jersey or going to the Longhorn -- which is what many breeders are doing to control this problem. Some bulls in artificial breeding companies of more popular beef breeds are known to sire small, trouble-free calves, and, if you are on an AI program, you can probably use this information. One recent trade paper suggested that such bulls should be thoroughly tested, with each having at least 500 birthweights available before consideration. That seems excessive.

How about the bulls for the cows that are in replacement herds? Again, for the commercial breeder, the standards Dr. Brinks sets forth are probably not obtainable. Nevertheless, the points emphasized should be taken into account, i.e., lower birthweights but increased maternal ability, weaning weights, and yearling weights. In some breeders' herds the number of calves raised (the most probable producing ability) and accurate birthweights may be available. In the Hesperus test, the most-probable-produc-

TABLE 1. INDIVIDUAL AND BREEDING VALUE SPECIFICATIONS OF THREE TYPES OF BULLS

	Individual record at 14 mo	Breeding value, %
All bulls:		
Breeding soundness exam	EBV 85 or greater[a]	105 or greater[a]
Scrotal circumference	36 cm or greater	
% Primary abnormalities	<10%	
% Total abnormalities	<25%	
Bulls for heifers (2 Herefords, 2 Angus)		
Calving ease	No assistance	105*[b]
Birth weight	<80 lb	105*[b]
Weaning weight	>500 lb	103[c]
Maternal (milk)	NA	105*[c]
Yearling weight	>1,000 lb	103[c]
Mature size (est.)	Goal	100[a]
Carcass cutability	<.3-in. probe	100[b]
Bulls for cows (3 Herefords, 3 Angus)		
Calving ease	No assistance	100[b]
Birth weight	<90 lb	100[b]
Weaning weight	>550 lb	108*[c]
Maternal (milk)	NA	108*[c]
Yearling weight	>1,050 lb	108*[c]
Mature size (est.)	Goal	100[b]
Carcass cutability)	<.3-in. probe	100[a]
Bulls for terminal cross (10 terminal sires)		
Calving ease	No assistance	100[b]
Birth weight	<100 lb	100 or less[b]
Weaning weight	>600 lb	High*[c]
Maternal	NA	NA
Yearling weight	>1,100 lb	High*[c]
Mature size	High	Prob. high[a]
Carcass cutability)	<.2-in. probe	105*[b]

* Most important values.
[a] Available in the future.
[b] Available from some breed association programs.
[c] Available from most breed association programs.

ing-ability (MPPA) values are given for the dams of station bulls but not for the cooperator herds. I would not like to buy a bull for producing replacement heifers without information on the dam's MPPA and the sire's daughters' MPPAs.

WHICH BREEDS?

The surge of interest we have seen in the introduction or reintroduction of exotic breeds is in no small part due to the interest in the rate-of-gain performance testing. Dr. Cartwright, at a recent commercial cow conference in Ft. Collins, indicated that selecting from among many breeds currently demands more emphasis than does selecting among bulls. Table 2 shows the ranking in productivity of breeds by traits; this data is taken from a 1984 report of the Meat Animal Research Center (MARC). Some information in table 3 was not included by MARC and represents only personal evaluation. There are many choices for selecting terminal-sire breeds that also provide productive brood cows with respect to age at puberty and milk production. These characteristics are fairly good in four of the breeds listed, which would probably serve as terminal-sire breeds as well: the Brown Swiss, Gelbvieh, Simmental, and Maine Anjou. These should give outstanding F_1 heifers for brood cows where nutritional conditions are favorable for cattle of their body size. If we are fairly certain that we are not going to save the crossbred F_1 heifers, then we might well consider Limousin, Charolais, or Chianina. Calf survival may be somewhat less in these breeds.

In the selection of bulls for the yearling heifers, the easy way is to go either to the Jersey or the Longhorn, except that the Jersey calf seems to have a lower survivability. We do not have concrete evidence on the Longhorn, but Dr. Hauser at Wisconsin says, "Those Guernsey calves just want to lie down and die, but the Longhorns are determined to live." (He did not comment on the Jerseys.) Some advertising for the Gyr suggests their ease of calving in 2 yr olds, and evidence from the old Louisiana experiments indicates that the Red Sindhi cross calves are very small and easy calving. The Red Sindhi are found in small numbers in Mexico and are not currently available in the U.S. to my knowledge. Using the Brahman, the Holstein, or any of the larger European breeds on those yearling heifers would not be advisable. Among the more traditional breeds, the Red Angus has a good reputation, and some Hereford and Angus bulls are acceptable for use on yearlings.

The toughest problem is how to handle the cow herd that is to supply the replacement females. Findings of Jenkins and others at MARC indicate that, for a given feed resource, the most efficient levels of total beef production are not obtained by incorporating the large terminal-sire breeds, but rather by indicating a more favorable relationship for the more moderate or even smaller breeds, such as the

Hereford-Angus crosses or the Jersey crosses. Despite this powerful experimental evidence, Bourdon at Colorado (and others in systems analyses) indicates that more total economic gain is obtained by use of the larger, heavier cows of good milk-producing ability, such as we have seen in many of the imported European breeds or even the Holstein. Brinks feels that a cow herd should have at least one-fourth dairy breeding (such as from Holstein or Brown Swiss).

TABLE 2. BREED CROSSES GROUPED IN BIOLOGICAL TYPE ON BASIS OF FOUR MAJOR CRITERIA[a]

Breed group	Growth rate and mature size	Lean to fat ratio	Age at puberty	Milk production
Jersey-X	X	X	X	XXXXX
Hereford-Angus-X	XX	XX	XXX	XX
Red Poll-X	XX	XX	XX	XXX
Devon	XX	XX	XXX	XX
South Devon-X	XXX	XXX	XX	XXX
Tarentaise-X	XXX	XXX	XX	XXX
Pinzgauer-X	XXX	XXX	XX	XXX
Brangus-X	XXX	XX	XXXX	XX
Santa Gertrudis-X	XXX	XX	XXXX	XX
Sahiwal-X	XX	XXX	XXXXX	XXX
Brahman-X	XXXX	XXX	XXXXX	XXX
Holstein	XXXX	XXX	XX	XXXXXX
Brown Swiss-X	XXXX	XXXX	XX	XXXX
Gelbvieh-X	XXXX	XXXX	XX	XXXX
Simmental-X	XXXXX	XXXX	XXX	XXXX
Maine Anjou-X	XXXXX	XXXX	XXX	XXX
Limousin-X	XXX	XXXXX	XXXX	X
Charolais-X	XXXXX	XXXXX	XXXX	X
Chianina-X	XXXXX	XXXXX	XXXX	X

[a] The number of "Xs" indicates increasing relative differences among breed groups for (1) growth rates and mature size, (2) lean to fat ratio, (3) age at puberty, and (4) milk production found in the Germ Plasm Evaluation Program at the Roman L. Hruska U.S. Meat Animal Research Center, Clay Center, Nebraska.

TABLE 3. ASSUMPTIONS RELATIVE TO BREED DIFFERENCES IN CALF SURVIVAL*

Crossbred cows by sire group	Calf survival due to: Calf genotype	Dam
Jersey	SS	SSSSS
Hereford-Angus	SSSS	S
Red Poll	SSS	SS
Devon	SSSSS	SS
South Devon	SS	SSS
Brangus	SSSS	SS
Santa Gertrudis	S	S
Sahiwal	SS	SSSSS
Brahman	SSS	SSSSS
Holstein	SS	SS
Brown Swiss	SSS	SSS
Gelbvieh	SSS	SS
Simmental	SSS	SSS
Maine Anjou	SSSS	SS
Limousin	SS	SSS
Charolais	SS	SSS
Chianina	SSS	SS
Tarentaise	SSSS	SSS
Longhorn	?????	????
Guernsey	?	????
Red Angus	SSS	S
Sindhi	??	?????
Gyr	SSS	

* Increasing survival ranked 1 to 5.
S indicates various data sources.
? indicates field observations.

HOW FAR NORTH CAN WE USE BRAHMAN F_1S?

Of the crosses at MARC, the Brahman F_1 cow has shown the highest cow production. This superiority has been observed in the southern states, including Texas, for many years. At Reno these cows also have been highly productive, but not so outstandingly. Older studies in Canada have indicated heavy weaning weights.

Informal observations indicate no predisposition to brisket disease at high altitudes. Late puberty makes it difficult to fit the Brahman-cross heifers into a 2-yr-old calving regime. But the F_1s calve easily with their first calf. Feedlot tests in winter suggest good gains but poor feed conversion. The jury is still out.

MAJOR PROBLEMS WITH DIFFERENT BREEDS

Commercial breeders find some of the following serious problems with the different breeds: dystocia, or calving difficulty (Charolais, Gelbvieh, Brahman, Simmental, and several others of the European exotics); fertility (Brahman and Santa Gertrudis); disposition (Brahman, Chianina, and Angus); cancer-eye, pink-eye (Hereford, Holstein, and Simmental); brisket disease (Angus and Holstein); milk production (Charolais, Limousin, and Chianina); age to puberty (Brahman, Sahiwal, Limousin, Charolais, Chianina, Brown Swiss, and Santa Gertrudis); calf loss (Santa Gertrudis, Brahman, Jersey, and Sahiwal). These problems are associated with conditions in the more temperate and(or) higher-altitude zones in the U.S.

GENETIC WEAKNESSES OTHER THAN DYSTOCIA

It seems doubtful that we are making any appreciable advances in livability of our cattle. Currently, dystocia is probably our largest, single, genetically-manageable cause of death loss. As Dr. Bellows indicated, we are losing 16% up to weaning and a high percentage of this loss is due to dystocia. In our area, we find that death losses in calf feeding are conservatively considered to be about 4% or, if we are feeding yearlings, about 2%. Periodically, the losses are much higher in certain groups of cattle and much of the reason seems to be unexplained. Important losses stemming from a genetic base are calf scours and shipping fever. Other weaknesses that cause great losses are brisket disease, prolapse, and bloating. If we were dealing with chickens or pigs, we might have diminished many of these disease problems by now through breeding. And, if I were a young researcher beginning in genetics today, I would surely attempt to design experiments to give the industry a more precise view relative to opportunities for increasing livability of cattle.

BREEDING FOR INTERMEDIATE CHARACTERISTICS

For many characteristics, the extreme levels of performance that are available in many of the cattle breeds are not optimum for commercial production. Some of the traits that seem certain to require an intermediate rather than optimum level are milk production, birthweight, mature size, disposition, body composition -- and, yes, even rate of gain. For many years to come it appears that the commercial cattleman will profit most by using different breeds that have contrasting or complementary characteristics to obtain optimum intermediate levels of production. We know some commercial breeders who have feed resources to utilize half-bred Holstein-Friesian cows without severe loss in fertility and calf crop. Others do not have the feed resources for the level of milk production that such cows are capable of without a marked drop in calf-crop production. For the producer of seed stock, there are inherent genetic limitations in selection as a tool to produce a herd with a high level of prediction for the intermediate expression of production. Years ago, Sewall Wright and Jay L. Lush indicated the methodology for breeding for the intermediate. In theory, selection only takes us to a plateau of 50% to 60% of animals with desired intermediate characteristics. To push beyond this ceiling, we must combine a considerable amount of inbreeding with this selection to increase the percentage of animals having the desired intermediate.

Dr. Bob Long recently said "I visualize, whether we like it or not, larger operations under sophisticated management that will use superior inbred strains of breeding stock of known combining ability in crossbreeding programs." The development of inbred lines seems to be something that eventually can be managed by the larger experiment stations. The results of the Line One Herefords at Miles City were spectacularly successful. In Colorado, with a number of inbred Hereford lines, the Prospector is currently the most popular. Unfortunately, when these lines were begun, there was very little interest in breeds other than the Hereford. There are a number of practical limitations to the use of inbred bulls, but an often overlooked advantage is that no matter what the characteristic, or its level of heritability, we can select from among the lines that have the desired expression of that trait. Who would have guessed in the early 1940s that we would, in the 1980s, be interested in something like pulmonary arterial pressure or calving difficulties? Line difficulties have been identified for these traits and bulls from within these lines can be used with the assurance that they will take the breeder in the direction that he wants to go.

Nearly everyone will concede that remarkable changes have been brought about in the past 50 yr through application of genetics to cattle breeding. I hope many of you will agree with many of us that the expression "You ain't seen nothin' yet!" will foretell progress into the next 50 yr.

SUMMARY

The commercial cattle producer can easily find the high-gaining, muscular, large-frame breeds of bulls for terminal-sire crossbreeding at central performance tests. These may meet up to 50% of his bull requirements. Terminal-cross heifers, while not meeting his requirements, may be purchased to expand other cow herds. That probably affects the total beef business negatively.

Because of calving difficulties and high losses at birth associated with terminal sires, bulls that meet 20% or more of bull requirements for breeding yearling heifers will have to come from breeds that have less acceptance in the feedlot, such as Longhorn, Jersey, Tarentaise, Gyr, or Red Sindhi, if available. These breeds will not be found in central performance tests. If AI is possible, there are identified bulls from more popular beef breeds that can be used on yearlings to reduce calving difficulty.

For breeding replacement heifers, search out those seed stock breeders located in your environment who can provide a ranking of their cows on most probable producing ability (MPPA). Try to buy sons of these cows that are sired by bulls with high MPPA daughters. It means buying according to seed stock breeders' recommendations more than to relative daily gain performance of bulls in a central testing station where MPPA values often are not available.

For many important traits, an intermediate level of performance is desired. Genetically, selection is a weak tool for increasing the percentage of animals with optimum intermediate phenotypes. This is a challenge to researchers and seed stock producers, which is not being resolved satisfactorily at present.

Looking ahead, we can encourage researchers to look into genetic resistance to diseases. The pulmonary arterial pressure (PAP) test of bulls already can help reduce susceptibility to high mountain disease or brisket disease. We urgently need research on breeding greater resistance to such economically important diseases as shipping fever and calf scours.

REFERENCE

Heinze, Lynn. 1984. Have you written out bull specifications? Colo. Rancher and Farmer. 38:4.

8

CROSSBREEDING BRAHMANS, EUROPEAN, AND SOUTH AMERICAN INDIGENOUS BREEDS

Dieter Plasse

INTRODUCTION

Beef cattle have been crossbred in the Latin American tropics for more than 20 yr. Most of the available data are from research projects in which Criollo (native *Bos taurus)* cattle have been crossed with different *Bos indicus* breeds.

Criollo cattle were introduced to Latin America by the invading Spaniards and Portuguese in the sixteenth century. During the present century, they have been absorbed by different *Bos indicus* breeds. Because of the danger that the Criollo might be eradicated, several experiment stations and producers have investigated their potential for crossbreeding.

The experimental breed combinations have included:

- *Bos indicus* x Criollo
- Charolais x *(Bos indicus* x Criollo)
- Santa Gertrudis x Criollo
- European x *Bos indicus*
- Gene pools and inter se matings
- *Bos indicus* x *Bos indicus*

This presentation analyzes published crossbreeding results by Plasse (1981, 1983). In the breed comparisons, the term "Criollo" refers to different lines of the native *Bos taurus;* "zebu" refers to all treated *Bos indicus* breeds; and "European" to *Bos taurus* breeds other than Criollo but of European origin. Comparisons are made according to breeding systems, which implies that certain genotypes are repeated in various comparisons.

CROSSBREEDING PROGRAMS IN TROPICAL LATIN AMERICA

Most Latin American beef cattle are produced on native grassland, and the existing extensive production systems (Plasse, 1976) are characterized by very small or very large herds, suboptimum nutritional levels, droughts and floods, diseases, and direct climatic stress. However, beef production is more intensive in some areas having improved pastures and a high level of management. Countries having crossbreeding programs aimed at increasing production under the more intensive conditions include:

- Bolivia. In 1962, a crossbreeding project was started in two herds of the Estancias Elsner Hermanos in the tropical lowlands of the Beni. This project is now under the technical supervision of the author; it includes 10,000 cows maintained on native grass, but has an advanced management and genetic program. A 4-mo breeding season is used and calves are weighed at weaning (8 mo) and again at 18 mo to 20 mo. Initially, the management program involved upgrading Criollo to zebu and rotational crossbreeding between these two breeds. The bulls were bred in two elite herds of Criollo and zebu. Some F_1 zebu x Criollo cows were bred by AI to Charolais (Bauer, 1968, 1973).

  In the second phase, F_1 zebu x Criollo were bred inter se, and a gene pool was established that contained those two breeds and Charolais. Both populations were closed genetically.
- Costa Rica. A reciprocal crossbreeding program between Criollo, Brahman, and Santa Gertrudis was initiated in 1960 at the experiment station of Turrialba, Costa Rica. The F_1 cows from this project were bred to Charolais bulls. Cattle were maintained on introduced pasture. A 3-mo breeding season was used and calves were weaned at 8 mo and calf weights recorded at that time. After a 140-day test on pasture, they were weighed again (Munoz and Martin, 1969).
- Colombia. In 1967, the Instituto Colombiano Agropecuario (ICA) initiated crossbreeding programs with three Criollo lines: Romosinuano, San Martinero and Blanco Orejinegro (BON) as well as zebu and Charolais. At three experiment stations (Turipana, La Libertad, and El Nus) cattle were kept on introduced pasture, the breeding season lasted 4 mo, calves were weaned at 9 mo, and postweaning weights were taken at 18 mo of age (ICA, 1976; Hernandez, 1981).
- Venezuela. Since 1965, a long-term cooperative research program has been conducted by the Universidad Central de Venezuela and the Fondo Nacional de Investigaciones Agropecuarias (FONAIAP) at the Calabozo experiment station in the tropical lowlands. The program includes pure breeding of each of two lines of Criollo and Brahman cattle, upgrading Criollo to Brahman and Santa Gertrudis, and rotational crossbreeding of Brahman - Criollo, Brown Swiss - Brahman, and Charolais - Brahman. Initially, Brahman cows also were bred to Red Poll bulls. Cows were run on native grass irrigated during the dry season. The management program includes a 4-mo breeding season, weaning and weighing at 7 mo, and

recording weight again at 18 mo. Between ages of 7 mo and 18 mo, half of the males of each genetic group were kept on native grass, while the other half and all females were run on introduced pasture (Plasse, 1981).

Experimental crossbreeding programs for inseminating grade zebu cows with semen of six European breeds are now being evaluated in private herds in the tropical lowlands of Venezuela. In another program, three types of F_1 European x zebu bulls are being used on grade zebu cows.

RESULTS OF CROSSBREEDING PROGRAMS

F_1 Zebu x Criollo

Growth traits. Table 1 shows the heterosis percentage for birth weight, preweaning average daily gain, weaning weight, postweaning average daily gain, and postweaning weight on pasture -- as well as the superior values for the F_1s as compared to the zebu. This latter comparison is very important, because the zebu is the best purebred alternative in tropical Latin America. Crossbreeding is of interest only if crossbreeds are sufficiently superior to this parent breed. In these comparisons, the data for the four countries agreed well (Plasse, 1983).

TABLE 1. HETEROSIS PERCENTAGE AND SUPERIORITY OF CROSSBREDS OVER PUREBREDS IN F_1 CRIOLLO X ZEBU IN TROPICAL LATIN AMERICA

Item	Heterosis (%)[a] (n = 2113)	Superiority of F_1 vs zebu (%)[a] (n = 11513)
Birth wt	9	12
Preweaning avg daily gain	12	6
Weaning wt	11	7
Postweaning avg daily gain	30	22
Postweaning wt	16	12

[a] Unweighted mean of published data from Turrialba, Calabozo, Turipana, La Libertad, El Nus, and Beni stations (Plasse, 1983).

Reproductive traits. Data available from the Calabozo project show considerable heterosis for age and weight at puberty in F_1 zebu x Criollo heifers. F1 heifers were 11%

younger and 11% heavier at puberty as compared to the mean of their parents; their advantage over zebu was 9% and 4%, respectively (215 observations). When calculating heterosis for pregnancy percentage of F_1 zebu x Criollo cows, a total of 53,123 records from Calabozo and Beni reflected values varying between 9% and 16%. However, in Calabozo, where Criollo cows have had low reproduction rates and the zebus had good rates, F_1 cows had the same values as did zebu. In Bolivia, where Criollo have had very high pregnancy rates and zebu have been inferior, F_1 cows were 15% and 20% superior to the zebus in two different evaluations. At Calabozo and Beni, Criollo cows nursing an F_1 calf had a lower pregnancy percentage, as compared to those nursing a purebred calf.

Growth of Calves of F_1 Zebu x Criollo Cows Bred to Criollo, Zebu, and Charolais Bulls

On stations at Beni, Turipana, La Libertad, El Nus, and Turrialba, F_1 zebu x Criollo cows were bred to Criollo, zebu, and Charolais bulls. The resulting weaning and 18-mo weights are given in coded form in table 2, where the weights for the zebus are transformed to 100 and the other groups are given as a deviation from the zebu breed. This same procedure is used in the following tables.

TABLE 2. WEANING AND 18-MO WEIGHT OF CALVES OF F_1 CRIOLLO X ZEBU COWS BRED TO ZEBU, CRIOLLO, AND CHAROLAIS BULLS COMPARED WITH PUREBRED CRIOLLO AND ZEBU (ZEBU = 100)

Breed[a]			
Sire	Dam	Weaning wt[b] (n = 20800)	18-mo wt[b] (n = 4256)
Z	Z	100	100
C	C	93	91
C	1/2Z 1/2C	105	101
Z	1/2Z 1/2C	114	98
CH	1/2Z 1/2C	122	113

[a] Z = Zebu, C = Criollo, CH = Charolais.
[b] Coded data (Zebu = 100) derived from publications from Beni, Turipana, La Libertad, El Nus, and Turrialba projects (Plasse, 1983).

There is a clear advantage of 22% for weaning and 13% for 18-mo weight for the triple-cross calves sired by Charolais bulls. A comparison with all other available data shows that this breed group has produced the relatively highest weaning weights in the Latin American tropics. In two locations, 18-mo weights of 3/4 Criollo and 3/4 zebu were inferior to zebu (Plasse, 1983).

Upgrading Criollo to Zebu

Data on upgrading Criollo to zebu have been published and are summarized in coded form for pregnancy rate, weaning weight, 18-mo weight, and carcass weight in table 3.

TABLE 3. COMPARISON OF DIFFERENT GENOTYPES IN UPGRADING CRIOLLO TO ZEBU (ZEBU = 100)[a]

Breed group[b]	Pregnancy, %[c] (n = 63825)	Weaning wt[d] (n = 29423)	18-mo wt[d] (n = 5568)	Carcass wt, steers[e] (n = 14705)
Z	100	100	100	
C	113	97	87	100
1/2Z 1/2C	118	105	110	116
3/4Z 1/4C	106	110	101	115
7/8Z 1/8C	84	104	97	
15/16Z 1/16C		99		

[a] D. Plasse (1983).
[b] Z = Zebu, C = Criollo.
[c] From Beni project.
[d] From Beni and Calabozo projects.
[e] From Beni. No zebu available; Criollo = 100.

In growth rate, the zebu has been superior to the Criollo; however, the Criollo pregnancy rate has been superior. The highest total production has been obtained by the F_1 cows and the 3/4 offspring. Pregnancy rate in the 7/8 zebu cows has dropped considerably probably because of heterosis. One explanation for the poor reproduction rate of the 7/8 zebu cows is that the purebred elite herds are under higher selection pressure and have better management.

These results confirm that high-grade zebu cattle in Latin America have low reproduction rates. According to table 3, a decrease in preweaning growth rate should be expected after the 3/4 zebu level, and a decrease in postweaning weight expected after the F_1. In carcass weight comparisons, F_1 steers were superior to Criollo by 16%; 3/4 zebu steers were superior by 15%.

There is no doubt that grade zebu populations lose all of the benefit shown by intermediate crosses between Criollo and zebu and, in some cases, their production may fall below that of Criollo cattle.

Rotational Crossbreeding Between Zebu and Criollo

Table 4 lists the coded pregnancy percentages, weaning weight, and 18-mo weights of genotypes for a rotational crossbreeding program between Criollo and zebu; the data comes from the Beni and Calabozo projects. F_1 cows have had the highest pregnancy rate; in 3/8-5/8 and 1/4-3/4 animals, pregnancy rates decreased. These two groups of reciprocals were much alike; however, all crossbreds had pregnancy rates above those of the zebus, although only the F_1 were superior to the Criollos. Weaning weights were highest in the 3/4 and 11/16 zebu, with low performance in 3/8 zebu calves. Neither of these results can be explained readily. At 18 mo, all crossbreds were at the level of the zebus -- except the 1/2 zebu, which were superior.

TABLE 4. COMPARISON OF DIFFERENT GENOTYPES FROM ROTATIONAL CROSSBREEDING PROGRAMS (ZEBU = 100)[a]

Breed group	Pregnancy, %[b] (n = 69321)	Weaning wt[c] (n = 29246)	18-mo wt[c] (n = 5571)
Z (Zebu)	100	100	100
C (Criollo)	112	97	89
1/4Z 3/4C	110	105	90
3/8Z 5/8C	108	94	101
1/2Z 1/2C	118	105	110
5/8Z 3/8C	107	108	101
3/4Z 1/4C	106	110	101
11/16Z 5/16C		110	

[a] D. Plasse (1983).
[b] From Beni project.
[c] From Beni and Calabozo projects.

Results of the large-scale rotational crossbreeding program at the Beni project have raised many questions. A small, well-controlled experimental group is being continued. It is hoped that the results of this experimental group and the final evaluation of the Calabozo project (where six generations from a rotational crossbreeding program will be available), will provide some answers.

Criollo x Santa Gertrudis Crossbreeding

The Turrialba project included reciprocal F_1 Criollo x Santa Gertrudis crosses, whose weights were superior to those of Brahman at birth, weaning, and 1 yr of age by 13%, 12%, and 9%, respectively. As compared to purebred Santa

Gertrudis, they were inferior by 5% at birth, and superior by 3% at weaning and by 1% at 1 yr.

In Calabozo, F_1 Santa Gertrudis calves out of Criollo cows were superior to Brahman by 22% at birth, by 9% at weaning, and by 8% at 18 mo. However, the pregnancy percentage of F_1 cows was 12% inferior to that of Brahman. Weights of calves of 3/4 Santa Gertrudis breeding were 5% superior at birth and 8% at weaning; they were 2% inferior to the Criollo at 18 mo.

When a Criollo x Brahman herd was upgraded by crossing to Santa Gertrudis, reproductive efficiency decreased as the level of Santa Gertrudis inheritance rose from 1/2 to 15/16. When the inheritance level of Santa Gertrudis became more than 3/4, the 18-mo weight was not affected.

Inter Se F_1 Criollo x Zebu and Gene Pool Charolais - Zebu - Criollo

In the Beni project, crossbred bulls have been used on crossbred cows in two programs. The first is an inter se mating at the F_1 (zebu x Criollo) level. In the second program, cows of 1/2 zebu - 1/2 Criollo, 3/4 zebu - 1/4 Criollo, and 1/2 Charolais - 1/4 zebu - 1/4 Criollo have been combined in one group and bred to bulls of the last named genotype. The comparative results obtained with the foundation cows and their progeny are summarized in coded form in table 5. F_1 zebu x Criollo cows bred to F_1 bulls have had a 20% higher pregnancy rate than that of zebu cows; their calves weighed 11% more at weaning and 5% more at 18 mo. Pregnancy percentage in the gene pool has been an average of 11% superior to that of zebus; weaning weight and 18-mo weight were 10% and 9% higher, respectively. In general, both the inter se and the gene pool foundation groups have shown a good production increase when compared to that obtained with pure zebu breeding. An expected production decrease in the second generation has not yet been evaluated.

Reproductive Efficiency of Purebred vs Crossbred Bulls

In the Beni project, 34,006 observations were made of purebred and crossbred bulls used with cows of the same breed composition. In three comparative analyses, crossbred bulls produced pregnancy rates 3%, 4%, and 9% higher than those of the purebred bulls.

F_1 European x Zebu

Growth traits. In recent years, zebu cows have been crossbred to European bulls in tropical Latin America. The first results from different breed combinations have been published and summarized by Plasse (1983) (table 6).

TABLE 5. COMPARISON OF DIFFERENT BREED GROUPS IN INTER SE AND GENE POOL PROGRAM IN BOLIVIA (ZEBU = 100)[a,b]

Breed[c]		Pregnancy, %	Weaning wt	18-mo wt
Bull	Cow	(n = 20788)	(n = 7217)	(n = 1907)[d]
Z	Z	100	100	100
Z, C	C	105	100	92
1/2CH 1/4Z 1/4C	1/2Z 1/2C	119	113	110
1/2CH 1/4Z 1/4C	1/2CH 1/4Z 1/4C	105	113	105
1/2CH 1/4Z 1/4C	3/4Z 1/4C	110	105	112
1/2Z 1/2C	1/2Z 1/2C	120	111	105

[a] D. Plasse (1983).
[b] Foundation generation.
[c] Z = Zebu, C = Criollo, CH = Charolais.
[d] Only small numbers of crossbreds involved.

TABLE 6. SUPERIORITY (%) OF F_1 EUROPEAN X ZEBU OVER ZEBU IN VENEZUELA (GROWTH TRAITS)[a]

		Location[b]							
		1			2	3		4	
Item	F_1 sired by[c]	RP	BS	CH	BS	S	M	BS	CH
Birth wt		-5	8	12	-1			1	4
Preweaning avg daily gain		0	7	7	12	10	5		
Weaning wt		-1	5	8	8	9	6		
Postweaning avg daily gain					9				
18-mo wt		2	11	9	6	10	10		

[a] D. Plasse (1983).
[b] Location 1 Calabozo, (n = 510).
2 Guarico, (n = 147).
3 Apure, (n = 417).
4 Barinas, (n = 372).
[c] RP = Red Poll, BS = Brown Swiss, CH = Charolais, S = Simmental, M = Marchigiana.

The Calabozo experiment station and technically supervised research programs on private ranches in the states of Guarico, Apure, and Barinas (Venezuela), have published data comparing calves of grade zebu cows sired by Red Poll, Brown Swiss, Charolais, Simmental, and Marchigiana bulls. The results are not consistent, however. When compared with pure Brahman, most F_1 crosses have shown from 6% to 9% superiority at weaning and around 10% superiority at 18 mo on pasture.

Limited unpublished data indicate that carcass weight of F_1 zebu x Simmental and zebu x Marchigiana finished on good quality pasture may be up to 19% superior to purebred

Brahman sired by progeny-tested bulls. Carcass quality also was improved.

Pregnancy percentage. When F_1 European x zebu cows raised reciprocal-backcross calves, their pregnancy percentage was lower than that of pure zebu (table 7). These results from Calabozo, however, might be affected by the low nutritional level under which these cows raised crossbred calves of high growth-rate potential (table 8).

TABLE 7. COMPARATIVE PREGNANCY RATES OF F_1 EUROPEAN X BRAHMAN COWS AND PUREBRED BRAHMAN (BRAHMAN = 100)[a]

Breed[b]		Pregnancy, %
Bull	Cow	(n = 795)
CB	CB	100
CH	1/2CH 1/2B	100
RB	1/2CH 1/2B	95
BS	1/2BS 1/2B	82
RB	1/2BS 1/2B	92

[a] D. Plasse (1983).
[b] B = Brahman, CH = Charolais, BS = Brown Swiss.

TABLE 8. GROWTH TRAITS OF RECIPROCAL BACKCROSSES OUT OF F_1 EUROPEAN X ZEBU COWS COMPARED TO BRAHMAN (BRAHMAN = 100)[a]

Breed group[b]	Birth wt (n = 581)	Weaning wt (n = 525)	18-mo wt (n = 497)
B	100	100	100
3/4CH 1/4B	117	117	100
3/4B 1/4CH	113	119	109
3/4BS 1/4B	114	125	101
3/4B 1/4BS	121	120	110

[a] D. Plasse (1983).
[b] B = Brahman, CH = Charolais, BS = Brown Swiss.

Reciprocal European x Zebu Backcross Calves

Results from the first phase of the rotational crossbreeding program in Calabozo are summarized in table 8.

Reciprocal backcross calves out of F_1 Charolais x Brahman and Brown Swiss x Brahman cows are compared to calves of commercial Brahmans. At weaning, the superiority of the backcrosses was 18% for the calves from F_1 Charolais and 23% for those from F_1 Brown Swiss cows. In each case, the 3/4 Brahman were better than the reciprocals. At 18 mo, only the 3/4 Brahman were superior to the purebreds.

Bos Indicus x *Bos Indicus*

Many producers in Latin America have claimed that crosses between different *Bos indicus* breeds are superior to purebreds; however, information from three publications does not support this supposition. When 2,101 offspring of Nellore cows in Paraguay were compared to that of F_1 Nellore x Brahman, no differences for weaning or 18-mo weight were detected. In the tropical lowlands of Venezuela, weaning weights were evaluated for 1,700 Brahman and F_1 Nellore x Brahman calves. These breed groups were similar in performance. At Apure, Venezuela, in comparisons of 1,438 weaning weights and 1,341 weights at 18 mo of Brahman, 1/2 Guzerat - 1/2 Brahman, inter se F_1 Nellore x Brahman, and 3/4 Nellore - 1/4 Brahman, the crossbreds weighed significantly less than did pure Brahman. The values were 4%, 6%, and 8% less for the respective genotypes at weaning, and 3%, 4%, and 5% less at 18 mo.

CONCLUSIONS

Results from many locations in tropical Latin America provide strong evidence that native or introduced pastures can provide for higher growth rates and (probably) better reproductive efficiency from crossbreds than from Criollo or grade zebus. The average superiority of F_1 zebu x Criollo over zebu has shown to be 12%, 7%, and 12% for weight at birth, weaning, and 18 mo of age, respectively. Weight and age at puberty of the F_1 females were superior to that of the zebu. The averages of the data reported show that F_1 cows had a 15% higher pregnancy rate than pure zebu. The magnitude of superiority in growth traits of the F_1 to the Criollo has been greater on the average than that of the F_1 to the zebu.

F_1 zebu x Criollo cows have weaned heavy calves, especially when they were bred to Charolais bulls. These calves weighed 22% more at weaning than did zebus. For terminal crosses, the data suggest that this is the best breeding option.

Upgrading Criollo to a higher percentage of zebu has not been successful and growth rate and reproductive efficiency have decreased considerably after the third generation. This finding is in agreement with past results.

Final conclusions cannot yet be drawn about the merit of rotational crossbreeding between the Criollo and the zebu. The mean reproductive efficiency and preweaning growth rates during the first three generations of such a program were superior to those of pure zebus. However, the 3/8 zebus had lower reproduction rates and weaning weight than did the purebreds. No improvement was evidenced in the 18-mo weight, except in the F_1 generation.

When F_1 zebu x Criollo cows were bred to F_1 bulls, reproductive efficiency was 20% above that of the zebus, and weaning and 18-mo weight of the calves were 11% and 5% higher, respectively. However, several generations must be evaluated before drawing firm conclusions about this system.

A gene pool was established in Bolivia that contained Criollo, zebu, and Charolais genes. The results of the foundation generation and its offspring have been very good, but final judgment must wait until data from several generations are available.

Crossbreeding several *Bos indicus* breeds has not resulted in the expected increase in growth rate. However, before the final conclusions can be made of the merit of this breeding system, reproductive efficiency and viability also must be evaluated. It might be, however, that faster improvement results from rigid within-breed selection.

In the majority of the cases, F_1 zebu x European calves of different breed composition have shown higher pre- and postweaning growth rate than have pure zebus. However, reproductive efficiency of the F_1 cows, when raising backcross calves, was lower than that of the zebu, perhaps because of the high lactation stress produced by the fast-growing calves. These backcross calves have weighed around 20% more at weaning than have Brahmans; but, at 18 mo, only the 3/4 Brahman have been superior. Under the conditions evaluated, these crosses do not seem to have been able to demonstrate their genetic potential because of nutritional deficiency. When F_1 zebu x Simmental and zebu x Marchigiana were finished on introduced and well-managed pastures, they had a 19% higher carcass weight as compared to that of pure Brahmans sired by progeny-tested superior bulls.

Final conclusions about these different systems and breed combinations using Criollo, zebu, and European cattle must wait until results are available on disease resistance, mortality, postweaning growth on different quality pasture, and slaughter characteristics. If genotypes of higher growth potential are to be produced, improvements must be made in pasture quality (at least for part of the production cycle) and in disease control. The revised publications and unpublished data from Venezuela indicate the advantage of strategic crossbreeding when certain breeds are combined in production systems designed for these genotypes.

REFERENCES

Bauer, B. 1968. Problemas de la cria de ganado vacuno de carne en el tropico de Latinoamerica (Bolivia) y medidas que hemos adoptado para solucionarlas. Proceedings, 2nd. Annual Conference on Livestock and Poultry in Latin America, Gainesville, FL.

Bauer, B. 1973. Improving native cattle by crossing with zebu. In: M. Kroger, T. J. Cunha, and A. C. Warnick (Ed.). Crossbreeding Beef Cattle. Series 2. pp 395-401. University of Florida Press, Gainesville, FL.

Hernandez, G. 1981. Las razas Criollas Colombianas para la produccion de carne. In: B. Muller-Haye and J. Gelman (Ed.). Recursos Geneticos Animales en America Latina. pp 52-76. FAO. Study: Animal Production and Health, No. 22. Rome, Italy.

ICA. 1976. Razas Criollas Colombianas. Instituto Colombiano Agropecuario. Manual de Asistencia Tecnica. No. 21. Bogota, Colombia.

Munoz, H. and T. G. Martin. 1969. Crecimiento antes y despues del destete en ganado Santa Gertrudis, Brahman y Criollo y sus cruces reciprocos. ALPA, Mem. 4:7.

Plasse, D. 1976. The possibility of genetic improvement of beef cattle in developing countries with particular reference to Latin America. In: A. J. Smith (Ed.). Beef Cattle Production in Developing Countries. pp 308-331. Edinburgh University Centre for Tropical Veterinary Medicine. Edinburgh, UK.

Plasse, D. 1981. El uso del ganado Criollo en programas de cruzamiento para la produccion de carne en America Latina. In: B. Muller-Haye and J. Gelman (Ed.). Recursos Geneticos Animales en America Latina. pp 77-107. FAO Study: Animal Production and Health, No. 22. Rome, Italy.

Plasse, D. 1983. Crossbreeding results from beef cattle in the Latin American tropics. Animal Breeding Abstracts. 51, 11:779.

9

DESIGNING PRODUCTION SYSTEMS TO OPTIMIZE NET MERIT AND PROFIT

H. A. Fitzhugh

Designing production systems to optimize net merit and profit -- this has to be the number one priority for every livestock producer! Since most livestock producers are experiencing real difficulty making a profit, does this mean that this design process is beyond the capabilities of the average producer? No! The truth is that major requirements for a successful design are experience, common sense, and a willingness to analyze before acting.

SYSTEMS APPROACH

Livestock producers intent on making a profit need to follow the same approach followed by other successful businessmen and decisionmakers. This "systems approach" is often associated with high technology, complicated models, and high speed computers. But it is more than tools and techniques; the systems approach is a strategy for the planning and management of complex systems, for identifying constraints, and for designing interventions that favorably change the system's productivity and efficiency.

Use of the systems approach is nothing new to livestock producers. With their basic livelihood at stake, they generally do their best to consider all the facts at their disposal before making decisions. Often, however, factors outside the production system -- such as climatic changes and government policies -- and beyond the control of the livestock producer determine profitability. In fact, often it is the policymaker, not the producer, who has failed to use the systems approach to the detriment of productivity and profitability.

The essence of the systems approach is included in four steps:

1. Characterization of the system resource inputs, product outputs, operational processes, and interactions among them
2. Analysis to identify constraints and practical options for their resolution

3. Design and evaluation of interventions to existing system
4. Implementation of proven, cost-effective interventions

Before considering how the systems approach applies to genetic improvement of livestock, it is instructive to consider the hierarchy of levels that make up agricultural systems. To some degree, producers have an opportunity to apply the systems approach at all these levels.

Global Level

Profitability of livestock production may be affected -- directly or indirectly -- by world-level economic and political factors. Countries such as New Zealand and Australia depend on exports of meat, milk, and wool to generate profits; others such as Japan depend on imports to feed their people. Producers can rarely have much impact at the global level; however, this does not lessen their need to anticipate and prepare for major shifts in supply and demand for their products.

National Level

At the national level, policies on prices, trade, and credit are formulated and political and economic interventions to livestock production are commonly made. Both individual countries and politically/economically linked groups of nations, such as the European Common Market, are included.

Historically, livestock producers have made little input at the national level. However, producer associations, such as the National Cattlemen's Association, increasingly attempt to influence the policymaking process to the benefit of their members.

Both producers and consumers benefit from publicly supported research and extension activities to improve production efficiency. Unfortunately, public funding for agricultural research has lagged far behind funding for other areas such as human health and defense. Producers have a direct vested interest in promoting increased support for research and extension at both the local and national levels.

Farm Level

The farm is the operational level at which producers make management decisions with respect to types of activities (crops vs livestock, ruminants vs nonruminants, cattle vs goats, beef vs milk), type and amount of investment in improvements in different activities, and marketing of products. Often the farm enterprise includes several different production systems incorporating different crops and livestock production activities.

Herd Level

The herd is the basic production unit consisting of three components: the breeding male, the breeding female, and the nonbreeding animals. The herd is the smallest unit for economic and biological analysis of livestock production systems because while these three basic components may require different inputs and yield different outputs, it is their combined requirements and outputs that determine productivity and efficiency. Also, interactions between these components (complementarity) affect systems efficiency. An example of complementarity is the mating of large sires to small cows to increase beef offtake while reducing costs of cow maintenance.

Subsystems below the herd level include the individual (traditionally known as the basic unit of selection), the organ, and the cell. While interventions can be made at these lower levels, measurement of the net biological or economic effects of these interventions on production efficiency will be made at the herd level or at higher levels.

CLASSIFICATION OF LIVESTOCK PRODUCTION SYSTEMS

Agricultural systems involving livestock were classified according to predominant agricultural activities (Winrock International, 1982). Three basic types of systems were identified:

1. Animal based. The animal component is the major, often only, source of production (food, fiber, etc.) in the farm system; ruminants predominate because the major source of nutrients is from grazing range or permanent pasture lands.
2. Mixed crop and animal. The animal component is an important, even essential, component of a balanced production system; the relative importance of crop and animal components varies widely among mixed systems in different regions.
3. Crop based. The animal component plays a minor, complementary, but not essential role relative to cropping component; examples include weed control and utilization of crop processing by-products by animals.

These classifications are not mutually exclusive by any means; examples of overlapping between mixed systems with animal-based or crop-based systems are common around the world. Complex stratified systems may combine animal-based and crop-based production systems at different stages, e.g., production of calves from beef herds on western range and the finishing of calves on grain produced by farmers who keep few, if any, livestock.

The first stage in the systems approach to improving any of these production systems is characterization, especially the identification of constraints.

CONSTRAINTS TO LIVESTOCK PRODUCTIVITY

Any of the components of livestock production systems -- resource inputs, production processes, and product outputs -- can be a constraint to system productivity. Alleviation of constraints is the implicit goal of any strategy to improve productivity.

The three general categories of constraints include:

1. Ecological: land, climate
2. Biological: livestock nutrition -- water, feed; livestock health -- disease, parasites, and predators; livestock genotype -- production and adaptation traits
3. Socioeconomic: labor availability and management skills; consumer taste/preference and disposable income; credit availability and cost; marketing infrastructure; and policies -- trade, prices, and land tenure

Generally little can be done to change ecological constraints. However, biological and socioeconomic constraints can often be resolved through a systems approach.

These constraints are listed as if they were discrete factors, each affecting livestock production independently. In fact, interactions among constraints are the rule, not the exception, with their effects often multiplicative rather than additive. One constraint may mask the effects of others. Thus, it is necessary to consider the total system so that multiple interacting constraints can be systematically resolved in order to achieve substantial improvement (Fitzhugh et al., 1978).

SYSTEMS APPROACH TO GENETIC IMPROVEMENT

Strategies for genetic improvement will be used to illustrate design of interventions to resolve constraints and improve productivity. Nature favors those genotypes that survive and thrive in a natural environment. Man, however, generally imposes a different definition of merit involving quantity and quality of meat, milk, and fiber. Often these performance preferences reduce fitness. The production environment must be modified to support livestock no longer able to fend for themselves. Thus, breeding and management strategies must be complementary.

Net Merit

Many traits influence livestock productivity (table 1). A successful breeding program pays attention to all these traits. Consideration must be given to the costs of inputs as well as to the value of outputs. High production levels must be balanced against adaptation to environmental stresses. Thus, the appropriate goal for genetic improvement is Net Merit.

TABLE 1. IMPORTANT TRAITS FOR LIVESTOCK PRODUCTION

Category	Traits
Fitness	Adaptations to environmental stress -- coat type, resistance to disease and parasites, neonatal survival, longevity, temperament Adaptability to environmental change
Fertility	Fecundity -- ovulation rate, fertilization rate, embryo survival Parturition interval -- postpartum interval to conception (postpartum anestrus, conception rate), gestation period Weaning rate -- maternal behavior, milk production, vigor of young Age at sexual maturity Male traits -- libido, semen quality
Size and efficiency	Growth and maturing rates Body weight Birth weight -- neonatal survival Slaughter weight -- meat yield Mature weight -- maintenance requirements Body composition -- edible disease Voluntary feed intake Composition of diet selected Efficiency of nutrient utilization for maintenance and production
Lactation	Days of lactation Amount and persistency of daily yield Composition of milk
Fiber	Weight and yield of fleece Fineness and uniformity of fiber diameter Strength of fiber

Theoretically, Net Merit includes all traits that affect livestock production. Each trait is weighted according to its relative genetic and economic value. Practically, many traits are too difficult or too expensive to measure. A systems approach implies that the definition of Net Merit is both practical and cost effective.

Genetic Decisionmakers

There are two general types of decisionmakers involved in genetic improvement, but both operate by determining which males mate which females.

Commercial producers. Here the mating decisions focus on improving profitability of meat, milk, and fiber production. Profitability may be increased by increasing quantity and quality of product but decreased by raising production costs. Often there are trade-offs, i.e., maximizing product or minimizing costs is not necessarily the most profitable strategy.

Commercial producers generally operate on a short-term horizon -- a few years from mating to slaughter in the case of cattle, a few months in case of swine, sheep, and goats. Thus, commercial producers must have flexible breeding programs that can be adjusted in response to short-term changes in supply and demand.

Seed stock producers. Their mating decisions should be directed to producing sires and dams that meet the requirements of commercial producers. However, the principal contribution of seed stock will not be meat and milk per se but the performance of their progeny.

A major opportunity for seed stock producers is to develop and maintain genetic variations that fit the needs of different production environments and different breeding programs (table 2). Whereas commercial producers often need to produce a uniform product to meet market needs, seed stock producers should produce the genetic variants that will allow commercial producers to mix, match, cross and combine genotypes.

Optimal Performance Levels

Both experience and common sense confirm that extremes in performance do not fit the needs of profitable livestock production. Over the past few decades productivity has suffered dramatically from emphasis placed on extreme levels of performance. Examples include:

- Beef cattle that were too big and slow maturing early this century became too small and early maturing by midcentury
- Swine that were too short and fat in the 1930s became too long and lean by the 1970s

The unfortunate, but all too human, tendency to follow trends to the extreme has led to these pendulum swings between undesirable extremes in performance levels.

TABLE 2. RELATIVE EMPHASIS TO BE PLACED ON BEEF CATTLE SEED STOCK FOR DIFFERENT COMMERCIAL PRODUCTION SYSTEMS[a]

Trait	Straightbred, synthetic, or rotational cross	Maternal line	Sire line
Fertility			
Male	+++	+	+++
Female	+++	+++	+
Fitness	+++	+++	+++
Calving ease	+	+++	+
Mature size	0	-	-
Growth rate	+++	0	+++
Milk yield	+	++	0
Lean yield	+	0	++

[a]Symbols indicate direction (+, increase; -, decrease; 0, no change or change not important); number of symbols indicates relative importance of change.

Intermediate levels are optimal for many traits. Often producers know this intuitively. Trends often start as an attempt to move away from an extreme level toward an intermediate one. Unfortunately, overreaction to an extreme with obvious undesirable qualities makes it difficult for producers to recognize when the desired intermediate level has been obtained. The consequence is that the intermediate is passed by enroute to the opposite extreme.

PARTNERS IN IMPROVING PRODUCTION SYSTEMS

The systems approach to genetic improvement calls for analysis of needs, for fitting genotypes to production environment, and for anticipating future market requirements. This analysis leads to design of interventions that will **probably** be profitable. However, the systems approach is a conservative strategy. Interventions should be carefully tested first before widespread application.

How should this testing be done and who should do it? This is an obvious role for publicly supported research agencies. Unfortunately, public funding for animal agriculture is low, less than .5% of the marketed value of animal products. If the research establishment is to protect producers from the risk of unproven interventions, producer (and consumer) support for research is needed.

As an example of how publicly supported research deals directly with production problems, consider the computer simulation studies that have documented the value of intermediate performance levels for body size and milk yield to beef production systems (Cartwright, 1979). In this case, evaluation of potential productivity of a wide range of genotypes depends on accumulation and analysis of extensive data using computers and mathematics beyond the resources of producers. However, the ultimate success of the research activity will be decided if the results are accepted and applied by producers. Thus, it behooves both scientists and producers to work together.

The consumer is the other essential partner in the systems approach to profitable livestock production. Consumers are increasingly concerned about animal products in their diet. In the more affluent societies, consumers worry that too much animal product in the diet may be detrimental to their health; for the majority of the world's poor, the consumer concern is just the opposite.

Producers following a systems approach will consider profitability for both the short- and long-term. Thus, the growing debate on the relative diet-health advantages/disadvantages of animal products for the consumer must be followed closely and breeding and management strategies made accordingly.

CONCLUSIONS

Livestock are an important -- often essential -- component of balanced agricultural systems. However, they compete with other components for a share of resource inputs and a profitable share of the market. Now, and increasingly in the future, Net Merit of livestock production must be evaluated in the context of the overall agricultural food and fiber system. Given the complexity of the problem, a systems approach is essential to the design of an effective management and breeding program that will lead the producer towards profitable livestock production. This systems approach will take into account the available resource base, consider the feasibility of any interventions, and measure the success of interventions involving the livestock component in terms of sustainable bioeconomic efficiency of the overall system. Emphasis will likely be on intermediate instead of maximal performance levels.

REFERENCES

Cartwright, T. C. 1979. The use of systems analysis in animal science with emphasis on animal breeding. J. Anim. Sci. 49:817.

Fitzhugh, H. A., H. J. Hodgson, O. J. Scoville, T. D. Nguyen, and T. C. Byerly. 1978. The Role of Ruminants in Support of Man. Winrock International, Morrilton, AR.

Winrock International. 1982. Livestock program priorities and strategies. Prepared for U.S. Agency for International Development. Winrock International, Morrilton, AR.

Part 2

REPRODUCTION OF CATTLE

10

REPRODUCTIVE TECHNOLOGIES FOR THE FUTURE

Roy L. Ax

INTRODUCTION

Though embryo transfer (ET) methods were used for years on an experimental basis, only in the last decade has ET become commercially widespread. It is estimated that 100,000 to 200,000 calves resulting from ET will be born annually from 1985 to 1990. The major boons to increasing ETs in the livestock industry were the introduction of prostaglandins and Synchro Mate-β for estrous synchronization and development of nonsurgical transfer methods. All of the newer biotechnology methods now depend upon ET for placing embryos into surrogate dams.

EMBRYO TRANSFER

The most important consideration for ET is to synchronize the estrous cycles of the donor animal and the recipients. Being able to store frozen embryos alleviates the need for synchronization because the embryo can be thawed and placed into the recipient at the correct stage of the cycle, 6 to 8 days after estrus.

Synchronization of estrus is accomplished by using prostaglandins to promote regression of a corpus luteum or employing Synchro Mate-β to mimic the presence of a corpus luteum. Either type of product is effective, and there are advantages and disadvantages with both.

Embryo transfer is performed as a single embryo or multiple-embryo procedure. For the latter, superovulation is required to induce multiple follicles to develop on the ovary, and this can be accomplished with follicle-stimulating hormone (FSH) injections on successive days in the middle of the cycle. Regression of the corpus luteum is induced by prostaglandins, the multiple follicles ovulate at estrus, and the eggs are subsequently fertilized. Herein lies the greatest puzzle in ET -- there is no accurate way to predict how many follicles will develop in response to FSH administration and, furthermore, the fertilization rates of ovulated eggs can vary tremendously. These phenomena pose problems in terms of deciding how many recipients will be required to receive the eggs at transfer.

In almost all cases embryos are now harvested by non-surgically flushing them out of the reproductive tract. They can be transferred surgically through a flank incision and deposited into the uterus, or they can be deposited in the uterus with an insemination catheter, which visually resembles artificial insemination.

The current practice of ET does not contribute significantly to genetic gain. To be effective, we must be able to progeny-test females using semen from many males for siring many offspring that can be evaluated statistically in performance testing programs. However, with the long generation interval in farm animals and low number of gametes ordinarily produced by a female compared to a male, by the time progeny-test data were compiled, a superior female would have died or been too old to produce many more offspring. The application of in vitro fertilization (IVF) will permit us to harvest many eggs in a short period of time so that contemporary comparisons can be made on offspring and enable females to contribute greatly to genetic improvement. IVF will be discussed in a later section.

Disease prevention is a major advantage of ET that needs to be aggressively promoted to expand international markets for livestock. Several studies have shown that when embryos harvested from infected reproductive tracts or directly exposed to infectious agents in culture were transferred to recipients, neither the offspring nor surrogate dams were serologically positive for the diseases. Another consideration is that offspring born to a surrogate dam will obtain immunity to local diseases in the colostral milk and adapt easier to a new environment than a mature animal that was shipped to a foreign country.

FREEZING OF EMBRYOS

We are all familiar with freezing of semen; however, freezing embryos poses a different problem because they are composed of masses of cells rather than being a single cell like a sperm. When freezing and thawing embryos, external cells are exposed to the environment before internal cells. This can cause cell-to-cell junctions to weaken and lead to disaggregation of the cell mass. The end result can be embryonic death. Extremely sophisticated thermal devices have been engineered that show promise for better control of the temperature for embryo survival. (The advent of microwave technology offers a novel approach to thaw the embryo from the inside out!) Preservation of whole organs such as kidneys and skin has provided researchers with some hints that have already led to significant improvements for frozen storage of embryos. Today, if embryos are frozen, thawed, and transferred to recipients, the success rate is 50% of that of fresh embryos.

SEXING OF EMBRYOS

Historically the sex of an embryo has been determined by doing a biopsy of the embryo and examining the chromosomal karyotype to determine whether the X or Y sex chromosome is present. A new approach will be to screen embryos for presence of the H-Y antigen on the embryonic surface. This antigen is expressed in males, but it is absent in females. Therefore, if antisera to the H-Y antigen is added to embryos, it binds to the cell surface. Addition of a fluorescent-labeled second antisera made against the primary antisera causes the male embryos to fluoresce, and they can be visualized under a fluorescence microscope and segregated from the females. Alternatively, individual embryos in test tubes could be evaluated for relative fluorescence in a fluorometer. This methodology will permit rapid screening for embryonic sex and avoids the hazards of microsurgical biopsies.

The ultimate use for sexing will be to determine the sex of an embryo followed by cloning to produce multiple copies of the embryo. Some have suggested that sex determination could be followed by transfers of twin bulls or twin heifers to a single recipient to reduce the likelihood of freemartin calves and to cut down the time and expense of maintaining recipients. However, it should be remembered that experiments to increase the percentage of twinning by transferring two embryos into a recipient produced dismal results. There is only a finite amount of caruncles for cotyledonary attachment, and twins compete for that space.

Sexing of sperm with a high degree of accuracy would enable us to make faster genetic progress compared to sexing individual embryos because we could skew the sex ratio of the entire population and end up with more female herd replacements. The embryo transfer industry will truly benefit when sexing of sperm becomes routine because producers will have the option of choosing the sex of the offspring without need for manipulating the embryos. A highly reliable method for sexing of sperm is now available, but the procedure kills sperm. In the future when methods are worked out that suggest that separation of viable sperm by sex is possible, we will be able to definitely confirm whether or not such methods are accurate.

IN VITRO FERTILIZATION

Sources of Oocytes

There are three considerations for sources of oocytes: 1) from the oviduct after ovulation, 2) large preovulatory follicles, or 3) small preovulatory follicles. To obtain viable oocytes, a surgical instrument called a laparoscope is used to visualize the oviduct or ovarian follicles. Options 1 and 2 above require superovulation treatments to yield enough oocytes for harvesting, and successes from superovulation are uncertain. Option 1 results in oocytes in the proper meiotic configuration for fertilization, but surgical trauma to the oviducts can result. With option 2, oocytes may or may not be in the correct meiotic configuration depending upon whether the follicles have been exposed to the ovulatory surge of gonadotropin. Qualitative assessment can be provided by examining the oocytes for expanded cumulus cells. If option 3 is used, the oocytes must be matured in vitro prior to in vitro fertilization.

Recovery rates are 50% to 70% for the number of oocytes harvested in relation to the number of follicles aspirated. Laparoscopies have been performed on animals repeatedly with no apparent harm. Adhesions can form, but they do not block access to oviducts or ovaries.

What is the advantage of maturing oocytes in vitro rather than setting up a superovulation program? The main advantage is that there is still a tremendous amount of uncertainty with superovulation procedures -- it is hard to predict how many follicles will develop in a particular animal injected with the hormones. By maturing the oocytes in vitro and examining them under a light microscope, only those oocytes that appear normal are used for subsequent in vitro fertilization. Another advantage is that about 20 follicles can be found on an ovary at any time -- even during pregnancy. Therefore, a female could lead a normal reproductive life and provide a continuous supply of oocytes for IVF.

Sources of Sperm

There are three potential sources of sperm: 1) epididymis, 2) fresh ejaculate, and 3) frozen extended semen. Epididymal collections provide concentrated samples of sperm that have not been exposed to seminal plasma and the decapacitating effects of seminal components. Fresh ejaculates offer concentrated specimens that are in seminal plasma, so sperm need to be removed from seminal plasma. Frozen extended samples are diluted and contain the cryoprotectants in the extender.

Sperm must undergo two processes prior to being able to fertilize an oocyte -- capacitation and the acrosome reaction. Sperm must reside in female reproductive tract secretions approximately 8 hr to become capacitated, which involves dilution from seminal plasma. Once capacitation has occurred, the acrosome reaction occurs within .5 to 1.0 hr in the presence of calcium. The acrosome reaction is a morphological change in the sperm head and is accompanied by activation of proteolytic enzymes to aid in digestion of vestments surrounding the ovum.

Our understanding of induction of capacitation and acrosome reactions in vitro is becoming clearer. High molecular weight polysaccharides termed glycosaminoglycans are extremely effective at promoting a significant occurrence of in vitro acrosome reactions using bull epididymal or washed ejaculated samples exposed for 22 hr or 9 hr, respectively. Exposure of epididymal sperm to seminal plasma for 20 min followed by washing and addition of glycosaminoglycans shortens the time for acrosome reactions to 9 hr, thus avoiding long-term cultures of sperm and the risk of loss of viability.

Glycosaminoglycans are found in secretions of the female bovine reproductive tract, follicular fluid, the intercellular matrix surrounding cumulus cells that exhibit expansion, and in the zona pellucida. Thus, it appears that Mother Nature has built in an "overkill" to guarantee that sperm are exposed to materials to prepare them for fertilization.

CONSIDERATIONS FOR IN VITRO FERTILIZATION

Cumulus expansion, acrosome reactions, and IVF are all temperature-dependent processes with 39°C optimal in cattle. Precise timing of these events and their interactions in terms of maximum success are current topics of research. That is, if oocytes are harvested from the oviducts or preovulatory follicles after the gonadotropin surge, they will not require time for cumulus expansion and nuclear maturation prior to the addition of sperm. Convenience dictates that oocytes harvested for in vitro maturation are cultured for 24 hr prior to addition of sperm.

An advantage of IVF as compared to fertilization in vivo is that fewer sperm are required per oocyte. As an example, consider a bull whose single ejaculate is extended to yield 500 services for AI. Since 20 to 30 million sperm are packaged per straw for AI, but only 200 to 300 thousand sperm are needed in a culture droplet containing 5 oocytes for IVF, one straw would provide enough sperm for 500 oocytes used for IVF. Or, that original ejaculate would contain enough sperm to fertilize 250,000 oocytes by IVF!

In embryo transfer we are trying to propogate offspring from elite cows bred to outstanding bulls. Imagine how IVF could extend the potential reproductive life of a bull if some of his semen were extended in lower concentrations and saved specifically for IVF. This might be a useful way to set aside some inventory from a bull as "insurance" in case he dies prematurely.

FUTURE APPLICATIONS OF IN VITRO FERTILIZATION

Application of IVF in the embryo transfer industry would have to be classified as prenatal but nearing parturition. It will have a short infancy with a sustained period of growth. Just as embryo transfer technology now finds itself as an invaluable tool for future applications in animal breeding, IVF will someday be in high demand so that clonal lines can be produced by nucleus transplants. In addition, for gene transfers to be feasible, genes would have to be inserted by microsurgery into a pronucleus prior to syngamy.

The embryo transfer industry and IVF lend themselves ideally to a symbiotic relationship. Veterinarians would harvest the oocytes and perform transfers of the embryos. Semen would be purchased from AI associations so that sperm from individual males could be used to fertilize individual oocytes. As IVF becomes more routine, AI companies may opt to freeze sperm in lower concentrations just for use in IVF. Oocyte cultures for in vitro maturation are straightforward and will not require elaborate facilities or equipment. Sexing of embryos by immunofluorescence necessitates a need for a fluorescence microscope or a fluorometer where individual embryos in test tube would be evaluated for relative fluorescence. The main obstacle that needs to be overcome before IVF can receive widespread commercial acceptance is a reliable means of culturing embryos from fertilization through the morula-of-blastocyst stage of growth.

EMBRYO CULTURE -- THE CURRENT LIMITATION FOR IN VITRO FERTILIZATION

When IVF is performed in humans, the early embryo is surgically placed back into the woman. In cattle, surgical deposition of early embryos into recipient oviducts would make the cost prohibitive and reduce chances for success. Research is desperately needed to develop a reproducible method for culturing embryos 6 to 8 days so that the embryos could be transferred as we do now. Thus, in the future, embryo transfer labs that use IVF will have to routinely culture embryos.

CLONING

Several laboratories have reported success at splitting embryos -- resulting in identical twins. The potential of this technology is that instead of 60 to 65 pregnancies out of 100 embryos, we could get 120 to 130 pregnancies from 200 half-embryos. Thus, the overall pregnancy rate from the original 100 embryos is potentially 120% to 130%. If we freeze the 200 half-embryos, our success would equal that of 100 fresh embryos. However, this procedure does not help us out in terms of identifying genetic superiority.

The method that holds greatest promise for cloning is that referred to as nuclear transfer. This procedure involves removing the 60 cells found in the blastocyst stage of embryonic growth, disaggregating the cells, and transferring the nucleus of each cell into a freshly fertilized egg from which the nuclear material was removed. In short order, 60 copies could be made, and each new blastocyst could be treated in the same manner to recycle the system ad infinitum. Any of the embryos could be frozen and saved for the future, until nuclear transfers could be performed again on a thawed embryo. This method would obviously permit us to obtain performance records on several copies and then make a decision to clone when a certain performance record is achieved.

Nuclear transfer could also be applied to insert the nucleus from one egg with the nucleus of another egg (gynogenesis) or the nucleus of one sperm with another sperm (androgenesis) into a freshly fertilized egg that has its nuclear material removed. Self-mating could also result from application of this technology so that highly inbred lines could be developed for subsequent cross-mating in order to obtain hybrid vigor.

CONCLUSION

Use of IVF in the ET industry will permit individual eggs from females to be fertilized with semen from different males. The offspring can subsequently be progeny-tested so that we can compute a more reliable genetic merit for the female as we now do for bulls in AI. This procedure will then let females make a significant genetic contribution to animal breeding programs. The main obstacle to overcome before IVF becomes commercially widespread is to develop reliable methods to culture embryos for 6 to 8 days. At that stage of development, embryos can be transferred into recipients or used for cloning, followed by transfer. A combination of these biotechnology methods offers exciting prospects for significant genetic improvement of livestock in a short period of time!

ACKNOWLEDGMENTS

The overview presented here would not have been possible without the research collaboration of Dr. David Ball, Ms. Mary Bellin, Dr. Neal First, Dr. Herb Grimek, Mr. Richard Handrow, Mr. Chin Lee, and Dr. Richard Lenz. Their research efforts have helped develop an understanding and a respect for limitations of IVF in cattle. This paper is a contribution from the University of Wisconsin College of Agricultural and Life Sciences.

REFERENCES

Ax, R. L. 1984. The potential of in vitro fertilization to the livestock industry. In: F. H. Baker and M. E. Miller (Ed.) Dairy Science Handbook, Vol. 16. pp 501-507. Westview Press, Boulder, CO.

Ax, R. L., G. D. Ball, N. L. First, and R. W. Lenz. 1982. Preparation of ova and sperm for in vitro fertilization in the bovine. Proc. 9th Tech. Conf. on Artif. Insem. and Reprod., Nat'l Assoc. of Anim. Breeders. pp 40-44.

Ax, R. L., R. W. Lenz, G. D. Ball, and N. L. First. 1983. Embryo manipulations, test-tube fertilization and gene transfer - Looking into the crystal ball. In: F. H. Baker (Ed.) Dairy Science Handbook, Vol. 15. pp 191-197. Westview Press, Boulder, CO.

Ball, G. D., R. L. Ax, and N. L. First. 1980. Mucopolysaccharide synthesis accompanies expansion of bovine cumulus-oocyte complexes in vitro. In: V. B. Mahesh, T. G. Muldoon, B. B. Saxena, and W. A. Sadler (Ed.) Functional Correlates of Hormone Receptors in Reproduction. pp 561-563. Elsevier-North Holland, NY.

Ball, G. D., M. E. Bellin, R. L. Ax, and N. L. First. 1982. Glycosaminoglycans in bovine cumulus-oocyte complexes: Morphology and chemistry. Molec. Cellul. Endocr. 28:113.

Ball, G. D., M. L. Leibfried, R. L. Ax, and N. L. First. 1984. Maturation and fertilization of bovine oocytes in vitro: A review. J. Dairy Sci. (In press).

Ball, G. D., M. L. Leibfried, R. W. Lenz, R. L. Ax, B. D. Bovister, and N. L. First. 1983. Factors affecting successful in vitro fertilization of matured bovine follicular oocytes. Biol. Reprod. 28:717.

Brackett, B. G., D. Bousquet, M. L. Boice, W. J. Donawick, J. F. Evans, and M. A. Dressel. 1982. Normal development following in vitro fertilization in the cow. Biol. Reprod. 27:147.

Brackett, B. G., Y. K. Oh, J. F. Evans, and W. J. Donawick. 1980. Fertilization and early development of cow ova. Biol. Reprod. 23:189.

Foote, R. H. 1984. New developments in embryo transfer and related technology. Holstein Science Report, Holstein-Freisian Assoc. of America, Brattleboro, VT.

Fulka, J., Jr., A. Pavlok, and J. Fulka. In vitro fertilization of zona-free bovine oocytes matured in culture. J. Reprod. Fert. 64:495.

Garner, D. L., B. L. Gledhill, D. Pinkel, S. Lake, D. Stephenson, M. A. Van Dilla, and L. A. Johnson. 1983. Quantitation of the X- and Y-chromosome-bearing spermatozoa of domestic animals by flow cytometry. Biol. Reprod. 28:312.

Handrow, R. R., R. W. Lenz, and R. L. Ax. 1982. Structural comparisons among glycosaminoglycans to promote an acrosome reaction in bovine spermatozoa. Biochem. Biophys. Res. Comm. 107-1326.

Iritani, A. and K. Niwa. 1977. Capacitation of bull spermatozoa and fertilization in vitro of cattle follicular oocytes matured in culture. J. Reprod. Fert. 50:119.

Lambert, R. D., C. Bernard, J. E. Rioux, R. Beland, D. D. Amours, and A. Montrevil. 1983. Endoscopy in cattle by the paralumbar route: Technique for ovarian examination and follicular aspiration. Theriogenology 20:149.

Leibfried, M. L. and N. L. First. 1979. Characterization of bovine follicular oocytes and their ability to mature in vitro. J. Anim. Sci. 48:76.

Lenz, R. W., R. L. Ax, H. J. Grimek, and N. L. First. 1982. Proteoglycan from bovine follicular fluid enhances an acrosome reaction in bovine spermatozoa. Biochem. Biophys. Res. Comm. 106:1092.

Lenz, R. W., G. D. Ball, M. L. Leibfried, R. L. Ax, and N. L. First. 1983. In vitro maturation and fertilization of bovine oocytes are temperature dependent processes. Biol. Reprod. 29:173.

Lenz, R. W., G. D. Ball, J. K. Lohse, N. L. First, and R. L. Ax. 1983. Chondroitin sulfate facilitates an acrosome reaction in bovine spermatozoa as evidenced by light microscopy, electron microscopy and in vitro fertilization. Biol. Reprod. 28:683.

Maxwell, D. P. and D. C. Kraemer. 1980. Laparoscopy in cattle. In: R. M. Harrison and D. E. Wildt (Ed.) Animal Laparoscopy. pp 133-156. Williams and Wilkins Co., Baltimore, MD.

Newcomb, R., W. B. Christie, and L. E. A. Rowson. 1978. Birth of calves after in vitro fertilization of oocytes removed from follicles and matured in vitro. Vet. Res. 102:461.

Shea, B. F., J. P. A. Latour, K. N. Bedireau, and R. D. Baker. 1976. Maturation in vitro and subsequent penetrability of bovine follicular oocytes. J. Anim. Sci. 43:809.

Singh, E. L., M. D. Eaglesome, F. C. Thomas, G. Papp-Vid, and W. C. D. Hare. 1982. Embryo transfer as a means of controlling the transmission of viral infections. I. The in vitro exposure of preimplantation bovine embryos to akabane, bluetongue and bovine viral diarrhea viruses. Theriogenology 17:437.

Singh, E. L., F. C. Thomas, G. Papp-Vid, M. D. Eaglesome, and W. C. D. Hare. 1982. Embryo transfer as a means of controlling the transmission of viral infections. II. The in vitro exposure of preimplantation bovine embryos to infectious bovine rhinotracheitis virus. Theriogenology 18:133.

Stringfellow, D. A., V. L. Howell, and P. R. Schnurrenberger. 1982. Investigations into the potential for embryo transfer from *Brucella abortus* infected cows without transmission of infection. Theriogenology 18:733.

Trounson, A. O., S. M. Willadsen, and L. E. A. Rowson. 1977. Fertilization and developmental capacities of bovine follicular oocytes matured in vitro and in vivo and transferred to the oviducts of rabbits and cows. J. Reprod. Fert. 51:321.

Wachtel, S. S., S. Ohro, G. C. Koo, and E. A. Boyse. 1975. Possible role for H-Y antigen in the primary determination of sex. Nature 257.

11

EMBRYO TRANSFER: PROBLEMS AND (OR) OPPORTUNITIES FOR THE PUREBRED INDUSTRY

Craig Ludwig

Over the past 20 yr, the role of embryo transfer (ET) has grown from that of being a research tool to becoming an economically important component of the purebred livestock industry. Nonsurgical embryo collection and transfer techniques developed in the 1970s opened the door for wide use of ET procedures within the registered segment of the beef cattle business. Whereas there were only 20 ET pregnancies in 1972, it is estimated that over 50,000 beef and dairy calves were born through ET in 1984.

Prior to ET, a cow was limited to about 8 to 10 calves over its productive lifetime. With so few calves, a cow's genetic influence on a herd or breed was almost meaningless. Herd genetic improvement was accomplished through the selection of sires over several generations. Now, however, ET procedures offer an exciting avenue for potential genetic improvement through female selection.

With the introduction of nonsurgical ET procedures, breeders of registered herds can envision groups of calves from one female and one bull that would look and perform exactly alike because they were full brothers and sisters. Potentially, through ET procedures, one donor dam could produce from 10 to 30 calves per year. This prospect launched a breeder search for genetically superior females within and across herds.

SELECTION OF DONOR DAMS

Breeders searching for genetically superior females soon discovered that the term "genetically superior" had a different meaning for each breeder. Originally, most breeders assumed that donor females for ET would be selected from producing cows on the basis of an analysis of their performance and pedigree information, such as growth breeding values for birth weaning, yearling weights, maternal breeding value, calving interval, structural soundness, and the acceptability or demand for any natural progeny the dam had produced. Breeder priority for any one of the traits varied; some put special emphasis on growth rate values while others selected more for maternal traits and factors indicating fertility. Regardless of the performance trait, pedigree acceptability of the donor dam has remained an important part of merchandising an ET program.

Selection of older cows with natural progeny and proven genetic superiority was the general rule in early stages of embryo transfer. Breeders who wanted to use ET to produce one great animal or a show-ring winner soon caused attention to focus on the donor's frame size and on the show-ring winnings of the donor's relatives. As these breeders searched for increased size and as the superior older cows fell victim to the generation interval, breeders began to select more and more young, virgin heifers as female-donor candidates. Many of these heifers had a minimum of performance data to indicate their genetic superiority or inferiority.

If "genetically superior" can be defined precisely as an index for all breeders, the success rate would improve for selecting the appropriate donor females for ET. However, everything from proven older cows to virgin heifers has been tried in transplant programs. We now know that some females should have simply remained in the herd and been allowed to produce calves under natural conditions.

SELECTION OF RECIPIENT DAMS

Recipient dam selection has become about as important to the ET breeder as is the selection of the donor dams. Soon after on-the-farm nonsurgical ET became a reality, marked differences were observed between full brothers and sisters. Breeders were quick to recognize that the different phenotypic makeup of ET calves was dependent on the kind of surrogate dam that raised the calf. More often than not, ET calves from large frame, heavy-milking recipient dams have greater birth, weaning, and yearling weights than litter mates raised from smaller, lower-milking recipient dams.

Again, to produce the one great animal or one show-ring winner, breeders selected recipient dams with a large-frame score potential and above-average milk production.

Maternal influences on calf performance and phenotypic makeup are not well understood and this lack of knowledge has prompted some associations to consider limiting recipient dams to the breed being transferred. Other associations have coded ET calves according to the breed makeup of the calf's recipient dam. Current research with split embryos to measure recipient dam effects on calf performance should help the industry to answer questions concerning maternal influence on calf performance.

CALF RECORDS FOR EMBRYO TRANSFER CALVES

Since use of ET began, breed associations have had problems dealing with the performance records of ET calves. Breeders investing money in an ET program wanted performance records for their own information, as well as for merchandising the ET calves. Almost without question, donor dams and service sires have been selected from animals above the breed average. Predicted breeding values for ET calves from such matings are usually above 100.

The real questions, genetically speaking, are 1) how greatly the ET calf's performance exceeds the breed average, and 2) are any records of performance on the calf valid measures of how it will perform as a parent.

Breed associations usually record actual birth weights, actual weaning and adjusted 205-day weights, actual yearling and adjusted 365-day weights for all ET calves. All breed associations, except one, either use no ratio or calculate a ratio of 100 on all individual-performance information for ET calves. Maternal and growth breeding values are calculated for ET calves when they are from donor dams and service sires that have past production and progeny records.

The individual performances of ET calves are not given ratios because of the wide variation on their records from one flush.

The data on one Hereford cow illustrate the performance variation in ET calves. The predicted breeding values for the donor dam after she produced seven natural calves were: maternal breeding value, 109; weaning growth breeding value, 114; and yearling growth breeding value, 110. Table 1 summarizes the calf performance data calculated over the first eight flushes of this donor cow.

Flushes 9 through 15 either produced single eggs or calves that were too young for recording of performance data. The first 8 flushes, however, demonstrate the individual variation in performance within each flush. Yearling weights within each flush for calves of the same sex varied by as much as 100 lb or more. The performance variation between male and female calves within each flush are quite distinct.

Within each flush, each calf had exactly the same predicted maternal breeding value, weaning growth breeding value, and yearling growth breeding value. Theoretically, when predicted breeding values are exactly the same, bull calves from the same flush should have equal values as sires, based on their progeny's performance.

In the case of bull calves 8511 and 8508 from flush number 3, their true genetic merit as parents can be measured through sire evaluation data analysis. Table 2 shows expected progeny differences (EPD) for birth, weaning, and yearling weights of these two bulls, along with their maternal breeding values.

TABLE 1. CALF PERFORMANCE SUMMARY FROM FIRST EIGHT FLUSHES OF A DONOR COW

Female ETs			Male ETs		
I.D.	Weaning wt, lb	Yearling wt, lb	I.D.	Weaning wt, lb	Yearling wt, lb
1st Flush (1978)					
8112	636	864	8115	714	1002
8113	626	837	8117	758	1151
8114	597	691			
8116	454	678			
2nd Flush (1978)					
9151	624	736	9160	576	1097
9153	634	926	9162	524	1095
			9164	568	1001
			9166	566	1058
3rd Flush (1978)					
(None)			8511	520	1028
			8509	495	-
			8508	493	983
4th Flush (1978)					
1 - Calf no data					
5th Flush (1979)					
07	512	768	04	658	1138
			06	795	1231
6th Flush (1979)					
(None)			704	717	-
			706	544	-
7th Flush (1980)					
126	507	718	122	447	1054
131	383	619	130	753	1185
8th Flush (1980)					
2	500	-	4	560	833
3	546	-	5	591	1085
			6	665	1180

TABLE 2. EXPECTED PROGENY DIFFERENCES OF TWO BULLS FROM SAME FLUSH BASED UPON PROGENY EVALUATION

Bull	Birth wt, lb		Weaning wt, lb		Yearling wt, lb		Maternal breeding value	
	EPD	ACC	EPD	ACC	EPD	ACC	%	ACC
8511	+3.9	.98	+31	.95	+68	.90	106.5	.82
8508	+1.3	.65	+15	.70	+33	.63	103.9	.70

For bulls 8511 and 8508, individual weaning and yearling weights did not differ significantly and their predicted breeding values from their own records for the four production traits were exactly the same. However, the progeny evaluation of the two bulls' true genetic merit (table 2) showed that they were quite different, although both were above average for the breed.

Table 2 reflects but one example of the variation that exists among ET calves in individual performance and performance as parents. In some cases, individual performances may vary greatly, but sire evaluation performance analysis may indicate that the individuals are almost exactly alike. Depending on your perspective, the records can be made to reflect almost any finding.

With 3183 ET calves on record, the American Hereford Association's (AHA) performance information shows differences between ET calves and natural calves. Table 3 shows the comparative performance of ET calves as compared to all natural calves processed through the AHA's total performance record program.

TABLE 3. INCREASE IN WEIGHT AND HEIGHT OF HEREFORD EMBRYO TRANSFER CALVES AS COMPARED TO NATURAL CALVES

Birth wt, lb		Weaning wt, lb		Yearling wt, lb		Yearling ht, in.	
ET females	ET males	ET females	ET males	ET females	ET males	ET females	ET males
+7.5	+6.8	+55	+78	+40	+112	+1.5	+2.0

In the Hereford breed, embryo transfer calves weighed more at birth, at weaning, and as yearlings, and were one frame score larger as yearlings than were natural born calves. We do not know how much of the increase in each of these four traits is due to genetics and how much is due to environmental influences, but it is reasonable to assume that both the environment and genetics are contributors to the increases. Based on progeny performances, ET bulls were superior to natural born bulls in terms of weaning and yearling weights. As a result of increased maternal breeding values, they contributed to greater milk production (table 4). The increase in growth rate of the progeny of ET bulls was accompanied by a 1.1 lb increase in birth weight that could be detrimental in some breeding programs.

TABLE 4. COMPARATIVE PROGENY PERFORMANCE FOR EMBRYO TRANSFER BULLS AND NATURAL SIRED BULLS IN 1984 HEREFORD SIRE EVALUATION

Bulls	Avg. no. of calves	Birth wt EPD, lb	Weaning wt EPD, lb	Yearling wt EPD, lb	Maternal breeding value
ET born	33	+1.2	+28.9	+50.7	103.4
Natural born	28	+0.1	+18.4	+33.9	101.6
ET advantage	+5	+1.1	+10.5	+16.8	+1.8

Of all 654 ET bull calves born from 1977 through 1981, 14.4% were used as herd sires in registered herds and the remaining 85.6%, or 560 head, were either sold into commercial herds or discarded and their genetic value is not known. However, we do know that their predicted breeding value was the same as that for the 94 ET bulls serving in registered Hereford herds. It seems likely that the ET bulls sold to the commercial breeders failed the visual appraisal examination of those who raised them and that of potential registered buyers.

We do not know if the registered herds benefited economically from using the 14.4% ET-born bulls as herd sires.

EMBRYO TRANSFERS -- SAVIOR OR VILLIAN

Embryo transplants have been a strong economic force in the registered cattle business over the past 7 yr; however, the genetic potential of ET is difficult to either prove or disprove. It appears to me that donor cows and service sires continue to be selected primarily for their potential to produce that one great animal that will sell at a very high price because of a winning show-ring performance. My opinion is that embryo transfers that are being done for genetic improvement in a herd can be counted on one hand.

At present, performance records on ET calves or donor dams are almost meaningless. In ET calves from the same flush, phenotypic variations are great and appear to be closely associated with the phenotypic value and milk production potential of the recipient dam. Genetically speaking (and in breeders' minds) ET appears to be a method to speed up genetic progress for selected traits. What most of us overlook is the genetic variation possible between full brothers and sisters. For example, if the total genetic merit of an animal is controlled by 10 pairs of genes (it is at least this many or more), there are 59,049 different genotypes possible from the same mating. Therefore, the chances of repeating the same genetic combination in different calves by using ET is only slightly better than that obtained by using natural service. Problems associated with the performance records on ET calves become even more substantial when we consider that their genetic variation is further masked by a completely artificial environment and lack of contemporary grouping.

SUMMARY

Theories for use of ET are sound. Many participating breeders using ET have found it exciting and advantageous to their merchandising program, while others have suffered dismal failures and economic loss. Embryo transfer as a genetic tool for herd and breed improvement depends on selection -- as do natural breeding programs. Masking the genetic potential of ET calves with environmental influences makes selection from ET calves difficult.

Thus, ETs have created much interest, but it has been difficult to measure and record performance records on traits that are economically important to the beef cattle industry. Embryo economics, merchandising, and the desire to produce a great animal or show-ring winner have dominated, or masked, the genetic gains.

12

POTENTIAL GENETIC IMPACT OF EMBRYO TRANSFER

H. A. Fitzhugh

Embryos -- collection, preservation, and transfer -- have been major news items in 1984. Most attention has been given to logistical and legalistic issues involving human embryos; however, livestock embryos have received a substantial share of attention as well. Great excitement and considerable publicity has been generated by cattle embryos selling for tens, even hundreds, of thousands of dollars.

Obviously, some livestock producers are profiting from embryo transfer (ET) today and many others hope to do so in the future. Not so obvious is the long-term genetic impact of ET. What will be the future genetic returns on the financial investments made today?

In order to answer this important question, the impact of embryo transfer will be considered with respect to:

- Genetic engineering, including sex selection, cloning, and gene transfer between species
- Conservation of rare genotypes
- Import/export and multiplication of exotic genotypes in new environments
- Genetic improvement of commercial meat and milk production

GENETIC ENGINEERING

All the technologies included here involve micromanipulation of embryos. Examples are cloning (accomplished by splitting of early-stage embryos to produce two or more identical siblings), biopsy of embryos to determine sex prior to gestation, and insertion of exotic genes or chromosomes into embryos. By their very nature, all these technologies involve removal of the embryo from the donor to allow in vitro micromanipulation followed by transfer of embryo to a recipient dam for gestation.

Some of the exciting potential advances in commercial livestock production from genetic engineering have been reviewed (Fitzhugh, 1984). These advances will likely have greatest impact when several technologies are combined. One example is sex control. This can be achieved by cloning embryos to produce multiple identical embryonic siblings (4, 8, or more), biopsying one sibling to determine sex of others, and then transferring only embryos from the clones of the desired sex.

Another dramatic advance involved insertion of copies of growth hormone genes (DNA) from rats into mouse embryos. The resulting mice grew much more rapidly than their normal litter mates. Research to increase productivity of commercial livestock species through similar interspecific DNA exchange is now being conducted.

CONSERVATION OF RARE GENOTYPES

Too often, potentially useful genetic variants are lost because they happen not to excel in the current production or market environment. But conditions can and do change. An example of a near loss is the Texas Longhorn. Other, even more dramatic, examples of loss of rare but potentially valuable genotypes exist around the world.

Part of the problem in conserving rare genotypes is the logistical difficulty and expense of maintaining populations of breeding stock sufficient in number to prevent serious inbreeding. Collection, freezing, and long-term storage of embryos provides a practical means for long-term conservation of potentially valuable, rare genetic resources (Smith, 1983). This procedure is akin to the well-established programs of international seed banks to preserve rare plant genotypes.

Smith (1984) estimated that rare animal genotypes could be conserved by maintaining an effective population size of only 25. This could be accomplished by preserving embryos from 25 unrelated pairs of parents. At this population size, inbreeding would increase by only 2%, the increase commonly occurring in approximately four generations for many domestic livestock breeds.

Highest priorities for conservation are the relatively poorly characterized genotypes in many developing countries. These genotypes have often evolved adaptations to disease and climatic stresses over many generations. Programs to conserve these potentially valuable genetic resources are now being planned (FAO/UNEP, 1983).

IMPORT/EXPORT OF EXOTIC GENOTYPES

Availability of commercially valuable genotypes may be restricted by distance and disease barriers. International transfer of animals has become increasingly expensive to the point that transportation costs often exceed the value of the animal. Transportation costs that prohibit long-distance transfer of breeding stock will be much less for embryos.

The major barriers, however, are very real concerns about the introduction of exotic diseases and their impact on a highly susceptible population. Foot-and-mouth disease, which is endemic in continental Europe, has limited the recent introduction of continental breeds to North America to those few that have passed the stringent quarantine restrictions.

Given the difficulties of exchange of genotypes between developed regions, it is not surprising that introduction of genotypes from developing regions is practically impossible. Thus, the American livestock producer cannot access potentially useful genotypes such as *Bos indicus* cattle from South Asia and Africa, prolific swine from China, and water buffalo from Southeast Asia.

The collection of embryos from dams under local quarantine in the country of origin and with transfer to recipient dams under quarantine in an importing country should preclude any transfer of disease. A related benefit is that embryos gestated by recipients adapted to local production conditions should benefit from acquired maternal immunities to local disease problems. Negotiations between the U.S. and other veterinary agencies are currently underway to implement the import/export of embryos.

GENETIC IMPROVEMENT OF PERFORMANCE

The reproductive technologies combined in artificial insemination have made possible major genetic improvements in the performance of commercial livestock, especially cattle (Foote, 1981). Claims have been made for potential of the same magnitude from embryo transfer. These claims have not been realized, nor is it likely they will be.

To understand why ET will not have the same genetic impact on commercial production as AI, consider a few basic principles governing the process of genetic improvement.

- The frequency of favored genotypes is increased when individuals with these genotypes contribute proportionately more progeny to the next generation. Both natural selection for fitness traits and artificial selection for man-preferred traits operate in this manner.
- Among commercial livestock species, individual males traditionally produce more progeny than do individual females. Therefore, selection primarily operates through the male side.

Artificial insemination increases the potential prolificacy of males much more than ET does for females. Using cattle as the example, individual bulls in natural service can quite easily sire 30 or more calves each year for 5 yr to 10 yr compared to a cow producing only one calf per year. Artificial insemination further increases the relative prolificacy of the male. Through AI, a bull can sire as many as 50,000 progeny per year. Even with repeated superovulation and embryo transfer under the best of circumstances, a cow can produce only a few hundred offspring per lifetime.

Consequently, the most rapid rate of genetic improvement -- even for female-limited traits such as lactation -- is through sire selection. Similarly, the genetic impact of world record dairy cows will be greatest as mothers of sires not as the mother of daughters -- even when the cows are superovulated.

These principles are illustrated in figure 1. Four pathways are considered in estimating potential genetic change in performance. Because fewer males are needed as sires, potential for genetic improvement is much greater through males than females, especially with AI. Consequently, the pathways SS and SD are generally the major source of genetic gain.

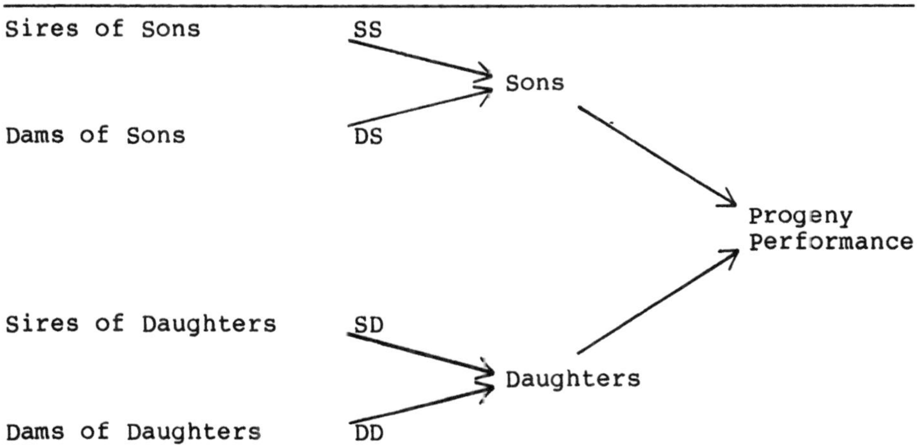

Figure 1. Four pathways for genetic improvement.

Embryo Transfer for Commercial Production

Van Vleck (1984) considered the benefits and costs of using ET to improve milk yield in a commercial dairy herd. Improvements from AI plus ET are compared to those possible from AI alone.

In an example for milk yield, the use of ET in addition to AI does substantially increase the intensity with which dams are selected (table 1). Thus, the contribution of the dam pathways to total gain in milk yield is increased (table 2). The net effect of ET over AI alone is projected to be an additional 76 lb of milk produced per cow per year.

TABLE 1. SOURCES OF SELECTION INTENSITY FOR MILK YIELD, PERCENTAGE OF POPULATION SELECTED

	Sires of		Dams of	
Technology	Sons	Daughters	Sons	Daughters
AI only	4	20	6	90
AI + ET	4	20	1	10

Source: Van Vleck (1984).

TABLE 2. GENETIC GAIN IN MILK YIELD, POUNDS PER YEAR

	Source of gain				
	Sires of		Dams of		
Technology	Sons	Daughters	Sons	Daughters	Total gain
AI only	95	62	67	7	231
AI + ET	95	62	90	60	307

Source: Van Vleck (1984).

Seventy-six pounds is a substantial increase. In addition, this annual gain is cumulative so that after 10 yr the average cow will be producing 760 lb more milk per year. Based on a profit of $.07 over feed costs per pound milk sold, this additional milk would have increased income by $53.20 per cow after the first 10 yr.

Unfortunately, the cumulative average cost of ET would greatly exceed the value of the additional milk. Per cow costs of ET range from $300 to several thousand dollars (Seidel, 1981). Van Vleck (1984) used the conservative cost for ET of $300 per cow. After 10 yr, $3,000 would be invested in ET to produce $53.20 worth of additional milk per cow.

Projected costs vs benefits for ET for a 100-cow herd are shown in table 3. Van Vleck (1984) projected that the value of increased milk would not equal the annual cost ($30,000) until year 60. Even with no interest charged on the investment in ET, the cumulative value of increased milk would not exceed the cumulative cost until about 120 yr after the program began.

Benefits relative to costs of ET for other commercial traits, such as weaning weight of beef cattle, are even less attractive (Van Vleck, 1984). Thus, we conclude that even though ET of commercial females may be technically feasible, it is not likely to be profitable unless costs per embryo transferred are reduced to approximately the current cost for AI ($10 to $15/cow).

TABLE 3. COSTS VERSUS BENEFITS FOR EMBRYO TRANSFER TO GENETICALLY IMPROVE MILK YIELD, $1,000

Year	Annual		Cumulative
	Costs[a]	Benefits[b]	profit (loss)[c]
1	30	<1	-29
2	30	1	-58
20	30	11	-488
50	30	27	-822
60	30	32	-826
100	30	53	-313
120	30	64	+261

Source: Van Vleck (1984).
[a] ET cost of $300/cow for 100-cow herd.
[b] Additional profit from selling a cumulative 76 lb annual increase in milk per cow.
[c] No interest charged on cumulative costs of ET.

Embryo Transfer for Producing Artificial Insemination Sires

Embryo transfer can, however, play a significant role in genetic improvement by intensifying selection of dams of bulls used in AI programs. Because AI sires can have such widespread influence, it can be profitable to utilize ET so that only the extreme top females serve as "bulldams."

In a hypothetical, but realistic, example presented by Van Vleck (1984), 50 beef sires would be needed to provide semen to breed .5 million cows per year. A testing program to improve weaning weight might involve 100 bulls per year from which the 50 AI sires would be selected. Without ET, approximately 300 cows would have to be bred to produce 100 bulls tested (taking account of expected conception and survival rates and sex ratios). Selection of the dams of sons (or bulldams) could be intensified through ET. Instead of choosing 300 cows to produce sons for testing, 50 cows could produce embryos for transfer to 300 recipients.

The impact of intensifying selection of "bulldams" is shown in table 4. Costs of ET would be measured against the benefits of the expected small increase in weaning weight (.7 lb per calf) multiplied across .5 million calves.

CONCLUSIONS

Notwithstanding short-term financial gains through tax savings, the economic value of embryo transfer will ultimately be measured from genetic improvements. Embryo transfer is essential to obtain genetic improvement from gene insertion, conservation of rare genotypes, and import/export of exotic genotypes. However, embryo transfer does not appear to be a cost-effective means of directly improving commercial traits such as milk yield and weaning weight. A reduction of 80% to 90% in current costs of embryo transfer appears to be necessary if this technology is to become a generally applied, cost-effective means of improving performance of commercial livestock. The exception may be the use of embryo transfer to intensify selection of the dams of sires used in large-scale, commercial AI programs.

TABLE 4. BENEFITS VERSUS COSTS FOR EMBRYO TRANSFER TO INCREASE SELECTION INTENSITY OF DAMS OF ARTIFICIAL INSEMINATION SIRES

Year	Genetic improvement in weaning wt/yr, lb	Increased value of .5 million calves, $1,000	Cost of ET, $1,000[a]	Cumulative difference, $1,000
1	0.7	105	90	15
2	1.4	210	90	135
3	2.1	315	90	360
4	2.8	420	90	690
5	3.5	525	90	1,125

Source: Van Vleck (1984).
[a] ET from 50 donors to 300 recipients at $300/ET.

REFERENCES

FAO/UNEP. 1983. Animal genetic resources conservation and management. Report of Joint Expert Panel. FAO, Rome.

Fitzhugh, H. A. 1984. Genetic engineering and commercial livestock production. In: F. H. Baker and M. E. Miller (Ed.). Beef Cattle Science Handbook, Westview Press, Boulder.

Foote, R. H. 1981. The artificial insemination industry. In: B. G. Brackett, G. E. Seidel, Jr. and S. M. Seidel (Ed.). New Technologies in Animal Breeding. pp 14-40. Academic Press, New York.

Seidel, G. E., Jr. 1981. Superovulation and embryo transfer in cattle. Science 211:351.

Smith, C. 1983. Estimated costs of genetic conservation of farm animals. In: Cryogenic Storage of Germplasm and Genetic Engineering. FAO, Rome (to be published).

Smith, C. 1984. Genetic aspects of conservation in farm livestock. Livestock Prod. Sci. 11(1):37.

Van Vleck, L. D. 1984. Genetic implications of embryo transfer. Proc. Beef Improvement Fed.

13

WHY DIFFERENT COWS HAVE DIFFERENT LEVELS OF CALVING DIFFICULTY

Rex M. Butterfield

Calving difficulty is a two-fold problem. Either the stockman has assisted the birth process until his cows are unable to calve unassisted, or he has been seduced by the call for muscular or larger-framed cattle.

The only way out of this problem in a breeding herd is to cull ruthlessly all those animals that are the subject of a difficult calving. Why some animals are difficult calvers may never be clearly defined as calving is such a complex process, but it is obvious that the quicker we "get back to nature" the sooner the problem will be minimized.

If we are to produce for the many markets the ideal carcass with maximum muscle, minimum bone, and optimum fat -- with that optimum at a very low level -- we need to search for more muscular cattle, despite the problems associated with muscularity.

Let us look at what has happened in the last century to change the body composition of cattle to serve a variety of needs and consumer demands. There was a time in Australian history, for example, when "beef" cattle were kept primarily for their hides, which is a long way from the current demand for red meat.

However, the first attempts to "improve" cattle were essentially to increase fatness -- not only to the pathological levels revealed in ancient pictures of British cattle but to the levels associated with "quality" beef in most non-European countries.

What did this selection for more fat do to cattle? If we look at the simple growth lines of muscle, bone, and fat, it is clear that the fattening phase was pushed onto younger and lighter cattle (figure 1).

This did not materially alter the composition of the calf at birth and was not a factor in increased calving difficulty.

Similarly, when the trend came for later-maturing cattle, the simple selection for this characteristic merely moved the fattening phase onto older and heavier cattle (figure 2). Once again the composition of the calf was not materially affected, and hence there was no increase in calving difficulty. However, smart breeders soon realized that the quick way to later maturity was not to look for late-maturing sires but to go straight to larger cattle. The breeder with the herd of unfashionably small cows sought the quick answer by acquiring the biggest possible bull. The disproportionate size of the bulls and the cows caused oversized calves and serious calving problems.

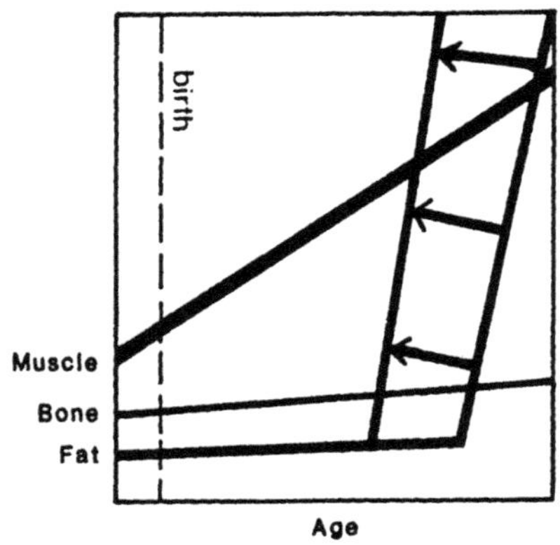

Figure 1. Selection for more fat through early maturity does not alter composition at birth.

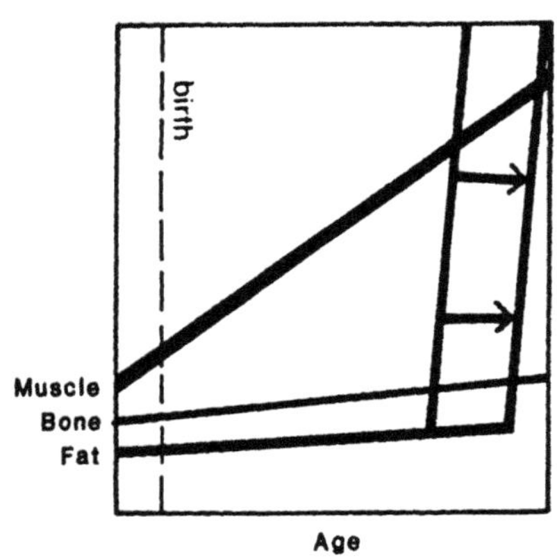

Figure 2. Selection for less fat through late maturity does not alter composition at birth.

Late maturity has been achieved through the combination of leaner and larger animals without the calving difficulty being increased.

The search for muscularity has caused the real problems. I remember seeing a quote, "Breed 'X' introduced calving difficulties to this country, but it hasn't taken the others long to catch up." Why? Because the trend has been towards more muscular animals.

When we select animals for increased muscle (figure 3), it is apparent that to increase the muscularity of our slaughter animals, we must also increase the muscularity of our calves. It is that simple. I have searched high and low for a breed, or a strain within a breed, where increased muscularity is not associated with more muscular calves. An unsuccessful search.

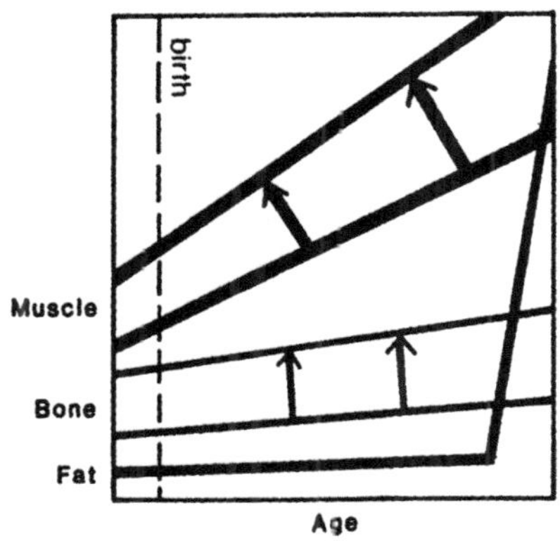

Figure 3. Selection for more muscle will alter composition of calf at birth by increasing muscle and usually the bone structure.

This does not mean that we should abandon attempts to increase the muscularity of our cattle, but rather that we should proceed with caution and accept **NO** level of calving difficulty. In other words, we should follow the philosophy of Tom Lassiter that there be no excuses. In this way, we would attempt to meet the requirements of the consumer without jeopardizing the easy calving that is so essential to economic beef production.

I have little doubt but that the consumers will try to push us too far either direction. The task ahead for cattle producers is to attempt to balance on the knife's edge. The alternative on either side of the knife is:

- A carcass that is a little less perfect than that sought by the trade but still profitable
- Cattle that produce the ideal carcass <u>but</u> require the maximum input of labor and veterinary skill to keep them alive and productive and, just maybe, highly profitable.

The choice is yours!!

14

THE RELATIONSHIPS BETWEEN BODY CONDITION, NUTRITION, AND PERFORMANCE OF BEEF COWS

Iain A. Wright

INTRODUCTION

In Northern Europe, the cost of feeding beef cows accounts for over 75% of the variable costs of suckled calf production, and over 70% of these feed costs are incurred by winter feeding. It is therefore obvious that any reduction in winter feed costs will have a large effect on the profitability of the enterprise. However, any reduction in feed inputs at that time may result in a penalty in terms of production; thus, before decisions can be made regarding the level of winter feeding, it is necessary first to understand the relationships between nutrition and performance and how these are modified by cow body condition.

The level of body condition has been used for years as an indicator of the nutritional status of livestock and has been used by stockmen to decide how much to feed their cattle. If cows are too thin, then feeding can be increased; if too fat, then feeding can be reduced. But what constitutes "too thin" or "too fat"? Until fairly recently there was no quantitative way of assessing or describing the level of body condition, but body-condition scoring now offers a method for describing the level of fatness of animals.

BODY-CONDITION SCORING

The Technique of Body-Condition Scoring

The technique of body-condition scoring originated in Australia for assessing fatness in sheep and was introduced into the United Kingdom for the same purpose. Since then it has been adapted for use in cattle and is now widely used by the extension services and by commercial breeders. The system used in the United Kingdom (Lowman et al., 1976) is easily learned and involves assessing the level of fat cover on two areas of the body: on the transverse processes of the lumbar region of the spine and around the tail head.

The depth of fat cover over the transverse processes of the spine is the major measurement, particularly in thin cattle. In fatter cattle (above condition score 3) the spine can no longer be felt and fat cover around the tail head is used to assess condition score. The animal is scored on a 0 to 5 scale, normally to the nearest 1/2 score. For more detailed experimental work, the nearest 1/4 score can be used. The description of each condition score is as follows:

Score 0. The animal is emaciated. No fatty tissue can be detected and the neural spines and transverse processes feel very sharp. Score 0 is very rarely encountered in practice.

Score 1. The individual transverse processes are sharp to the touch and easily distinguished.

Score 2. The transverse processes can be identified individually when touched, but feel rounded rather than sharp.

Score 3. The transverse processes can only be felt with firm pressure and areas on either side of the tail head have some fat cover.

Score 4. Fat cover around the tail head is easily seen as slight mounds, soft to the touch. The transverse processes cannot be felt.

Score 5. The bone structure of the animal is no longer noticeable and the tail head is completely buried in fatty tissue.

Condition Score and Body Fatness

Information taken from the records of cows that have been slaughtered at different body-condition scores has established the relationship between the fat content of the body and condition score (Wright and Russel, 1984). This relationship for three cow types is shown in figure 1.

As can be seen, body-condition score is a good indicator of the level of fatness. There are, however, differences between the breeds in the relationship and we now know that this is a result of the way in which different breeds deposit fat in the different fat depots in the body. These differences become more pronounced when cows have more than about 15% fat in their live weight. The Friesian, and dairy breeds in general, deposit more fat internally and less externally. Body-condition scoring only assesses the level of subcutaneous fat, and so at any given level of condition score the Friesians will have a higher level of total fatness because of their greater quantity of internal fat. Of the three breeds shown in figure 1, the Hereford x

Friesian has the lowest proportion of internal fat; thus at any given body-condition score the cross has the least total fat in the body.

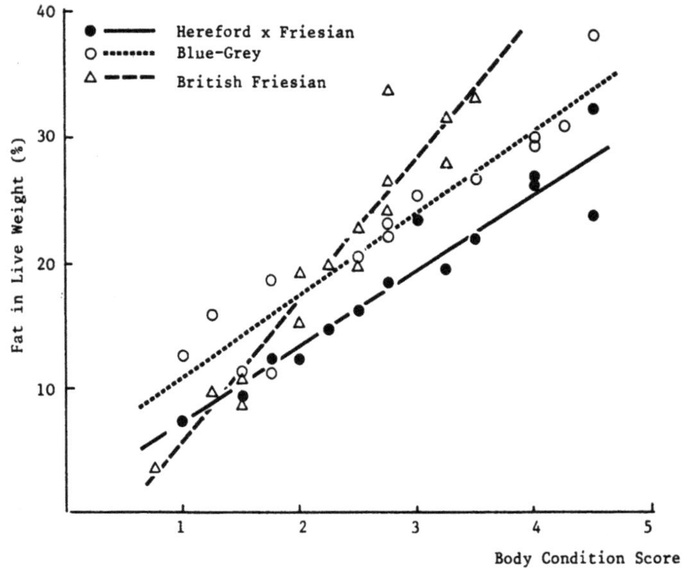

Figure 1. Relationships between fatness and body-condition score.

FEEDING IN LATE PREGNANCY

Level of Feeding

A number of studies have shown that cows can utilize their body reserves of energy by mobilizing body tissue, principally fat, to sustain fetal growth when their energy intake falls below their requirements. In one experiment (Russel et al., 1979), cows that were all in good body condition (body-condition score 3 or above) 3 mo before calving were fed a range of energy intakes of approximately 30 megajoule (MJ) to 80 megajoule (MJ) of metabolizable energy (ME) per day for the last 12 wk of pregnancy. This range corresponded to 65% to 170% of the maintenance requirements for nonpregnant cows.

The effect of energy intake on calf birth weight is illustrated in figure 2. Maximum birth weights resulted from energy intakes of about 120% of nonpregnant maintenance, and it appeared that reductions in birth weight were caused by very low or very high energy intakes. The lower levels of feeding reduced calf birth weight by 10% to 15%. This level of reduction in birth weight is perfectly acceptable; and, in fact, if a large breed of bull, such as the

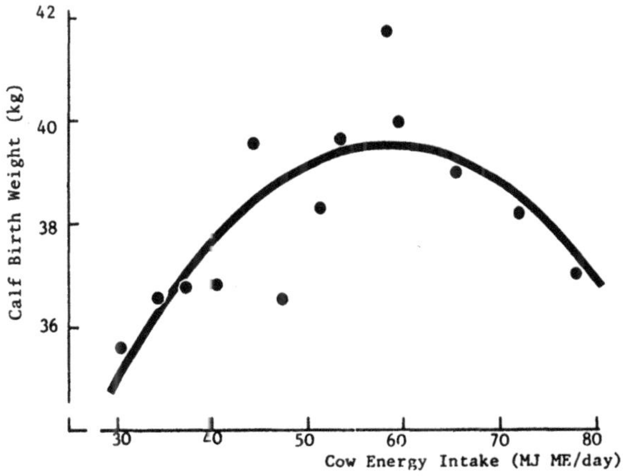

Figure 2. Relationship between calf birth weight and energy intake of cows in late pregnancy.

the Charolais or Simmental is being used, then a reduction in birth weight is a positive advantage because of the associated reduction in calving difficulties. The association between calf birth weight and the incidence of calving difficulties is well known and has been documented in the U.K. by the Meat and Livestock Commission (figure 3).

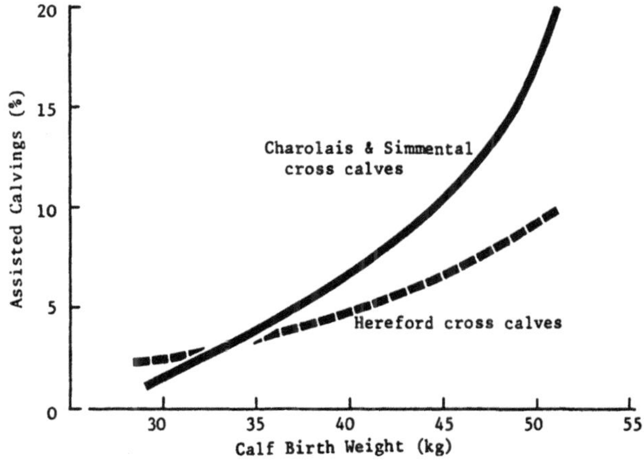

Figure 3. The relationship between calf birth weight and the percentage of assisted calvings.

Lower levels of feeding in pregnancy will result in thinner cows at calving and this too will reduce calving difficulties. As figure 4 shows, there is considerable advantage in ensuring that cows are below condition score 3 at calving.

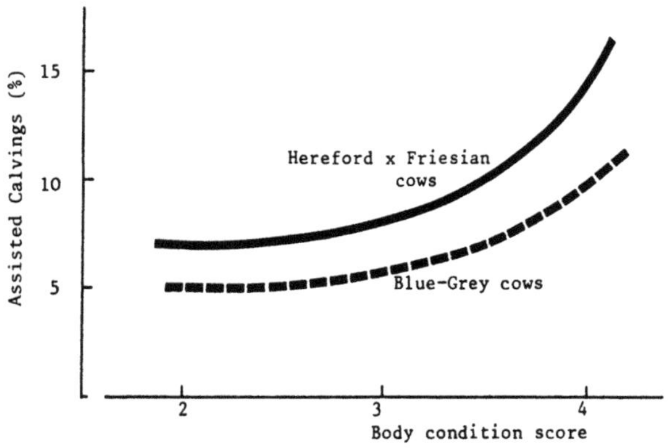

Figure 4. The effect of cow-condition score of calving difficulties.

In the experiment described above, the cows were well fed during lactation, and the level of milk production and calf growth rate were unaffected by the feed intake during pregnancy, but there was a large effect on cow weight change during lactation. Those cows that had been fed less during pregnancy gained more weight after calving than those that had been fed the higher levels. By 22 wk after calving there was no difference in weight.

Pattern of Feeding

In another experiment, some cows were fed flat rate (i.e., the same amount of feed was given each day throughout late pregnancy); others had their feed increased monthly in proportion to the energy demand for fetal growth, but the same total amount of feed was given to both groups. The pattern of allocation of feed had no effect on calf birth weight or the subsequent performance of the cows or calves, so it seems that the pattern of feeding is unimportant and, for ease of management, that flat rate feeding is perfectly acceptable in late pregnancy.

FEEDING IN LACTATION

Cows can continue to utilize their body reserves during lactation. Hodgson et al. (1980) fed Hereford x Friesian and Blue-Grey (Whitebred Shorthorn x Galloway) cows about 40 MJ ME daily for the last 3 mo before calving. After calving in February-March and until turn-out (a period of 55 days), the cows were fed one of two levels: approximately 100 MJ or 65 MJ of ME per day. At turn-out in mid-May until weaning in September, they were rotationally grazed.

The results of this experiment are given in table 1. During early lactation the cows on the higher level of feeding produced 9.5 kg of milk per day and maintained their body condition, while those on the lower level of feeding produced 8.1 kg of milk per day and lost .4 units condition score, clearly demonstrating the ability of beef cows to sustain reasonable levels of milk production by drawing upon their body reserves.

At turn-out, there was a large increase in milk production from all cows. As illustrated in figure 5, this increase was greater in the cows from the lower feeding level, and they consistently produced more milk throughout the summer. This increase in milk yield that occurs at turn-out has been found to occur at all stages of lactation. For example, in another experiment, cows that had calved at two different times of the year were turned out to grass in mid-May, at 5 mo or 2 mo after calving. Although the milk yield (indoors) of the cows that were at a more advanced stage of lactation was only 5.2 kg/day compared to 7.2 kg/day in the cows that had recently calved, the increase in yield at turn-out was similar at both stages of lactation (4.2 vs 3.9 kg/day).

TABLE 1. PERFORMANCE OF COWS AND CALVES FED AT TWO LEVELS IN EARLY LACTATION

	Feed level in early lactation	
	65 MJ/day	100 MJ/day
Winter Performance		
Cows		
Condition score at calving	2.6	2.4
Milk yield, kg/day	8.1	9.5
Condition score at turn-out	2.2	2.4
Calves		
Live weight gain, kg/day	.77	.93
Weight at turn-out, kg	80	86
Summer Performance		
Cows		
Milk yield, kg/day	11.7	10.8
Herbage intake, kg organic matter/day	14.1	13.6
Condition score at weaning	2.9	2.8
Calves		
Live weight gain, kg/day	1.17	1.14
Weaning weight	219	223

Source: Hodgson et al., 1980

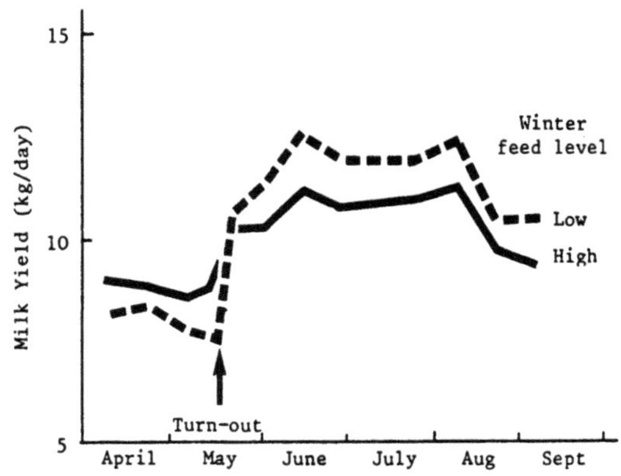

Figure 5. Milk yield of cows fed high or low levels in early lactation.

This large increase in milk yield at turn-out is a consequence of the sudden nutritional boost that occurs when cows are turned out to pasture. The higher milk yield of the cows that were previously on lower feed levels (table 1 and figure 5) results from these cows having slightly higher herbage intakes (14.1 kg as opposed to 13.6 kg organic matter/day). In addition to higher milk yields, the increased intake also results in greater weight gains and body-condition recovery, so that by weaning time there is no difference in the body-condition score of cows subjected to different feeding levels in early lactation (see table 1).

Despite the slightly lower growth rates in winter of the calves suckling cows on the lower energy intakes, and their lower live weight at turn-out, there was no difference in their weight by weaning time.

THE ENERGY VALUE OF CONDITION SCORE CHANGE

The relationship between body-condition score and the energy stored in the body (Wright and Russel, 1984) allows the quantification of the dietary-energy equivalent of body-condition change. Each unit loss of body-condition score will supply the equivalent of 3200 MJ dietary ME. Each unit gain in body-condition score needs about 6500 MJ dietary ME.

This information can be used to ration cows according to body condition and is best explained by an example. Suppose a cow is at condition score 3 in early November and is to calve at condition score 2 1/2 in early April, i.e., to lose 1/2 unit of body-condition score over the last 5 mo of pregnancy.

(EXAMPLE)

	MJ ME
Maintenance (52 MJ for 150 days)	7800
Energy cost of pregnancy for the last 150 days	1800
Total energy requirements	9600
Loss of 1/2 unit condition score will supply	1600
Therefore diet must supply the difference	8000

The 8000 MJ ME for 150 days is equivalent to 53 MJ ME/day; thus if 53 MJ ME/day are fed for the last 5 mo of pregnancy the cow should calve at condition score 2 1/2.

CONCLUSIONS

Body-condition scoring offers a very satisfactory method of assessing the level of body condition. If cows are scored regularly at some or all of the following times -- calving, turn-out, mating, weaning, and housing -- then the herd or groups of cows in similar body condition can be managed in such a way as to get them to the appropriate level of body condition for the next stage in the production cycle.

It appears that there is an opportunity for making substantial savings in the expensive winter feeding of beef cows. Cows can draw heavily on their body reserves, both pregnancy to sustain fetal growth and during lactation to sustain acceptable levels of milk yield. Relatively low levels of feeding do not result in unacceptably low levels of performance, provided cows are in suitable body condition at the start of the period of undernutrition. Indeed, there may be considerable biological as well as economic advantages to be gained by restricting the feed intake of beef cows at certain times.

To a large extent, the nutritional management of beef cows can be regarded as the manipulation of the cow's body condition. In general, body condition can be allowed to fall considerably in winter when the provision of feed is expensive. This can only be done when there is the opportunity to regain the lost condition the following summer when pasture provides a relatively cheap source of feed.

REFERENCES

Hodgson, J., J. N. Peart, A. J. F. Russel, A. Whitelaw, and A. J. Macdonald. 1980. The influence of nutrition in early lactation on the performance of spring calving suckler cows and their calves. Anim. Prod. 30:315.

Lowman, B. G., N. A. Scott, and S. H. Somerville. 1976. Condition scoring of cattle. East of Scotland College of Agriculture, Bulletin No. 6 (revised edition).

Russel, A. J. F., J. N. Peart, J. Eadie, A. J., and I. R. White. 1979. The effect of energy intake during late pregnancy on the production from two genotypes of suckler cow. Anim. Prod. 28:309.

Wright, I. A. and A. J. F. Russel. 1984. Partition of fat, body composition and body condition score in mature cows. Anim. Prod. 38:23.

15

NUTRITION FOR OPTIMUM REPRODUCTION IN BEEF CATTLE

Rodney L. Preston

INTRODUCTION

The most important stage in any animal production system is reproduction. Optimum reproduction is probably the major determinate of efficiency in beef production because of the large overhead that maintenance of the brood cow represents in the total feed required to produce beef.

Beef cows have only one real function -- to produce calves. They may be pleasing to look at and give a large deduction on income tax, but these attributes won't last very long without calves. Therefore, the object of any beef cow feeding program is to maximize the chance that each breeding female will have a calf each year. Feeding costs, however, must be minimized to maximize profits, but this doesn't mean to not feed anything.

Immediately one thinks of trace minerals and vitamins as magic diet ingredients that will assure good reproductive performance in cows. These, however, are not the most important considerations. Energy is most important because it is the nutrient that has the greatest affect on reproduction and also because it is the most expensive to supply -- probably 80% of the total feed cost.

Four critical feeding periods in the female's life cycle that relate to good reproductive performance in beef cattle are when she is a 1) heifer prior to breeding, 2) bred heifer, 3) lactating heifer, and 4) cow.

Three of these periods relate to heifers -- the most important period for proper nutrition. Before discussing feeding during each of these four periods, we must understand the role of proper nutrition -- especially energy -- on subsequent reproductive performance.

First, it may be instructive to compare the total daily nutrients of the cow-calf and feedlot phases in a life-cycle approach to cattle nutrition (figure 1). Here one sees that beef cows, even those nursing a calf, require a much lower concentration of energy (TDN), protein, calcium, and phosphorus in their diet as compared to either stocker or finishing cattle. Thus, higher-quality feeds should be used for stocker and finishing cattle and lower-quality feeds for the beef cows, but do not make the mistake of paying more attention to the nutrition of stocker and finishing cattle than to beef cows. This kind of thinking often leads to overnourishing the finishing cattle and underfeeding the brood cows.

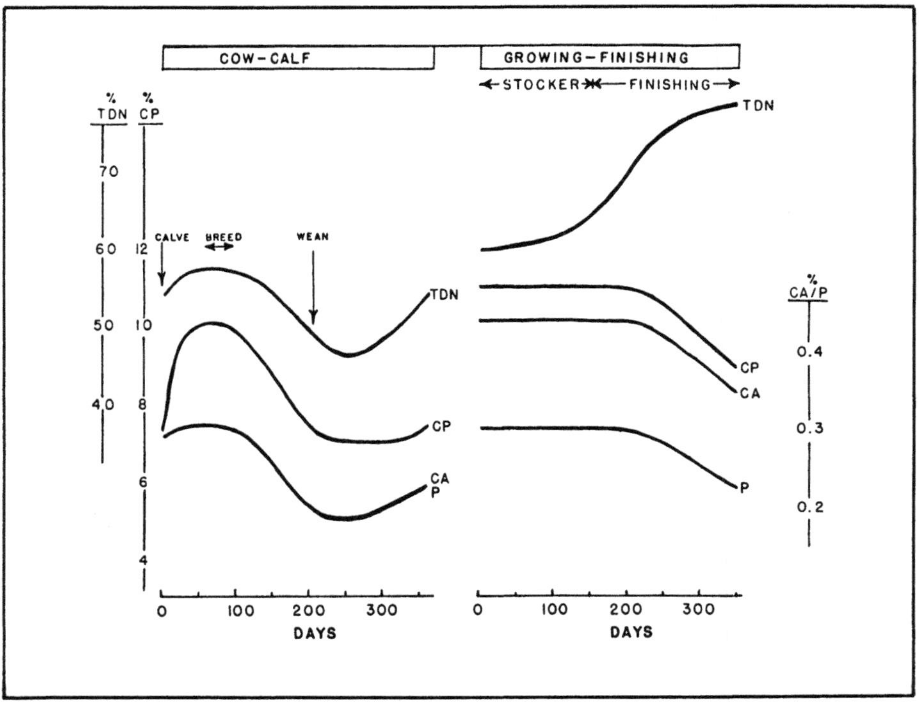

Figure 1. Life cycle of cattle nutrition.

Cow size has received a lot of rhetoric since the exotic breeds were introduced into the U.S. Large-sized cows obviously require more feed than small-sized cows. Research on this subject has shown, however, that if calves from various sizes of cows are carried to the same degree of finish, the total efficiency of the beef produced -- including feed for the cow and feed to grow and finish the calf -- remains the same across a wide range of mature cow weights. The weight to which calves must be fed to achieve the same degree of finish, however, increases at a rate parallel to the increase in mature size. Therefore the question of cow size resolves itself to one of carcass size rather than the differential efficiency of cows of various mature sizes.

HEIFER PRIOR TO BREEDING

Profit potential is of great importance when related to cow size and amount of nutrition larger-sized heifers require to reach puberty. There seems to be increasing difficulty in getting heifers from larger-sized cows to settle and calve as 2 yr olds. Increasingly I hear about heifers having their first calf as 3 yr olds in spite of overwhelming research showing a feed savings and increased total productivity of heifers calving as 2 yr olds. Perhaps the major problem of ranchers is providing the quality of diet required for larger-sized heifers to reach target weights by 385 days of age. This level and type of nutrition is necessary to ensure that they reach puberty, have two chances at becoming pregnant, and calve an average 21 days ahead of the rest of the cow herd.

Figure 2 shows three factors that affect the age at which heifers reach puberty. First, heifers wintered to gain 1 lb/day reached puberty at a younger age but at a heavier weight than heifers that gained 1/2 lb/day. Therefore, weight alone does not determine when heifers reach puberty. One can calculate from these data, however, that heifers reach puberty when they weigh about 60% of their mature cow weight.

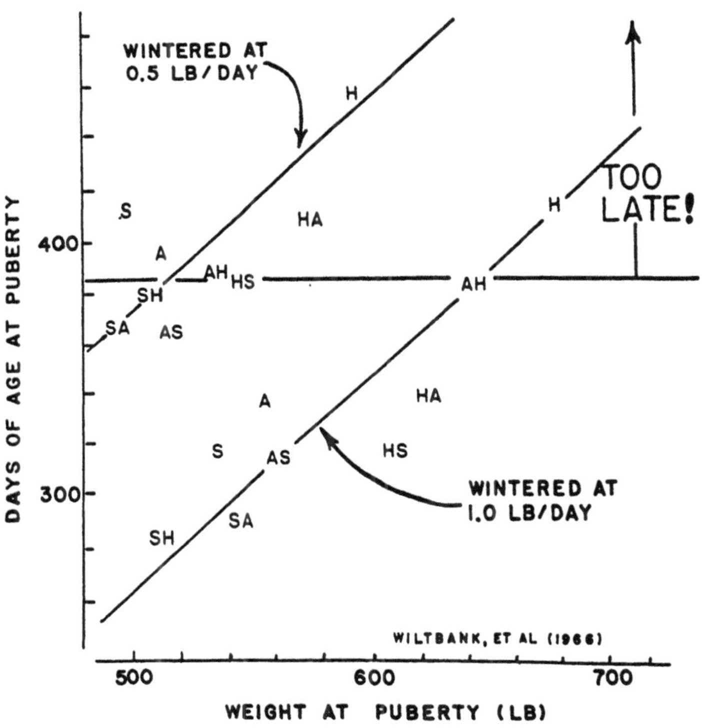

Figure 2. Feeding, weight, and heterosis effects on age at puberty.

Second, between straight bred heifers (Hereford, Angus, Shorthorn) and two-way crosses of all three breeds, the crossbreds reached puberty at a slightly, almost undiscernibly, younger age than did the straight breds. Therefore, straight breds and crossbreds appear to reach puberty about the same age.

The third, and most important item, is that the lower level of feeding during the winter enabled only half of the groups to reach puberty by 385 days of age, whereas all but one group on the higher winter diet reached puberty by this time. Why the 385-day puberty goal?

385	days, maximum days at puberty
42	days for two breeding cycles
282	days for gestation
21	days to rebreed one cycle before the older cows
730	days or 2 yr of age

The larger the heifer size, the faster the rate of gain required to reach the target weight, which requires a greater energy concentration in the diet. This is illustrated in table 1. A producer is going to be dollars ahead to feed 10% more TDN to get large-sized heifers to a target weight during the 6 mo between weaning and 385 days of age than to feed the 55% additional TDN required to carry these heifers an additional year before breeding. In my opinion, this is the major reason why the producer should be concerned about cow size and the feed requirements to get heifers to puberty by 385 days of age using the feed resources available. If your primary feed resource is low quality pasture, range, and roughage and you cannot or will not feed some grain, then smaller cows must be used.

TABLE 1. ENERGY AND PROTEIN REQUIREMENTS FOR REPLACEMENT HEIFERS TO ACHIEVE TARGET WEIGHTS BY 385 DAYS OF AGE

Mature cow size	Weaning wt, lb	Target at puberty, lb	ADG[a] required, lb	Requirement[b] TDN, %	CP, %
Large	495	730	1.3	64	10.0
Medium	430	610	1.0	60	9.7
Small	365	530	.9	58	9.5

[a]Daily gain required to reach desired body weight at puberty (385 days of age).
[b]Dry matter basis.

BRED HEIFER

What is the feeding objective for the bred heifer? To produce her fist calf? Wrong! If she is bred, the chances are that she will have a calf in spite of how you feed her. Your objective now is to have her cycling within 67 to 92 days after calving.

First of all, everyone knows that the cow's age affects the time required to show heat or estrus after calving. This is shown in figure 3. Thus a higher percentage of 5-yr-old cows come in heat within 40 to 60 days than 4-yr-old or 3-yr-old cows. If we are to keep cows on an annual cycle, they must begin to show heat by 70 days after calving; 90% of the older cows will be in heat by this time but only 65% of the first-calf heifers will be in heat by 70 days. This is why heifers should be bred 21 days before the older cows to allow more time for them to return to estrus after having their first calf.

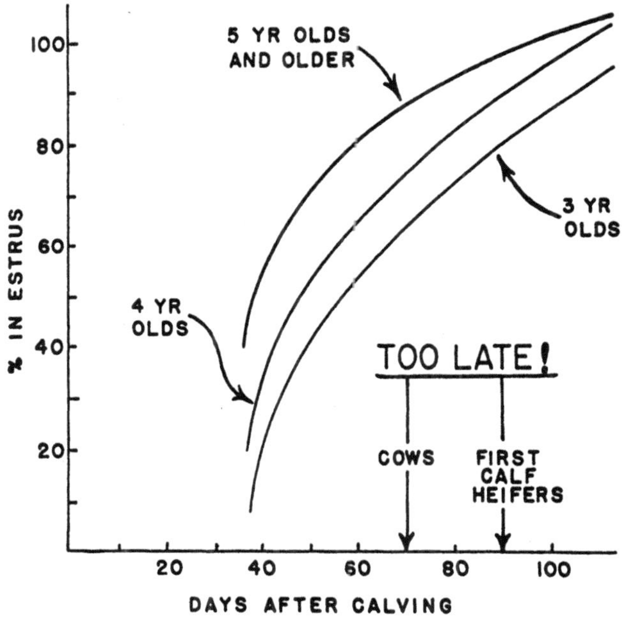

Figure 3. Age of cows at calving and estrus after calving.

Body condition at calving is one of the most important factors affecting the return of cows to estrus after calving. In figure 4, 70% to 85% of all cows in moderate to good condition at calving will return to estrus within 70 days, whereas only 55% of the cows in thin condition will be in heat by this time.

Since first-calf heifers are still growing while their calf is developing, it is vital that first-calf heifers continue to gain weight during their pregnancy and calve with good to moderate body condition for returning to estrus within 67 to 92 days after calving.

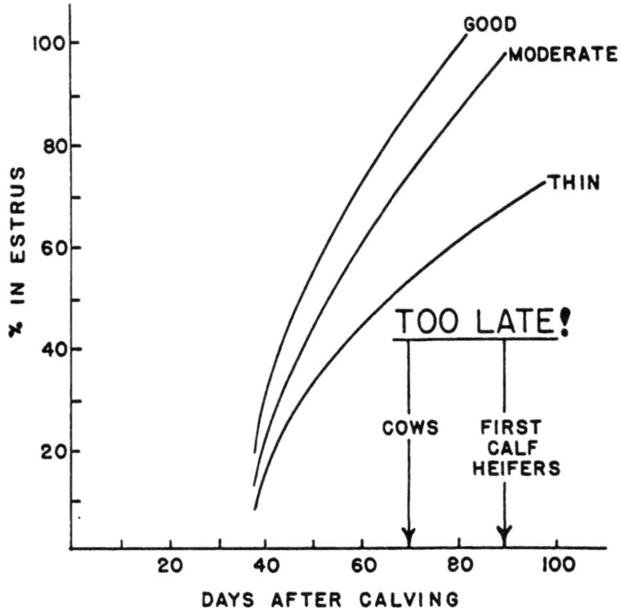

Figure 4. Body condition at calving and estrus after calving.

Table 2 gives requirements of pregnant heifers for ensuring proper body condition at calving. As with growing-replacement heifers, larger type cattle require a higher percentage TDN than smaller type heifers.

TABLE 2. ENERGY AND PROTEIN REQUIREMENTS OF PREGNANT HEIFERS

Mature cow size	Wt at breeding, lb	ADG[a] required, lb	Requirement[b] TDN, %	CP, %
Large	750	1.1	61	9.4
Medium	610	.9	58	9.0
Small	530	.7	56	8.8

[a]Daily gain required to reach desired body weight and condition at calving.
[b]Dry matter basis.

Crude protein requirements for the bred heifer are nearly the same for all sizes but lower than that required for the growing heifer. While an overall average daily gain is shown, most of this gain should occur during the last third of pregnancy when the size of the calf is increasing rapidly.

LACTATING HEIFER

The feeding objective for the lactating heifer is the same as for the bred heifer, namely to have her cycling within 67 to 92 days after calving. However, we must provide sufficient energy and protein not only for milk production but also for some additional growth, although this will be small during the heavy lactation period.

The most important factor affecting the return to estrus of lactating heifers and cows is their weight change after calving. Table 3 shows that a small weight increase between calving and breeding time greatly increases the chances of cows showing heat to become pregnant within the desired time. Cows that were losing weight were coming into heat later and a significant number had not shown estrus within the required 92 days to assure a calf 1 yr later.

TABLE 3. WEIGHT CHANGE AFTER CALVING AND PREGNANCY RATE

Wt change after calving	Pregnant From 1st service	Pregnant After breeding 20 days	Pregnant After breeding 90 days	Cows not showing heat
Losing wt, %	43	29	72	14
Gaining wt, %	60	57	82	0

In table 2 and table 4, the percentage TDN requirements for the lactating heifers are lower than for the pregnant heifer. Lactation stimulates higher energy requirements and so there is a greater energy intake -- even with a lower TDN concentration. Protein requirements, however, are higher.

TABLE 4. ENERGY AND PROTEIN REQUIREMENTS OF LACTATING HEIFERS

Mature cow size	Wt during 1st lactation, lb	Requirement[a] TDN, %	Requirement[a] CP, %
Large	1,080	59	10
Medium	900	57	10
Small	780	55	10

[a]Dry matter basis.

COWS

The major feeding objective with cows is to produce a calf each year and increase the average herd weaning weight. Relative to heifers, nutrient requirements for cows are lower and represent the best place to utilize low quality and inexpensive feed sources. Cows, however, also have critical nutrient periods.

One critical period for the cow is at calving, as shown in figure 4. Thin cows at calving have delayed estrus and breed late. Also, cows that lose weight between calving and breeding are less likely to become pregnant (table 3). Body condition of cows during the postpartum interval greatly influences their return to estrus. If cows do not have an estrous cycle by 70 days after calving, it's too late. Cows that do not cycle in time are either culled or kept until the next breeding season. Both alternatives represent considerable expense to the producer. Table 5 shows that cows must carry 15% to 20% body fat to assure that they begin to cycle 70 days after calving. Feeding cows at or above their energy requirements after calving and during lactation will help shorten the postpartum interval. At calving time, cows must have a moderate to good body condition and then gain some weight if they are to return to estrus in the time required to remain in the herd.

TABLE 5. ROLE OF ENERGY INTAKE AND BODY FAT ON THE POSTPARTUM INTERVAL IN BEEF COWS

	Percentage of energy requirement		
Fat, %	85	100	115
	------------No. of days-------------		
10	100	97	86
15	75	73	61
20	60	58	46
25	54	52	40

Source: S. J. Bartle, R. L. Preston, and J. R. Males (1982).

What are the energy and protein requirements of cows during pregnancy, early lactation prior to breeding, and late lactation? With the exception of early lactation, energy and protein requirements are quite low (table 6). In fact, preventing cows from becoming too fat during pregnancy is also important since cows that are too fat after calving may not settle as well as cows in moderate condition.

One can conclude that in the cow-calf production cycle, in replacement heifers, and in cows immediately before and after calving, the animal needs more nutrition than it can receive from low-quality roughages, pastures, or ranges.

TABLE 6. ENERGY AND PROTEIN REQUIREMENTS OF MATURE BEEF COWS[a]

Productive stage	Requirement[b]	
	TDN,%	CP,%
Pregnancy	52	7
Early lactation		
0 to 3 mo	57	10
Late lactation		
4 to 7 mo	54	7

[a]All mature sizes.
[b]Dry matter basis.

HERD ENERGY MANAGEMENT

In the management of breeding beef cattle, energy needs determine the feedstuffs to be used. Let's summarize these energy needs in relation to the energy content of roughages.

Table 7 summarizes the TDN requirements of breeding beef cattle. The cattle can be divided into three management groups: 1) growing and pregnant heifers that require a much higher concentration of TDN, 2) lactating heifers that require a lower concentration, and 3) cows that require the lowest concentration of energy. Table 7 shows relatively higher energy need for the growing and pregnant heifers that will develop into larger mature cows rather than into smaller-sized cows.

TABLE 7. TOTAL DAILY NUTRIENT REQUIREMENTS OF REPRODUCING BEEF CATTLE

Productive stage	Mature cow size		
	Large	Medium	Small
Growing replacement heifers	64	60	58
Pregnant heifers	61	58	56
Lactating heifers	59	57	55
Pregnant cows	52	52	52
Lactating cows			
Early, 0 to 3 mo	57	57	57
Late, 4 to 7 mo	54	54	54

Table 8 divides these cattle into energy management groups based on their energy requirements. Data from table 8 indicate that in any given herd, heifers should be fed separately from mature cows until they wean their first calf because of their different energy requirements.

TABLE 8. ENERGY MANAGEMENT OF BEEF CATTLE

TDN,%[a]	Management groups
60 or more	Large-type growing and pregnant heifers
55 to 60	Medium- and small-type growing and pregnant heifers, all lactating heifers, and cows in early lactation
54	Cows in late lactation
52	Pregnant cows and cows in late lactation

[a]Dry-matter basis.

How do producers manage roughage sources to meet these energy needs? Roughages can be conveniently divided into four categories (table 9) to match these energy groups so that cattle producers can tailor their roughage supplies to the energy need of their cattle. Thus, roughages listed as high quality are better suited for heifers especially if these heifers could develop into large mature cows. Poor quality roughages should be fed to mature cows immediately after weaning their calves.

TABLE 9. ROUGHAGE GROUPS ACCORDING TO ENERGY CONTENT

Roughage quality	TDN,%[a]	Examples
High quality	60 or more	Dehydrated alfalfa Early spring pasture Wheat pasture Corn silage
Average quality	55 to 60	Alfalfa hay Immature grass hays Summer pasture
Low quality	52 to 55	Mature grass hays Fall range grass
Poor quality	48 to 52	Straws and stover Dormant range grass

[a]Dry-matter basis.

What about those times when high-quality roughages are not available or drought reduces the total amount of roughage available for feeding? This is where judicious use of concentrates can be beneficial. In fact, many times the use of protein supplements with low-quality roughages gives better results because of the energy they supply rather than the protein they contain.

With the threat of drought, there is interest in supplemental feeds, including hay, to carry beef breeding stock. In evaluating potential supplemental feeds, the use of concentrates should not be overlooked in this situation.

Adequate mineral supplies are important to successful reproduction. Salt is required and should be supplied free choice. Many roughages, especially poor-quality roughages, are deficient in phosphorus. Since phosphorus plays a vital role in the reproductive process of cattle, producers should be certain that adequate phosphorus is provided. Trace minerals are important, especially zinc and iodine; these are generally contained in trace mineralized salt.

During periods of rapid, lush growth of pasture grasses, a problem of acute tetany -- grass tetany -- can develop, especially in older, lactating cows. Apparently tetany is an acute deficiency of magnesium that requires immediate treatment with intravenous magnesium salts to save the affected cow. Acute tetany can be prevented or greatly lessened by feeding magnesium oxide in the salt or mineral mix at a level of 5% to 10% at the time that grass tetany may be a problem. An intake of 30 g of magnesium oxide/cow/day is required to prevent this condition.

Vitamin A, generally, is associated with the green color of forages, of practical importance for reproducing beef cattle. Badly weathered or bleached forages and forages stored for more than one season cannot be relied upon as a source of vitamin A. Adequate vitamin A levels are especially important during gestation and early lactation to assure strong, vigorous calves that are better able to resist scours and other early calfhood diseases. Cows should be provided adequate vitamin A, especially during the last third of pregnancy, by providing bright green forage, dehydrated alfalfa (1/3 lb/head/day), or a supplement containing vitamin A at a level of 30,000 units/cow/day. If the diet of pregnant cows is deficient in vitamin A, an injection of vitamin A during the last third of pregnancy can be very beneficial.

SUMMARY

While beef cattle, in general, and beef cows in particular, can be productive on diets that are relatively low in nutrient concentration, they can still suffer nutrient deficiencies, especially at critical times during their life cycle. Nutrient requirements at these critical times are presented. If nutrients are deficient, reduced productivity and efficiency are evident from rather severe deficiency symptoms. More often, however, with marginal deficiencies, productivity is reduced somewhat and may or may not come to the attention of the cattle producer. Since these marginal deficiencies generally go uncorrected, they are more costly in the long run. Energy intake is a major determinant of reproductive efficiency and is related to the size of the female. Energy and protein needs have been presented in relation to the differences in the mature size of the cattle.

REFERENCES

Bartle, S. J., J. R. Males, and R. L. Preston. 1984. Effect of enrgy intake on the postpartum interval in beef cows and the adequacy of the cow's milk production for calf growth. J. Anim. Sci. 58:1068.

NAS-NRS. 1984. Nutrient Requirements of Beef Cattle. (6th Ed.). Natl. Academy Press, Washington, D.C.

Wiltbank, J. N., K. E. Gregory, L. A. Swiger, J. E. Ingalls, J. A. Rothlisberger, and R. M. Koch. 1966. Effects of heterosis on age and weight at puberty in beef heifers. J. Anim. Sci. 25:744.

16

HORMONE THERAPY FOR CYSTIC AND REPEAT-BREEDER DAIRY COWS

Roy L. Ax and C. N. Lee

INTRODUCTION

Twenty-three percent of all dairy cows are culled from the herd because of reproductive problems. Only low milk production culls more cows (Call, 1978). Problems in reproduction are costly because they result in lower milk production per day of herd life, increase semen and feed costs, and reduce genetic gain within a herd.

Two common reproductive problems are cystic ovarian degeneration and repeat-breeder cow syndrome. Ovarian follicular cysts are follicles greater than 2.5 cm in external diameter that fail to ovulate and subsequently persist for 10 or more days. Estimates show that the incidence of cystic follicles can be as high as 20% in a dairy herd (Eyestone and Ax, 1984). Though cystic follicles are common in the early postpartum period, in many cases recovery is spontaneous. Cystic follicles are also more frequently diagnosed in higher-producing cows, older cows, and cows with parturient abnormalities, e.g., retained placenta, milk fever, dystocia, etc.

Repeat breeders are cows that require three or more services per conception, have a normal interestrus interval, and normal reproductive-tract condition. The incidence of repeat-breeder cow syndrome for a given herd can be as high as 25% (Maurice, 1982).

Both of these reproductive problems -- cystic follicles and repeat-breeder syndrome -- increase days open, calving interval, and average days in milk that are uneconomical in a dairy enterprise. Therefore, it is the objective of this discussion to identify some causes and solutions for these troublesome reproductive problems.

CYSTIC OVARIES

The causes of cysts in cows remain unknown. However, data indicate that the hypothalamic-pituitary-ovarian axis is affected (Eyestone and Ax, 1984). Cysts can be induced by passive immunization with antiluteinizing hormone (LH) antiserum (Nadaraja and Hansel, 1976). Administration of estrogens late in the luteal phase can also result in cystic follicles (Eyestone and Ax, 1984). There are biochemical differences between the fluid from follicles that is obtained from cystic and noncystic cows (Ax et al., 1982), but a mechanism to explain the error in development of a cyst needs to be experimentally established. Studies are now needed that center on causing cows to become cystic so that predisposing factors associated with cysts can be positively identified.

Results from a 40-yr survey of reproductive records of cows in our herd at the University of Wisconsin-Madison showed that there was no seasonal effect on cystic ovaries (Ax et al., 1984). The incidences of cysts increased with age, twinning rates, and postpartum abnormalities. Table 1 shows the frequency of cysts in cows with retained placenta and metritis. As the severity of milk fever increased, so did the probability that a cow would subsequently be diagnosed as cystic.

TABLE 1. CYSTIC OVARIES FOLLOWING RETAINED PLACENTA AND METRITIS

	Percentage of cystic ovaries	
	No	Yes
Retained placenta	5.1	6.5[a]
Metritis	5.1	10.4[a]

Source: R. L. Ax, R. U. Peralta, W. G. Elford, and A. R. Hardie (1984).
[a] $P<.05$

Results from another study using commercial herds in southern Wisconsin showed that approximately 60% of the cysts were diagnosed between days 30 to 60 postpartum and 86% within 120 days postpartum (Ax et al., 1984). Cystic cows had a 26-day longer calving interval than their noncystic herdmates, and the difference was significant ($P<.05$). Conception rates for cystic cows receiving various treatments and those not treated did not differ. However, cows diagnosed as cystic that were treated with gonadotropin releasing hormone (GnRH), human chorionic gonadotropin (hCG), or manual rupture had fewer days open following treatment than those cows not treated (table 2). While there was no advantage in days open among treatments, treated animals overall had 17 fewer days open than those not treated.

TABLE 2. DAYS OPEN (MEAN ± SEM) FOR CYSTIC COWS AFTER VARIOUS TREATMENTS[a]

Treatment	N	Days to conception following treatment
GnRH (100 µg)	69	54.6 ± 4.7[b,d]
hCG (10,000 IU)	51	53.7 ± 5.5[b,d]
Manual rupture	18	52.4 ± 8.9[c,d]
No treatment	32	70.1 ± 8.0

Source: R. L. Ax, R. U. Peralta, W. G. Elford, and A. R. Hardie (1984).

[a] N = cows.

[b] $P<.05$ compared to no treatment

[c] $P<.1$ compared to no treatment

[d] Did not differ among treatments

Only 14% of cows diagnosed one time as cystic repeated the condition in a later lactation (Ax et al., 1984). That finding suggests that treatment of a cow diagnosed as cystic, followed by careful attention at later calving, should provide the option for a cow to be a productive unit in a herd.

REPEAT-BREEDER COWS

Repeat-breeder cows are not sterile since they will ultimately conceive after repeated breedings. These animals are cyclic, have normal estrous cycles and a majority will settle when bred by a bull. The potential for maximum profits and genetic improvement is lost by repeat breeding and use of a bull.

The repeat-breeder syndrome is one of the most poorly understood subjects in bovine infertility. Management, nutrition, infection, genetics, hormonal imbalances, artificial insemination (AI) techniques, and semen quality are some of the factors cited as plausible causative agents for this problem. In recent years more attention has been focused on specific causes of repeat breeding, such as embryonic mortality and the fertilization process. Studies (Ayalon, 1984) have shown that heavy losses due to embryonic death occur around 6 to 7 days after breeding (table 3). This is the time when the morula is developing into the blastocyst and has just entered the uterine horn. Embryonic mortality occurring before day 17 of a cycle would allow for normal interestrous intervals as seen with repeat-breeder cows. Repeat breeders had higher levels of antibodies against seminal antigens in their genital secretions (Bhatt et al., 1979). Identification of additional specific factors that contribute to the repeat-breeder syndrome continue to pose a major research challenge to animal scientists.

TABLE 3. EARLY FERTILITY LOSSES IN DAIRY CATTLE AFTER BREEDING[a]

	Normal cows		Repeat breeders	
Time of slaughter (days)	No.	%	No.	%
4-5	22/25	88[b]	20/25	80[b]
6-7	10/12	83[b]	5/12	42
8-10	13/18	72[b]	9/18	5

Source: N. Ayalon (1984).
[a] Number of normal embryos/embryos found.
[b] $P<.05$

The incidence of repeat-breeder cows varies from herd to herd and year to year. The incidences reported in the scientific literature (Ayalon, 1984; Maurice, 1982) varied between 5% and 18% (table 4).

TABLE 4. INCIDENCES OF REPEAT BREEDERS IN DIFFERENT COUNTRIES

Country	Total cattle	Repeat breeder, %
Sweden	5,744	10.2
USA	5,858	15.1
USA	5,000	18.0
England	5,844	11.6
Israel	4,811	5.0
Israel	3,314	12.2

Source: R. Zemjanis (1980); N. Ayalon (1984).

In herds with 50% conception rates at first service, about 25% of the cows will be presented for three or more services, and these can be considered as repeat breeders.

While some of the causes of infertility could be corrected by improving management techniques, such as heat detection, proper handling of semen, and proper AI technique, others are more complex. Folman et al. (1973) showed a marked relationship between the level of nutrition, body weight changes, and plasma progesterone (P) concentrations. Boice (1979) had reported that 14% of postpartum cows fail to produce sufficient P to support pregnancy should it occur. Recent studies (Lee et al., 1983) showed that intramuscular injection of 100 micrograms of GnRH at the time of breeding improved conception rates in first-service postpartum as well as repeat-breeder cows (table 5). The mechanism of GnRH-induced fertility seems to be via the route of higher P production (Lee et al., 1984).

TABLE 5. PERCENTAGE OF CONCEPTION RATES FOR REPEAT-BREEDER COWS INJECTED WITH 100 μg GnRH OR 2 MILLILITERS PHYSIOLOGICAL SALINE AT BREEDING[a]

	GnRH	Saline
Pregnant	72.9 (135)	47.8 (77)
Nonpregnant	27.1 (50)	52.2 (84)
Total	185	161

Source: C. N. Lee, E. Maurice, R. L. Ax, J. A. Pennington, W. F. Hoffman, and M. D. Brown (1983).
[a] N = ( ).

In a previous review, Jainudeen (1965) reported an 8% to 11% heritability of repeat-breeding conditions. Infections by microorganisms are other possible causes for repeat breeding. The reduction of multiple services per conception would require the combined efforts of sound management and good herd health programs following established genetic principles and hormone therapy when appropriate.

ACKNOWLEDGEMENTS

This paper is a contribution from the University of Wisconsin College of Agricultural and Life Sciences.

REFERENCES

Ax, R. L., R. U. Peralta, W. G. Elford, and A. R. Hardie. 1984. Surveys of cystic ovaries in dairy cows. In: F. H. Baker and M. E. Miller (Ed.) Dairy Science Handbook, Vol. 16. A Winrock International Project. Westview Press, Boulder, CO. 16:205.

Ax, R. L., G. D. Ball, R. W. Lenz, and N. L. First. 1982. Preparation of ova and sperm for in vitro fertilization in the bovine. Proc. 9th Tech. Conf. on Artificial Insem. and Reprod., Nat'l. Assoc. of Animal Breeders. p. 40.

Ayalon, N. 1984. The repeat breeder problem. Proc. 10th International Congress of Animal Reprod. and Artif. Insem. Urbana, IL.

Bhatt, G. N., K. K. Vyas, S. K. Purohit, and P. R. Jatkar. 1979. Studies on immunoinfertility in repeat breeder cows. Indian Vet. J. 56:184.

Boice, M. L. 1979. The relationship between estrous behavior, ovarian activity and milk progesterone in postpartum Holstein cows. MS thesis. University of Illinois, Urbana-Champaign.

Call, E. P. 1978. Economics associated with calving intervals. In: C. J. Wilcox et al. (Ed) Large Dairy Herd Management Symposium. Univ. Press of Florida, Gainesville, FL. p. 190.

Eyestone, W. H. and R. L. Ax. 1984. A review of some endocrine and physiological factors associated with cystic follicles in the bovine. Theriogenology. (In press).

Folman, Y., M. Rosenberg, Z. Herz, and M. Davidson. 1973. The relationship between plasma progesterone concentration and conception in postpartum dairy cows maintained on two levels of nutrition. J. Reprod. Fert. 34:267.

Jainudeen, M. R. 1965. The "repeat-breeder" cow: A review. Ceylon Vet. J. 13:10.

Lee, C. N., E. Maurice, R. L. Ax, J. A. Pennington, W. F. Hoffman, and M. D. Brown. 1983. Efficacy of gonadotropin-releasing hormone administered at the time of artificial insemination of heifers and postpartum and repeat breeder dairy cows. Amer. J. Vet. Res. 44:2160.

Lee, C. N., J. K. Crister, and R. L. Ax. 1984. Endocrine changes in cows injected with gonadotropin-releasing hormone at first service. 79th Ann. Mtng. of Amer. Dairy Sci. Assoc. Abstract. p. 128.

Maurice, E. 1982. Use of gonadotropin-releasing hormone to improve fertility in repeat breeder dairy cows. MS thesis. University of Wisconsin, Madison.

Nadaraja, R. and W. Hansel. 1976. Hormonal changes associated with experimentally produced cystic ovaries in the cow. J. Reprod. Fertil. 47:203.

Zemjanis, R. 1980. "Repeat-breeding" or conception failure in cattle. In: D. A. Morrow (Ed.) Current Therapy in Theriogenology. Saunders Publ. Co. Philadelphia, p. 205.

17

THE RATIONAL USE OF ARTIFICIAL INSEMINATION IN TROPICAL BEEF CATTLE

Dieter Plasse

INTRODUCTION

The intensive use of artificial insemination (AI) in dairy and dual-purpose cattle has been one of the principal forces underlying the dramatic production increases and genetic progress in more highly industrialized temperate-climate countries in recent decades. In beef cattle, however, artificial insemination has been used to a lesser extent.

In tropical Latin America, only Cuba has an efficient national AI program that includes a large part of her cattle population. Statistics from other Latin American countries indicate that less than 5% of the national cow herd is now artificially inseminated.

This paper identifies the reasons for the limited use of AI in tropical Latin America; points out the advantages of this technique; and discusses alternatives for its strategic use in genetic programs.

The discussion is based on a research project that has been in progress 7 yr and that presently covers seven private herds with 5,000 cows, of which 3,500 are in AI programs. This project is cofinanced by the Consejo Nacional de Investigaciones Cientificas y Tecnologicas de Venezuela (CONICIT). Most of these farms are located in the Venezuelan lowlands under extreme tropical conditions. All herds have a genetic program that includes progeny testing and use reference sires *(Bos indicus)* across herds.

REASONS FOR LIMITED USE OF ARTIFICIAL INSEMINATION

Several reports from tropical Latin America (Troconiz et al., 1973; Walfenzao and Albers, 1973; Brouwer, 1975; Morales et al., 1976) and our own unpublished data show that reasonably good results may be obtained by artificially inseminating tropical beef cattle in Latin America if the program is intelligently planned and strategic use is made of the technique. The question then arises: Why is AI not yet widely used?

Some of the main reasons for the poor results in conception rate and, consequently, the discontinuation of the AI programs are:

- **Wrong objectives.** Until now, AI programs in Latin America have mainly been organized and conducted by people trained in reproductive physiology. All too often the AI technique has been presented merely as a means of reproduction; such an approach does not make clear that 1) AI conception rates are, in general, lower than those obtained with natural service (NS) and 2) this disadvantage can be counterbalanced by other specific advantages (discussed later). Clear objectives have seldom been stated for the strategic use of AI in genetic programs.
- **Insemination of the whole herd.** Producers often have been told that the only alternatives with AI are to inseminate the whole herd or do nothing. The whole-herd approach resulted in low conception rates and, consequently, frustration of the producer. The strategic use of AI in that part of the herd that is best prepared to conceive has seldom been attempted.
- **Lack of efficient management programs.** Artificial insemination has frequently been introduced before technical developments (basic management, sanitary, and nutrition programs) were sufficiently advanced to ensure satisfactory results.
- **Lack of trained personnel.** The success of an AI program depends, to a large extent, on the interest, knowledge, and ability of the inseminator as well as that of his collaborators and his supervisor. Most Latin American countries lack this kind of personnel.
- **Lack of tropical technology for AI.** Only in recent years has the reproductive physiology of *Bos indicus* begun to be understood, and it is now well established that it is different from that of *Bos taurus* cattle. However, AI is still being used along the lines developed in the temperate zone with *Bos taurus* cattle.
- **Lack of organization.** There has been a lack of good regional or national organization of AI programs (which also need to be under genetic supervision). In most of the cases, these programs have been too dependent on the importation of semen from developed countries, and such semen often has been lacking in genetic and physiological quality.

ADVANTAGES OF ARTIFICIAL INSEMINATION

Since AI conception rates usually are lower than those obtained with natural service, (and even lower in tropical cattle, apparently), AI must have other advantages if it is to be used. Such advantages may include:

- **More accurate proof of a bull's breeding value.** If bull selection is done in a single herd, the individual's record is normally the selection criterion. However, because overall heritability estimates are .29 for weaning weight and .45 for postweaning weight on pasture in the American tropics (Plasse, 1978a, 1979), such breeding value evaluation may not be sufficiently accurate and it might be desirable to progeny test the bull. However, in natural service, only 15 to 20 offspring can be expected from a sire each year, whereas AI has almost no such limit. With AI, high accuracy can be achieved in the estimation of breeding value by using the bull's own record and the result of a progeny test (sequential selection). Thus, with a larger number of progeny, a high accuracy in breeding value estimation can be achieved. This is important for bulls to be used intensively in AI programs.

  If bulls are to be selected across various herds (a more efficient practice with larger herds), at least partial use of AI is necessary to compare the bulls by means of a reference-sire system (Anonymous, 1981).
- **More efficient use of superior genetic material.** A proven, genetically superior bull that is used only in natural service will have a very limited number of offspring during his life; but the number of offspring is almost without limit when artificial insemination is used. Thus, the genetic impact of a proven bull can be very important.
- **Crossbreeding with unadapted *Bos taurus* bulls.** Recently, crossbreeding commercial *Bos indicus* populatons with *Bos taurus* breeds has been effective in the Latin American tropics under certain circumstances. However, poor adaptation of the imported bulls to the tropical environment has been a negative aspect in such programs. Artificial insemination can solve this problem if the benefit from crossbreeding is higher than the original cost of AI and its usually lower conception rate.

ALTERNATIVES FOR REPRODUCTION PROGRAMS USING ARTIFICIAL INSEMINATION

The design of an AI program must be adjusted to the production system of a specific herd. The strategic use of AI must emphasize:

- The achievement of maximum total reproductive efficiency of the herd. In tropical Latin America, which is characterized by low reproductive efficiency in beef cattle (FAO, 1977), the objective must be to have satisfactory results in the whole herd.

- The establishment of clear objectives that ensure optimal opportunities for the above-mentioned advantages.

To be successful, such strategies usually must combine AI and natural services. A limited breeding season must be established, with no more than two or three inseminations per cow, and followed by natural service (NS) if necessary. Based on this principle and practical experience, five options can be suggested for an AI reproduction program:

1. AI in nonlactating cows and well-developed heifers. (NS optional.)
2. AI in nonlactating pluriparous cows (that have calved early before the breeding season) and heifers. (NS optional.)
3. AI in the whole herd (2 to 3 services), except the youngest heifers and the cows in their first lactation, which would go directly to NS. AI would be followed by natural service.
4. AI in the whole herd (2 to 3 services), no exceptions. Followed by NS.
5. AI in the whole herd, without NS.

Option 1 could be used in herds just starting an AI program or that are being run under very extensive conditions. Option 5 is for herds with high reproductive efficiency and intensive production systems.

STRATEGIC USE OF ARTIFICIAL INSEMINATION IN PURE BREEDING

Individual Herds

Registered or commercial purebred herds might use AI according to one of the five alternative reproduction options with the objective of introducing semen from proven, genetically superior sires into part or all of the herd, followed by progeny testing for their own replacement bulls. The whole herd need not be inseminated. Good reproductive efficiency and good genetic progress can be obtained when only heifers and nonlactating cows are bred artificially to superior bulls; sons are used later in natural service in the rest of the herd.

A large commercial herd might find another application for partial use of AI to produce bulls from that herd. For this purpose, the herd should be thought of as a pyramidal structure that uses the best 10% to 20% of the cows for AI, with the remainder bred by NS to selected sons of this AI program (figure 1). Thus, by using a small number of cows for AI, the breeder can produce the bulls he needs and assure genetic progress in the commercial herd. Where good quality commercial cows and semen of proven bulls are available, this method may turn out to be better and cheaper than that of buying so-called registered bulls that offer only paper certificates but no production records.

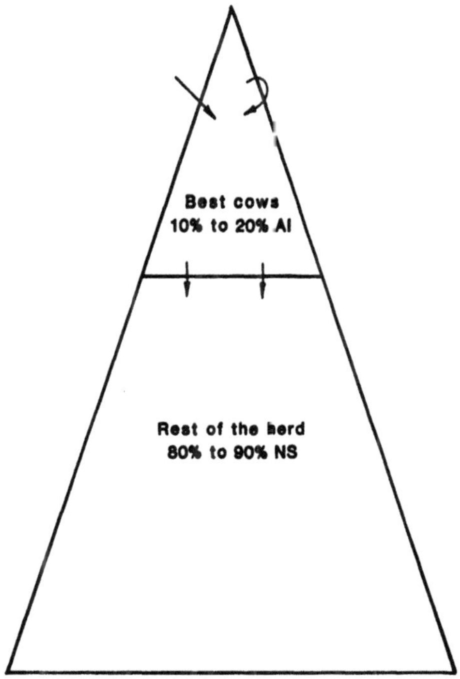

Figure 1. Population structure in a herd where bulls are produced by 10% to 20% of the best cows in the AI program and selected sons are used on the rest of the herd for natural service (NS).
↓↲Flow of bulls and semen.

Another variation of such a program is to use heifers and nonlactating cows for AI and produce the bulls required for the rest of the herd. For example, a herd with 60% weaning and culling of all open cows except those in first lactation, will provide about 33% of the herd as heifers and nonlactating cows to be candidates for the AI program needed to produce the bulls.

Cooperative Programs

The full impact of AI can be achieved only if it is used within a genetic program and in a large population. Although many large herds are found in the Latin American tropics, most cattle production units are below the size needed to assure a self-sufficient and effective genetic program using AI. Small herds have little sire selection. However, if many small herds are combined into a large population and a common genetic program is designed for using AI strategically, maximum selection can be achieved.

If a number of herds are incorporated in an integrated genetic program and reference sires are used across herds, replacement bulls can be selected from all herds in a sequential selection scheme, using their relative individual record on the farm and their progeny test in several herds. Central semen banks can thus be established. Such a cooperative program would require:

- Incorporation of well-organized farms
- A cooperative spirit among the producers
- Production testing
- Genetic design for the whole population
- Technical supervision of the genetic program
- Use of reference sires
- Availability of a computer service
- Statistical and genetic analyses of the results of the progeny test
- Establishment of a central semen bank

A reference sire is a progeny-tested bull with a greater breeding value as compared to those of other reference sires on a number of farms. These bulls would be used by AI in all herds of the cooperative program and the offspring of other bulls would be compared to their progeny. Thus, the best sires could be selected across herds after having been proven in a progeny test that they entered after being selected on their own record in the herd in which they were born. All cooperative herds would use production testing, and such schemes for beef cattle in tropical Latin America have been outlined (Plasse, 1982). Bulls could also be used in natural service, which makes it possible for small producers to participate in the cooperative progeny test and selection.

Cooperative programs need technical supervision and good cooperation among the cattlemen and between the cattlemen and the technical team. Such cooperation is now needed among more progressive producers, if the Latin American countries are to lessen their dependence on semen importation. The present economic conditions, and the frequently questionable quality of semen on the international market, strongly indicate the need for such cooperation.

STRATEGIES FOR USE OF ARTIFICIAL INSEMINATION IN CROSSBREEDING

Although crossbreeding with Criollo *(Bos taurus)* and Zebu *(Bos indicus)* cattle has proved to be beneficial in the Latin American tropics (Plasse, 1983), the use of modern European breeds in specific crossbreeding systems with *Bos indicus* also has had good results. However, the use of imported bulls in these programs has been problematic because of their poor adaptation to the tropics. Our experience shows that under many conditions where crossbreeds are superior to *Bos indicus*, the use of *Bos taurus* bulls is not feasible and AI must be used.

Several European breeds in the Venezuelan lowlands are now used in AI of nonlactating cows and well-developed heifers. The males are later moved for fattening to areas with improved pastures. The first results have been very good (Hoogesteijn et al., 1983a,b; Plasse et al., 1983) and other breeds are being tested for this purpose.

In larger herds, when good quality pasture is available and used inefficiently by pure Zebu, a strategic AI program in part of the herd could produce F_1 steers with hybrid vigor and with slaughter weights up to 20% greater than those of *Bos indicus*.

Semen of *Bos taurus* also might be used in rotational crossbreeding programs with *Bos indicus* bulls for the establishment of gene pools or in three-breed crosses (for instance, inseminating F_1 Criollo x Zebu cows with Charolais bulls).

REGIONAL OR NATIONAL GENETIC PROGRAMS WITH STRATEGIC USE OF ARTIFICIAL INSEMINATION

Maximum genetic progress in the beef cattle population of a country or a region can be achieved only by development of an integral structure and genetic program.

In most of the Latin American countries, the national or regional population (within a breed) is made up of different strata:

- Registered purebred herds
- Commercial cows of high genetic quality
- Commercial cows of lower genetic quality

The numbers of registered cows in Mexico (Plasse, 1978b) and Venezuela (Plasse, 1980) are estimated to be insufficient to produce the numbers of bulls needed, if minimum selection is practiced. In the future, most countries should consider including production-recorded commercial cows in the national bull-producing herd. The national or regional population would then have at least four strata:

- Registered cows in AI (producing bulls)
- Registered cows in NS (producing bulls)
- Commercial elite cows in NS (producing bulls)
- Commercial cows in NS (not producing bulls)

Figure 2 shows the approximate percentage of cows in each stratum.

Taking into account the current reproductive efficiency of Latin American beef cattle populations, about 10% of the cows should be assigned for the production of replacement bulls at the beginning of the program. Thus, 3% to 5% of all cows could be artificially inseminated; a goal that would be easy to achieve at the beginning.

However, the present situation in Mexico and Venezuela suggests that such percentages might be unrealistic; a less efficient but more realistic plan suggested (Plasse, 1978b; 1980) might assign only 3% of all cows to systematic bull production. This plan would not guarantee optimum selection

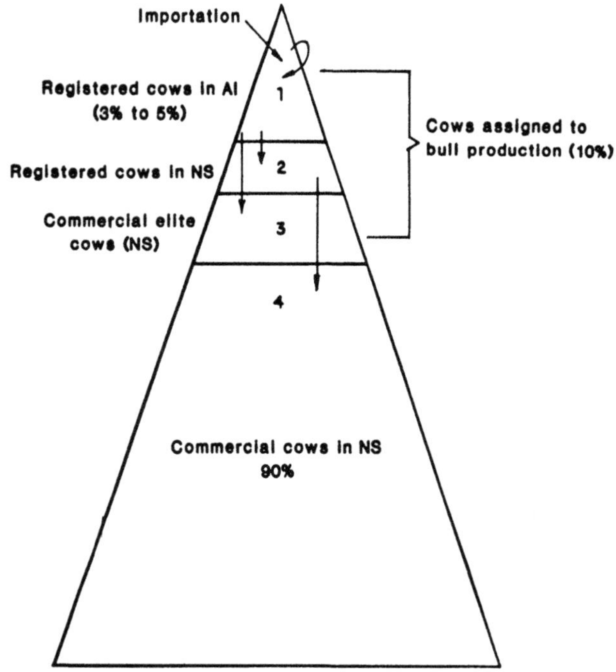

Figure 2. Desirable population structure to initiate a national or regional genetic program in Latin America.
↓↻Flow of bulls or semen.

progress, but would be better than the present situation where only about 1% of all cows are in registered herds that, officially, produce the replacement bulls for the national population.

Although I am not suggesting that the present national beef cattle populations in Latin America could use such a program, a beef improvement program might begin with a small regional plan or, as shown at the top of the pyramid (figure 2), by using registered cows. The use of AI, combined with a production and progeny testing program in 10% to 20% of the registered cows, would not be difficult to achieve. The sons of this program would be used later on in the rest of the registered herds that would produce the sires for the commercial bull-producing herds. These sires would (together with the registered population) produce the bulls for the commercial cows. In this way, a small but well-planned AI program incorporated in a national or regional scheme could benefit all strata, even the commercial population. Thus, genetic progress in the majority of the commercial cattle at the bottom of the pyramid would result from the use of highly selected bulls at the top. Because some commercial cattle have high genetic value, a limited genetic flow from the bottom to the top might be beneficial, also.

RESULTS OF REPRODUCTIVE PROGRAMS WITH COMPLETE OR PARTIAL USE OF ARTIFICIAL INSEMINATION

Drawing from the five options I previously recommended for an AI reproduction program, I have listed below some selected results of our program as used strategically in private herds. For more than 7 yr, more than 25,000 cow-years have been accumulated from programs where AI is used as the only method of reproduction, or is used along with natural service. (Numbers given in brackets after the mean in each option refer to the range between extreme years.)

Option 1 has been used in two herds. In a registered herd maintained under relatively intensive conditions, 646 two-year-old heifers were bred artificially over a 9-yr period. They conceived with 1.9 services per conception for a pregnancy rate of 80% (76% to 92%). The 241 dry cows had an 88% pregnancy rate, with 1.6 services per conception.

Option 2 was used under extensive conditions in the tropical lowlands, where 3,455 dry commercial Zebu cows and heifers were inseminated with semen of *Bos taurus* bulls according to option 1. Over a 7-yr period, the mean pregnancy rate from AI was 77% (60% to 86%) and another 10% (8% to 14%) from clean up (Zebu) bulls totaling 87% (73% to 96%). There were 1.7 services per conception.

Option 3 was used only recently in three herds, and it is hoped that it will slightly improve total reproductive efficiency under extensive conditions.

Option 4 was used for 4 yr in two herds of *Bos indicus* cows. In a total of 8,125 cow-years, 54% of all cows conceived from AI and 16% from natural service, resulting in a total pregnancy rate of 70%. (In these areas the usual mean calving percentage is estimated to be around 40%.)

Option 5 involved AI in the whole herd, without natural service on two ranches. One of the herds has Guzerat cattle under very extensive conditions, with low pasture quality. In a 3-yr period, and for 643 cow-years, a pregnancy percentage of 61% (52% to 68%) was achieved with 2.0 services per conception.

This option was used also in a registered Brahman herd, under intensive conditions with good pasture supply, and in a climatically favorable area. AI was used for 5 yr (without bulls) and an 82% (76% to 85%) pregnancy was achieved in 1,117 cow-years. There were 1.8 inseminations per conception.

CONCLUSIONS

Artificial insemination is a useful technique when used strategically in genetic programs for beef cattle. Several options might be used to combine AI with natural service to obtain optimum results in reproductive efficiency.

In the Latin American tropics, the fùture development of crossbreeding programs and efficient selection programs will depend on the rational use of AI. The application of this method, however, requires carefully considered objectives.

Our data and experiences show that an 80% pregnancy rate can be achieved by integral use of AI under favorable tropical conditions. Under extensive tropical conditions in Latin America, a 70% or greater pregnancy rate might be achieved, if AI is used together with natural service and an adequate management program.

REFERENCES

Anonymous, 1981. Guidelines for uniform beef improvement programs. USDA - FED Program Aid 1020.

Brouwer, J. R. 1975. Problemas de inseminacion artificial en el ganado en America Latina. 9a. Conferencia Anual de Ganaderia y Avicultura en America Latina. Mem. 1-29.

FAO. 1977. Informe de la consulta de expertos para el mejoramiento de la eficiencia reproductiva del ganado vacuno en America Latina. FAO, Maracay, Venezuela.

Hoogesteijn, R., D. Plasse, O. Verde, P. Bastidas, y R. Rodriguez. 1983a. Peso al nacer de becerros de vacas Brahman comercial y toros Brahman, Simmental y Marchigiana en Apure. III Congreso Venezolano de Zootecnia. Mem. F28.

Hoogesteijn, R., O. Verde, D. Plasse, P. Bastidas, y R. Rodriguez. 1983b. Peso al destete y mortalidad predestete de becerros de vacas Brahman comercial y toros Brahman, Simmental y Marchigiana en Apure. III Congreso Venezolano de Zootecnia. Mem. F29.

Morales, J. R., C. Iglesias, A. Menendez, J. Dora, y H. Chavez. 1976. Resultados de los servicios de inseminacion artificial (IA) en hembras bovinas de las razas de leche y carne, en el clima subtropical de Cuba. Rev. Cub. Reprod. Anim. 2:60.

Plasse, D. 1978a. Aspectos de crecimiento del *Bos indicus* en el tropico Americano. (Primera parte.) World Rev. Anim. Prod. XIV, 4:29.

Plasse, D. 1978b. Modelos geneticos que utilizan la inseminacion artificial para la Asociacion de Ganaderos de Criadores de Cebu de la Republica de Mexico. V Ciclo Internacional de Conferencias sobre Ganaderia Tropical, Villahermosa, Mexico 24-27 de Octubre de 1978. Mem. (In press.)

Plasse, D. 1979. Aspectos de crecimiento del *Bos indicus* en el tropico Americano. (Segunda parte.) World Rev. Anim. Prod. XV, 1:21.

Plasse, D. 1980. Modelos geneticos basados en inseminacion artificial en ganado de carne. ALPA Mem. 15:17.

Plasse, D. 1982. Performance recording of beef cattle in Latin America. World Anim. Rev. 41:11.

Plasse, D. 1983. Crossbreeding results from beef cattle in the Latin American tropics. Anim. Breeding Abstracts. 51, 11:779.

Plasse, D., R. Hoogesteijn, O. Verde, P. Bastidas, y R. Rodriguez. 1983. Peso a 18 meses de becerros de vacas Brahman comercial y toros Brahman, Simmental y Marchigiana en Apure. III Congreso Venezolano de Zootecnia. Mem. F30.

Troconiz, J., A. Hernandez, M. Quintero, D. Navarro y O. Silva. 1973. Evaluacion de un programa de inseminacion artificial en ganado de carne en un hato llanero. In: R. Sosa, H. Welcker, y R. Salom (Ed.) Ganaderia en los Tropicos. Caracas, Venezuela.

Walfenzao, M. and E. Albers. 1973. Inseminacion artificial con semen congelado en un rebano Cebu. In: R. Sosa, H. Welcker, and R. Salom (Ed.). Ganaderia en los Tropicos. Caracas, Venezuela.

18

CHANGING REPRODUCTIVE PERFORMANCE IN BEEF COW HERDS

J. N. Wiltbank

Reproduction in a beef cow herd is a fragile thing, easy to disrupt and difficult to re-establish. The beef cow has priorities. Her first priority is survival, her second priority is the survival of her calf, and her third priority is reproduction. This means the first two priorities must be met before the third can be accomplished. It is difficult, with today's economy, to have good reproduction without excessive cost. The purpose of this paper is to briefly outline those ingredients necessary for good reproduction to occur in a beef cow herd, show methods that can be utilized to predict reproduction performance, outline methods for determining the economic feasibility, and set up methods whereby reproduction can be improved in your cow herd.

NECESSARY INGREDIENTS FOR GOOD REPRODUCTION

Most beef herds contain many nonproducers such as dry cows, replacement heifers and bulls, and many cows that wean light calves. For an example of nonproducers, let's look at a beef herd containing 100 cows (table 1). In addition to the 100 cows, there would be 15 replacement heifers and 5 bulls. If 90 cows weaned a calf, there would be 30 nonproducing animals in this herd. Even larger numbers of nonproducing animals and greater costs are noted when only 70 or 80 calves are produced.

TABLE 1. NONPRODUCERS IN A 100-COW HERD[a]

No. calves weaned	Dry cows	Replacement heifers	Bulls	Nonproducers No.	Nonproducers %	Cost per calf
90	10	15	5	30	25	$333
80	20	15	5	40	33	375
70	30	15	5	50	42	428

[a]$250 per animal carrying cost.

The number of nonproducers must be reduced to make production of calves economically feasible. The cost of keeping nonproducers is as great or greater than the cost of keeping producers, thus the cost of producing a calf is higher because of the large number of nonproducers.

Calves that are light at weaning will not pay the costs of keeping the cow. As an example, consider calves weaning at different weights (table 2). It does not take a mathematician to recognize the value of the heavy calf.

TABLE 2. WEANING WEIGHT AND NET RETURN

Weaning weight, lb	Gross return at $.70/lb	Cost of keeping a cow, $	Net return, $
500	$350	250	100
450	315	250	65
400	280	250	30
350	245	250	-5
300	210	250	-40

Calves wean light because they are born late in the calving season or do not grow, or both. As an example, look at table 3.

TABLE 3. WEANING WEIGHT AS INFLUENCED BY TIME OF BIRTH AND AVERAGE DAILY GAIN

Day of calving	Average age weaning, days	Average daily gain, birth to weaning, lb		
		2.25	2.0	1.75
0-20	220	565	510	455
21-40	200	520	470	420
41-60	180	475	430	385
61-80	160	430	390	350
81-100	140	385	350	315
101-120	120	340	310	280
121-140	100	295	270	245

Look at the differences in weaning weights in this herd. Calves varied from 565 lb to 245 lb. The late calves were light even when they gained 2.25 lb a day. You just cannot leave calves on the cow and expect the calves to continue to gain. Calves stop growing when grass dries up and milk production stops in the cow. If calves are to wean at heavier weights, they must be born early in the calving season, and they must have the genetic ability to grow and the necessary nutrients to grow. A cow must wean at least 350 lb of calf to pay her own costs. When you consider paying the cost of nonproducers, each cow must wean a considerably heavier calf.

To get the complete cost picture, consider the net returns when costs of heavy calves and nonproducers are analyzed together (table 4).

TABLE 4. INFLUENCE OF NONPRODUCERS AND WEANING WEIGHT ON POUND OF CALF WEANED AND NET RETURN IN A 100-COW HERD

Calves weaned in 100-cow herd	Total animals in herd	Non-producers	Lb of calf weaned per animal[a]			Net return per animal[b]		
			500	400	300	500	400	300
90	120	30	375	300	225	12	-40	-92
80	120	40	333	267	200	-17	-63	-110
70	120	50	292	233	175	-46	-87	-128

[a]Average weaning weight per calf.
[b]Calves at $.70/lb and $250 carrying cost.

If the herd is to make money, the number of nonproducers must be kept low and the average weaning weight must be high. For cows that weaned calves averaging 500 lb, the pounds of calf weaned per animal in the herd varied from 375 lb to 292 lb because the pounds of calf weaned must be averaged out over a lot of nonproducers. Most of the figures on net return are negative. The returns are positive only for the herd in which cows weaned 500 lb of calf, and in which there were only 30 nonproducers. Now look at table 3 and see how many calves weighed 500 lb or more -- only those earliest-born calves that gained 2 lb or more a day weighed over 500 lb.

A worthwhile goal for herd reproduction is to have 75% to 80% of the cows calving in the first 20 days of the calving season, with 95% of the cows calving within a 60-day season. When 80% of the cows calve in the first 20 days, a calving rate of 95% within 60 days is relatively easy. Consequently, our attention in this paper will be centered on achieving an 80% pregnancy rate in 20 days. If 80% of the cows are going to calve in the first 20 days of the calving season, then 95% to 100% must show heat in the first 20 days of the breeding season and 80% to 85% must become pregnant on first service. To achieve this, we must have precise management control in a beef cow herd. We cannot just hope. Everything must be done correctly, with attention given to details.

METHODS FOR PREDICTING PERFORMANCE

The number of cows becoming pregnant early in the breeding season is determined by the following formula:

Cow in heat 1st 20 days of breeding	x	Cows becoming pregnant from 1st service	=	Cows pregnant 1st 20 days of breeding season

The rate at which cows become pregnant from first service is a combination of cow fertility and fertility of the bull. Therefore, three things must be accomplished if 75% to 80% of the cows are to become pregnant during the first 20 days of the breeding season.

1. Ninety-five percent to 100% of the cows must show heat the first 20 days of the breeding season.
2. Cow fertility must be high.
3. Cows must be bred by a fertile bull.

These three factors are not additive, but multiplicative. In other words, the formula is:

Cows in heat 20 days	x	Cow fertility	x	Bull fertility	=	Cows pregnant 20 days

not:

$$\frac{\text{Cows in heat 20 days} + \text{Cow fertility} + \text{Bull fertility}}{3} = \text{Cows pregnant 20 days}$$

Poor performance in one area cannot be averaged out. If one factor is low, then the ultimate goal of cows pregnant early in the breeding season will be low. Three examples can illustrate the importance of this concept.

	Cows in heat 20 days, %	x	Cow fertility, %	x	Bull fertility, %	=	Cows pregnant 20 days, %
Ex. 1	95	x	95	x	95	=	86
Ex. 2	65	x	95	x	95	=	59
Ex. 3	65	x	95	x	60	=	37

In example 1, only 86% of the cows are pregnant, even though all factors are 95%. In this equation you do not average the factors but they are multiplied, consequently, 95% x 95% x 95% = 86%.

In example 2, only one factor -- in heat in 20 days -- is low, but note that even though the other two factors are high, only 59% of the cows are pregnant. The percentage pregnant will be lower than the lowest factor. In example 3, two of the factors are low; consequently, only 37% of the cows are pregnant early in the breeding season. The proportion pregnant can be no higher than the lowest factor. These examples illustrate that to achieve good reproductive performance all factors must be high.

A system was designed to cause most of the cows to show heat in the first 20 days of the breeding season and to have a high conception rate at first service. This system was called the O'Connor system.

O'Connor System

The O'Connor management system was first put into practice on Mr. Tom O'Connor's ranch near Victoria, Texas. The reproductive performance in a small group of cows was noted to be exceptionally high (table 5).

TABLE 5. REPRODUCTIVE PERFORMANCE IN A HERD AT O'CONNOR RANCH

21 days	42 days	63 days	84 days
80	87	87	93

A large proportion of the cows became pregnant in a short period because 1) all cows in this group calved at least 30 days prior to the start of the breeding season; 2) cows were in moderate or good body condition at calving time; 3) cows were gaining weight for 3 wk prior to the start of the breeding season and for the first 3 wk of the breeding season; 4) calves were removed from cows for 48 hr at the start of breeding season; and 5) cows were bred to fertile bulls.

The number of cows involved were small, therefore an experiment was designed at Brigham Young University to further test the concepts of this management system and compare pounds of calf weaned with a control group. The work was done cooperatively on a ranch at Elberta, Utah. Mr. Dale Jolley was the manager. Two hundred and thirty cows were checked for pregnancy in October. An attempt was made to divide the cows into groups by stage of pregnancy. The cows had been exposed to bulls for 5 mo; consequently, some cows were only 35 to 40 days pregnant at the time of pregnancy examination.

Cows selected to be in the O'Connor management group were all early-calvers (estimated to calve 30 days before the start of the breeding season), while cows in the control group were expected to calve within a 120-day period that extended into the breeding season. The control group contained the same percentage of early-calving cows as were found in the original 230 head. Cows were scored for body condition and divided so that both groups had cows in similar conditions. Most cows in both groups were in moderate or good body condition at calving time. Cows in the O'Connor management group were full-fed corn silage, starting 2 wk before breeding, and were continued on this ration for the first 3 wk of breeding. Calving started in the last of January and bulls were turned in with the cows on April 22. All bulls were evaluated for fertility 4 wk before the start of the breeding season. All bulls turned with the O'Connor group had testicles larger than 32 cm in circumference and had 70% or more normal sperm.

Calves were removed from cows for 48 hr and the bulls were placed with the cows at the time of calf removal.

Thirty-three of the eighty-nine cows in the O'Connor management group showed heat within 24 hr after calf removal. Twenty-five days after the start of the breeding season, 95% of the cows in the O'Connor system had been bred. This increased to 98% by 46 days after the start of breeding, while in the control group only 72% had shown heat after 46 days of breeding (table 6).

TABLE 6. REPRODUCTIVE PERFORMANCE AT ELBERTA, UTAH, USING THE O'CONNOR SYSTEM

	Cows managed under		
	O'Connor system	Control system	Difference
No. cows	89	86	
Showing heat after breeding, %			
25 days	95	59	36
46 days	98	72	26
Pregnant after 1 breeding	80	50	30
Calves, %			
After 20 days	80	28	52
After 40 days	91	52	39
After 60 days	99	72	27
After 120 days	99	93	8

Conception rate at first service was 80% in the O'Connor group as compared to 50% in the control group. Eighty percent of the cows managed under the O'Connor system calved during the first 20 days of the breeding season, in contrast to 28% in the control group. Ninety-one percent of the O'Connor cows had calved within 40 days, while only 52% of the controls had calved. It was 120 days before 91% of the control group of cows had calved. The losses following pregnancy check are shown in table 7.

TABLE 7. LOSSES OF EMBRYOS OR CALVES FOLLOWING PREGNANCY DIAGNOSIS

	O'Connor system	Control system	Difference
No. cows bred	98	91	--
Cows pregnant, %	99	93	6
Cows calved, %	98	91	7
Live calves born, %	95	87	8
Calves weaned, %	92	78	14

The O'Connor group had 8% more calves alive at birth; this was largely because 6% more cows were pregnant in the O'Connor group. Losses in pregnancy due to calving were similar; however, the losses following calving differed markedly between the two groups. Six percent of the O'Connor cows lost calves from birth to weaning compared to 13% in the control cows. Losses were higher in late-calving cows, mostly as a result of scours. The O'Connor cows weaned 14% more calves than did the control cows, which can be attributed to a higher pregnancy rate and a lower loss of calves from birth to weaning.

Five management principles for comparing fertility in cow herds can help ensure that a high percentage of cows become pregnant in a short period of time: 1) a 60-day breeding season; 2) a nutrition program to ensure that all cows are in moderate or better body condition at calving; 3) a nutrition program to make certain cows are gaining weight for 3-wk period prior to breeding and in the first 3 wk of breeding; 4) a method for removing calves for a 48-hr period at the start of the breeding season; 5) a program for evaluating bulls for potential fertility.

ECONOMIC FEASIBILITY

The next question -- does it pay? In the Brigham Young University study, calf weaning weight was determined and additional costs were known (table 8). The additional costs were $1,095 for cows in the O'Connor System. An additional 9,016 lb of calf were weaned or approximately 90 lb of calf per cow bred. It was estimated that income was increased $5,410 in a 100-cow herd when calves sold for $.60/lb.

Reproduction can be improved, but each thing must be done correctly. Handling cows, as outlined in the O'Connor system, will ensure that 90% to 95% of the cows will show heat in the first 20 days of the breeding season, with a high conception rate at first service.

The important question is, "Will the O'Connor method work on your ranch and make you money?" To answer this question, you must first determine the reproductive performance and the pounds of calf weaned on your ranch. To help achieve this, we can look at some ways to measure reproductive performance and to predict pounds of calf weaned. There are two important components of reproduction; number of calves weaned and calf size. Differences in calf size are mainly the result of calf age. Thus, time of birth is important. Thus, to measure reproduction in a cow herd, we must know the number of calves and have an estimate of birth date.

Information necessary to estimate the pounds of calf weaned includes an estimate of the birth date, the weaning date, the average daily gain, and an estimate of birth weight. Number of days to grow is calculated from the weaning date and birth date. Let's assume that a calf has a weaning date of October 16 and an estimated birth date of March 3; it has 227 days to grow. This calculation can be made easily using the Julian calendar, which lists the days in numerical order for two years (figure 1). As an example, January 1 is 1, December 31 is 365, May 15 is 135. In calculations used here, March 3 is 62 while October 16 is 289: 289 - 62 = 227 days.

TABLE 8. ESTIMATED ECONOMIC VALUE OF THE O'CONNOR MANAGEMENT SYSTEM IN A 100-COW HERD

	Cows managed under		
	O'Connor system	Control system	Difference
Additional cost, $			
Feed	910		910
Labor	60		60
Semen evaluation	125		125
Total	1,095		1,095
Production			
Calves weaned, %	92	78	14
Age at weaning, days	253	229	24
ADG, lb	1.52	1.43	.09
Weaning weight, lb	488	460	28
Total lb weaned	44,896	35,880	9,016
Return at $.60/lb	$26,938	$21,528	$5,410
Additional cost	$1,095		$1,095
Increase in net return	$25,843	$21,528	$4,315

Worksheet #1 (see Worksheet Appendix) can be used to calculate weaning weight; in this example, four calves were born 20 days apart. The average daily gain can be estimated if average weaning weight is known and an estimate can be made of average calving date. In this example, the average weaning weight of all calves is 460 lb. It was estimated that 50% of the calves were born by April 10. The calves were weaned October 16 so that the calves had 189 days to grow (289 - 100 = 189). The calves are estimated to weigh 80 lb at birth.

The ADG is calculated in this manner:

Avg weaning wt - Birth wt = Gain, birth to weaning

460 - 80 = 380

Gain = Avg daily gain

380/189 = 2.0 lb/day

This figure was used in calculating weaning weights.

Calving time and age of cow: More early-calving cows will show heat during the first 20 days of the calving

No.	1	2	3	4	5	6	7	8	9	10	11	12
Day	January	February	March	April	May	June	July	August	September	October	November	December
1	1	32	60	91	121	152	182	213	244	274	305	335
2	2	33	61	92	122	153	183	214	245	275	306	336
3	3	34	62	93	123	154	184	215	246	276	307	337
4	4	35	63	94	124	155	185	216	247	277	308	338
5	5	36	64	95	125	156	186	217	248	278	309	339
6	6	37	65	96	126	157	187	218	249	279	310	340
7	7	38	66	97	127	158	188	219	250	280	311	341
8	8	39	67	98	128	159	189	220	251	281	312	342
9	9	40	68	99	129	160	190	221	252	282	313	343
10	10	41	69	100	130	161	191	222	253	283	314	344
11	11	42	70	101	131	162	192	223	254	284	315	345
12	12	43	71	102	132	163	193	224	255	285	316	346
13	13	44	72	103	133	164	194	225	256	286	317	347
14	14	45	73	104	134	165	195	226	257	287	318	348
15	15	46	74	105	135	166	196	227	258	288	319	349
16	16	47	75	106	136	167	197	228	259	289	320	350
17	17	48	76	107	137	168	198	229	260	290	321	351
18	18	49	77	108	138	169	199	230	261	291	322	352
19	19	50	78	109	139	170	200	231	262	292	323	353
20	20	51	79	110	140	171	201	232	263	293	324	354
21	21	52	80	111	141	172	202	233	264	294	325	355
22	22	53	81	112	142	173	203	234	265	295	326	356
23	23	54	82	113	143	174	204	235	266	296	327	357
24	24	55	83	114	144	175	205	236	267	297	328	358
25	25	56	84	115	145	176	206	237	268	298	329	359
26	26	57	85	116	146	177	207	238	269	299	330	360
27	27	58	86	117	147	178	208	239	270	300	331	361
28	28	59	87	118	148	179	209	240	271	301	332	362
29	29	-	88	119	149	180	210	241	272	302	333	363
30	30	-	89	120	150	181	211	242	273	303	334	364
31	31	-	90	-	151	-	212	243	-	304	-	365

No.	1	2	3	4	5	6	7	8	9	10	11	12
Day	January	February	March	April	May	June	July	August	September	October	November	December
1	366	397	425	456	486	517	547	578	609	639	670	700
2	367	398	426	457	487	518	548	579	610	640	671	701
3	368	399	427	458	488	519	549	580	611	641	672	702
4	369	400	428	459	489	520	550	581	612	642	673	703
5	370	401	429	460	490	521	551	582	613	643	674	704
6	371	402	430	461	491	522	552	583	614	644	675	705
7	372	403	431	462	492	523	553	584	615	645	676	706
8	373	404	432	463	493	524	554	585	616	646	677	707
9	374	405	433	464	494	525	555	586	617	647	678	708
10	375	406	434	465	495	526	556	587	618	648	679	709
11	376	407	435	466	496	527	557	588	619	649	680	710
12	377	408	436	467	497	528	558	589	620	650	681	711
13	378	409	437	468	498	529	559	590	621	651	682	712
14	379	410	438	469	499	530	560	591	622	652	683	713
15	380	411	439	470	500	531	561	592	623	653	684	714
16	381	412	440	471	501	532	562	593	624	654	685	715
17	382	413	441	472	502	533	563	594	625	655	686	716
18	383	414	442	473	503	534	564	595	626	656	687	717
19	384	415	443	474	504	535	565	596	627	657	688	718
20	385	416	444	475	505	536	566	597	628	658	689	719
21	386	417	445	476	506	537	567	598	629	659	690	720
22	387	418	446	477	507	538	568	599	630	660	691	721
23	388	419	447	478	508	539	569	600	631	661	692	722
24	389	420	448	479	509	540	570	601	632	662	693	723
25	390	421	449	480	510	541	571	602	633	663	694	724
26	391	422	450	481	511	542	572	603	634	664	695	725
27	392	423	451	482	512	543	573	604	635	665	696	726
28	393	424	452	483	513	544	574	605	636	666	697	727
29	394	-	453	484	514	545	575	606	637	667	698	728
30	395	-	454	485	515	546	576	607	638	668	699	729
31	396	-	455	-	516	-	577	608	-	669	-	730

Figure 1. Julian Calendar.

season than will late-calving cows, and fewer young cows will show heat than will older cows. An example can help in understanding this concept (table 9).

TABLE 9. CALVING TIME AND HEAT, FIRST 20 DAYS OF THE BREEDING SEASON

Time of calving	Young cows, %	Mature cows, %
First month	79	94
Second month	44	69
Third month	5	10

Hypothetical Ranch as Example

To help understand the concepts of the O'Connor system, a hypothetical ranch will be used. The breeding system last spring started May 8 and ended July 27. The cows were checked for pregnancy in September and the calving dates were estimated as shown in Work Table #1 (see Work Table Appendix).

The body condition of the cows was estimated at the time of the pregnancy diagnosis. The problem will be concerned with the effect of expected calving date on pregnancy rate. In this problem, we will estimate that the 80 cows are in moderate body condition. They are expected to calve February 14 to May 14. Twenty cows calved between February 14 and March 5. Breeding will start May 14, and cows will be bred for 60 days. A step-by-step solution to the problem is outlined below.

PROBLEM 1

1. Use Work Sheet #2 for post-partum cows.

2. Divide the expected calving dates into 20-day intervals. **Example:** February 14 and March 5, March 6 to March 25, etc.

3. Calculate the average calving date for each group. **Example:** If calving dates are between February 14 and March 5, the average calving date would be February 24. Then find the Julian date (i.e., February 24 = 55, March 16 = 75, etc.).

4. Record the number of cows in each calving group on the work sheet for the first, second, and third 20 days. **Example:** February 14 to March 5 = 20 cows calved.

5. Look up the Julian date for start of calving. **Example:** May 15 = 135. Calculate and record the Julian date at

the end of the first 20 days of breeding. **Example:** 135 + 20 = 155. Also 135 + 40 = 175 or the end of the second 20 days or 40 days of breeding. 135 + 60 = 195 or the end of the third 20 days or 60 days of breeding.

6. Find difference between average calving date and date end of 20 days of breeding. **Example:** 155 - 55 = 100. Do the same for the end of 40 and 60 days. 175 - 55 = 120; 195 - 55 = 140.

7. Look up the expected percentage to be cycling at the indicated days post-partum in Work Table #2 under moderate and record. **Example:** 100 days post-partum, 100% are cycling.

8. Calculate expected number of cows cycling. **Example:** 20 cows x 100% = 20 cow cycling.

9. Assume a 60% conception rate.

10. Calculate expected number of cows pregnant. **Example:** 20 x 60% = 12.

11. Calculate for second 20 days of breeding (June 5 through June 24).

12. Look up those cycling at 120 days (100% x 20 = 20).

13. Subtract cows pregnant to previous breeding.

14. Assume conception rate of 60%.

15. Calculate pregnant (8 x 60% = 5).

16. Calculate expected number of cows pregnant; figure for third 20-day period. Remember, "previously pregnant" involves 2 breedings. In this example -- 12 + 5 = 17.

17. Repeat procedure for each 20 days of expected calving period.

You can now estimate the calving date for cows next year. Cows becoming pregnant in the first 20 days of the breeding season (average date is May 25, day 145), will calve March 3 (average breeding date [145] + length of gestation [282] = 427, or March 3). The average calving date for cows bred in the second 20 days is March 28 and in the third 20 days, April 12. Worksheet #3 is designed to help you to estimate the average weaning weight of calves born to cows that calved the first year, February 14 to March 5, March 6 to March 25, etc.

To calculate the estimated pound-of-calf-weaned and gross return, you first use Worksheet #2 to find the number of cows pregnant. For example, during calving period 2-24, 12 cows became pregnant in the first 20-day period, 5 in the second 20 days and 2 in the third 20 days. Enter these findings on Worksheet #3 under cows pregnant first 20 days, second 20 days, and third 20 days. Next, use Worksheet #1 to calculate average weaning weight. Enter average weaning weight and multiply by number of calves; calculate the total pounds of calf weaned and gross return per group of cows; note the cows that make the most money!

Body condition is important in determining the proportion of cows showing heat and becoming pregnant. Many cows in thin body condition do not become pregnant. In one study in Florida, the proportion open varied from 77% in very thin cows to 5% in those in good body condition (table 10).

TABLE 10. RELATIONSHIP BETWEEN BODY CONDITION AND PREGNANCY RATE IN FLORIDA

	Very thin	Thin	Slightly thin	Moderate	Good
No. of cows	115	545	564	344	234
% open	77	49	27	14	5
Early calvers, %	5	15	19	40	56

Only 5% of the thin cows calved early compared to 56% of the cows in good body condition. The main reason thin cows do not become pregnant or calve late is that the proportion of cows showing heat is delayed in cows in thin body condition. In Work Table #2, the proportion of cows that have shown heat by 60 days after calving is much higher among cows that are in good body condition (91%) as compared to those in moderate (61%) or thin (46%) condition. Even at 100 days after calving, only 70% of the cows in thin body condition had shown heat.

Problem 2 involves the same ranch and provides an example of the relationship between body condition, calving date, heat and pregnancy.

PROBLEM 2: CHECKLIST

1. Use all cows on Work Table #1
2. Use Work Table #2
3. Estimate that breeding begins May 15; continue for 40 days
4. Assume 60% conception rate
5. Use Cow Worksheet #4
6. Be sure to calculate pounds of calf weaned and gross return per cow

Two approaches can be used to keep cows in moderate body condition. First, cows should be observed carefully for 1 mo or 2 mo before calves are scheduled to be weaned.

If cows are thin, then calves should be weaned right away. This will give cows a few months of good feed before the quality of the forage declines. Calves are probably growing at a slow rate because of low-quality feed available. Consequently, you can wean the calves and put on more feed for faster growth.

With the second approach, cows can be sorted by body condition at weaning time. Cows should be sorted for body condition using a rating of from 1 (thinnest) to 9 (fattest). A scoring sheet describing this method is found in the figure 2. Decisions on feeding should then be made. The amount of weight gain needed to change body condition must be kept in mind. Work Table #3 is included to help with this. Examples of how to use Work Table #3 are found in the following exercises:

1. How many pounds does a cow have to gain if she has a score of 5? __________
2. What does this gain consist of? __________
3. How much weight does a cow have to gain if she has a score of 4? __________
4. What does weight gain consist of? __________
5. To improve a body condition score of 1, how much weight does a cow have to gain? __________
6. If a cow scores a 3 and is 100 days from calving, how much weight does she have to gain each day? __________ What if she is 180 days from calving? __________
7. What does she need to gain each day to score a 5 at calving time? __________
8. What methods can we use to make certain each cow scores a 5 at calving time? __________

Next consider the weight changes needed by a cow following calving to change body condition (see Work Table #4):

1. The body condition for a cow at breed time is a _____.
2. The weight gain needed to improve body condition one score is _____ lb.
3. The ADG needed by a cow calving 60 days before start of breeding and scoring a 3 is _____.

The level of feed required to make ADG is shown in Work Table #5:

1. Does a cow suckling a calf or a pregnant cow require more feed to make a weight change? __________
2. Is the level of feed for a cow scoring a 3 the same as for a cow scoring a 5? __________
3. Is it easier to change body condition before or after calving? __________
4. Are the costs of changing body condition before and after calving comparable? __________
5. What do you need to do to change body condition? __________

Thin	1.	Poor - starving - bordering on inhumane - survival questioned during stress.
	2.	Very thin - poor milk production - chances for rebreeding slim to none. Some fat present along backbone but no fat cover over ribs.
	3.	Thin - lowered milk production - poor reproduction. Fat along backbone and slight amount of fat cover over ribs.
Moderate	4.	Borderline - reproduction bordering on inadequate. Some fat cover over ribs.
	5.	Moderate - minimum necessary for efficient rebreeding - good milk production - generally good overall appearance. Fat cover over ribs feels spongy.
	6.	Moderate to good - milk production and rebreeding very acceptable. Spongy fat cover over ribs and fat beginning to be palpable around tailhead.
Good	7.	Good - fleshy - maximum condition needed for efficient reproduction.
	8.	Fat - very fleshy - unnecessary - no advantage in rebreeding from having cows in this condition. Cow has large fat deposits over ribs, around tailhead, and below vulva.
	9.	Extremely fat - extremely wasty and patchy - may cause calving problems. Cow extremely over-conditioned.

Figure 2. Recommended scores for evaluating body condition in beef cattle.

Each year is different; cows are different. You must assess the body condition of your cows, the forage available, and then put together a plan so that cows will score a 5 or 6 at calving time. Don't ignore the problem and think it will go away. Thin cows will come back to haunt you next year; they will be open or calve late.

<u>Suckling</u>

The length of interval from calving to first heat is 20 days to 42 days longer in cows suckling calves than in milked cows (table 11). Methods that might be used to shorten the interval of calving to first heat include weaning the calves early or decreasing the frequency of suckling.

TABLE 11. EFFECT OF SUCKLING ON THE INTERVAL FROM CALVING TO FIRST HEAT

Type of cow	Suckled, days	Nonsuckled, days	Difference, days
Holstein	58	38	20
Milking Shorthorn	94	64	30
Beef	73	31	42

Flushing and 48-hr calf removal can be helpful in improving reproductive performance. Neither practice alone is as beneficial as a combination of the two. This principle is evidenced in a study conducted at Howell's in South Texas with first-calf cows that were slightly thin (a score of 4) at calving time (table 12).

TABLE 12. PREGNANCY RATES FOLLOWING CALF REMOVAL AND FLUSHING (HOWELL'S)

	Control	Fl[a]	Cr[b]	Fl + Cr
No. cows	18	21	21	21
Pregnant, %				
21 days	28	14	38	57
42 days	56	52	62	72
63 days	72	76	62	86

[a]Flushed with 10 lb of corn for 2 wk before breeding and first 3 wk of breeding.
[b]Calf removal for 48 hr at start of breeding.

The pregnancy rate was highest in the group that used both the practices of flushing and calf removal. Flushing cows for 3 wk before breeding was associated with increased pregnancy rate, also.

Cows with a score of 4 or more will respond beautifully to a little extra feed for 3 wk or so prior to breeding, <u>if</u> the calves are removed for 48 hr when the bulls are placed in the breeding pasture. (Note what happened again at Howell's with flushing alone, compared with flushing and calf removal.)

How do you get cows to gain a little weight just prior to breeding? Grain is one way; a good pasture with some dry matter is another. However, you cannot expect a cow to gain weight on just a little, short, green grass; that kind of grass is 90% water. Get good hay, grain, on find a pasture that has some good growth or you will be disappointed.

Removing calves for 48 hr can be a problem in some situations. The best way to remove them without involving extra labor is to remove calves for 24 hr and then work the calves (brand, castrate, etc.). After working them, keep calves from their mothers until the end of the 48-hr period. Calves must not nurse for 48 hr to get maximum results.

Work Table #6 indicates the kind of results that can be obtained with flushing. (Note that there is little or no change in thin cows showing heat.) Problem #3 is intended as an example to show the benefits of 48-hr calf removal and flushing.

PROBLEM 3

With the cows used in Problem #2, remove the calves for 48 hr and flush the cows. Use Work Table #6 to estimate cows showing heat. Calculate cows pregnant after 20 days of breeding. Use 60% conception rate. Compare results to Problem #2. Enter on Worksheet #5.

Fertile bulls must 1) produce adequate amounts of sperm, 2) a large proportion of the sperm produced must be normal, 3) the bull must have the desire and ability to deposit the sperm in the cow. A good measure of semen production is scrotal circumference. It can be measured quickly and easily with a tape. Available data indicate that bulls with a scrotal circumference of less than 30 cm have reduced fertility. Ten percent to 15% of the bulls in most breeds have little or no desire to breed. Simple reliable tests for determining these bulls in all herds are yet to be developed, although tests for bulls that have been handled regularly have been developed and are reliable.

The effect of selecting bulls for semen quality was demonstrated recently at the King ranch. Semen from 79 bulls was collected and evaluated; 27 of the bulls with 80% or more normal sperm were selected and placed with 675 cows. Another 26 bulls were selected as a representative sample of the original group of bulls, and were placed with 655 cows. The pregnancy rates after 120 days of breeding were 93% in the selected group and 87% in the controls. In a second study, bulls with 80% or more normal sperm were compared with 1) bulls with 70% or more normal sperm and

2) control bulls. Five percent fewer cows bred to the control bulls became pregnant than those bred to the bulls selected for semen quality.

Bulls should be evaluated each year. Semen quality will improve in certain bulls from the first semen collection to the second. If a bull has poor semen, collect a second time immediately. Evaluate, and if semen is still poor, collect the bull 3 wk or 4 wk later. Then make a decision. Do NOT compromise. DO NOT use a bull with poor semen (table 13).

TABLE 13. BULLS SELECTED FOR SEMEN QUALITY AT KING RANCH

	Control	80% +	70% +
Multiple sire, 1980[a]			
No. exposed	572	656	
Pregnant, %	87	93	
King multiple sire, 1981			
No. exposed	1,179	522	769
Pregnant, %	85	90	91

[a]Four bulls per 100 cows.

Cow fertility is affected by two factors: the length of time from calving to breeding and weight change near breeding. Conception rate at first service increases as the interval from calving to breeding increases up to 40 days postcalving. By 50 days after calving, cows generally have reached optimum conception rates. This means higher conception rates at first service in early-calving cows.

Cows losing weight after calving have a lower conception rate than do cows gaining weight; 43% of the cows losing weight conceived on first service, compared to 60% of the cows gaining weight.

The effect of days postcalving, weight change following calving, and selecting bulls for fertility is shown in Work Table #7.

Problem #4 is designed to show the effect of this on pregnancy rate.

PROBLEM 4

Selected cows from the hypothetical ranch have been used and are shown in Work Table #8. Calves were removed for 48 hr. Use Work Table #7 for conception rates; Work Tables 2 and 6 for heat rates. Breed 20 days.

There is one other important ingredient you need to get from where you are to where you want to be. Calving should be early in the calving season. The length of the breeding season is an important factor in determining pregnancy rate. As you know, late-calving cows have smaller calf crops than early-calving cows. The only reliable methods for making sure that cows calve early in the calving season is to have a short breeding season. Results shown here indicate that the breeding season should not last more than 60 days.

Shortening the breeding season from 150 days or even from 90 days to a 60-day season may present a cash-flow problem. The first year the breeding season is shortened there could be fewer calves for sale, thus some suggestions are in order as to how you can minimize this problems. A first step is to get an estimate of how many calves were dropped in each of the weeks of the calving season. This estimate should then be compared to the breeding season dates to ascertain when cows were being bred. Next, an estimate should be made of the amount and quality of forage available in different months of the year. We have discussed the nutrient requirements of cows. A breeding season should be selected so that nutrient requirements of cows match as nearly as possible the available forage supply. The present calving pattern should be compared with the desired calving pattern; then you can make intelligent changes.

Sometimes the breeding season can be shortened so as to produce only small losses in calf numbers during the first year. At other times, rather drastic changes must be made. There are two possible methods: 1) A plan is developed in which the breeding season is shortened 2 wk to 4 wk per year. (A heifer development program where heifers are bred only 45 days is an important part of this program and must be implemented or the plan will not work.) 2) A plan is developed in which cows are bred in a fall and spring program. Forage supply must be carefully evaluated in this type of program; calf numbers may actually be increased.

An example follows of how the breeding season might be shortened from 150 days to 60 days (table 14). Thirty replacements per 100 cows are added each year for 3 yr. To have these 30 replacement heifers calving in a 45-day period, 35 heifers would be bred and open heifers culled. The cost per animal in the herd is increased from $250 to $270. The net return is changed from a debit of -$39 to +$49. This is assuming a 90% calf crop each year. Generally when you have a long calving season, the calf crop is lower.

This particular method resulted in an increase in revenue, but it would be necessary to find a place to carry an extra 25 heifers each year for 3 yr; thus, the practice might not be feasible. Such a practice could be implemented by checking cows for pregnancy and culling open and late-calving cows. Using this system, the number of cows replaced would be determined by the number of pregnant replacement heifers available to be placed in the herd.

TABLE 14. CHANGING LENGTH OF CALVING SEASON

Expected day of calving	1st year	2nd year	3rd year	4th year	5th year
1-20	10	30	50	70	75
21-40	10	20	20	25	20
41-60	10	10	20	5	5
61-80	20	20	10		
81-100	20	20			
101-120	10				
121-140	5				
141-150	5				
Total no. pregnant	100	100	100	100	100
No. replacements saved	35	35	35	12	12
Pregnant replacements placed in herd	10	30	30	30	10
Cost per animal, $	250	270	270	270	250
Calf crop weaned	90	90	90	90	90
Animals per 100 calves	127	135	135	135	127
Lbs calf weaned per animal	281	308	336	352	393
Net return, $	-39	-39	-18	-6	-45

WORK TABLE #1

Estimated calving date	No. of cows	Body condition: Thin	Moderate	Good	Average calving date	Julian date
February 14 - March 5	60	20	20	20	February 24	55
March 6 - March 25	60	20	20	20	March 16	75
March 26 - April 14	60	20	20	20	April 5	95
April 15 - May 4	60	20	20	20	April 25	115

WORK TABLE #2
BODY CONDITION AT CALVING AND HEAT AFTER CALVING

Body condition at calving	No. of cows	Days after calving: 30%	40%	50%	60%	70%	80%	90%	100%	110%	120%
Thin	272	0	19	34	46	55	62	66	70	75	77
Moderate	364	10	21	45	61	79	88	92	100	100	100
Good	50	12	31	42	91	96	98	100	100	100	100

WORK TABLE #3
WEIGHT GAIN NEEDED PRIOR TO CALVING

Body condition	Desired body condition at calving	Weight gain: Calf and fluids membranes	Fat or muscle	Total	Days to weaning	ADG needed
5	5	100	0	100	100	1.00
					120	.83
					140	.71
					160	.62
					180	.55
					200	.50
4	5	100	80	180	100	1.80
					120	1.50
					140	1.28
					160	1.12
					180	1.80
					200	.90
3	5	100	160	260	100	2.60
					120	2.20
					140	1.86
					160	1.62
					180	1.44
					200	1.30
2	5	100	240	340	100	3.40
					120	2.83
					140	2.43
					160	2.12
					180	1.89
					200	1.70

WORK TABLE #4
WEIGHT GAIN NEEDED FOLLOWING CALVING

Body condition at calving	Desired body condition at breeding time	Weight gain needed	Days calving to breeding	ADG
5	5	0	80	0
			60	0
			40	0
4	5	80	80	1.0
			60	1.33
			40	2.00
3	5	160	80	2.00
			60	2.67
			40	4.00
2	5	240	80	3.00
			60	4.00
			40	6.00

WORK TABLE #5
FEED NEEDED TO MAKE CHANGES

	Pregnant cow		Cow nursing a calf[a]		
ADG	Alfalfa hay, lb	Corn	Alfalfa hay, lb		Corn
0	12	0	31		0
.5	15	0	29	+	5
1.0	17	0	19	+	15
1.5	21	0	17	+	20
2.0	25	0			
2.5	21	5			
3.0	18[b]	9[b]			
3.5	13[b]	15[b]			

[a] Where no data are shown it is not feasible for cow to eat enough energy.
[b] It is not probable that a dry cow will eat this much dry matter.

WORK TABLE #6
BODY CONDITION AT CALVING, FLUSHING, CALF REMOVAL FOR 48 HOURS AFTER HEAT

	30	40	50	60	70	80	90	100	110	120
Thin	0	19	40	55	65	72	76	80	85	85
Moderate	45	61	79	88	92	100	100	100	100	100
Good	42	91	96	98	100	100	100	100	100	100

WORK TABLE #7
CONCEPTION RATE AS INFLUENCED BY WEIGHT CHANGE AFTER CALVING, FERTILE BULLS, AND TIME FOR CALVING

	Untested bulls, days after calving			Bulls tested for potential fertility,[a] days after calving		
	30	30-60	Over 60	30	30-60	Over 60
Losing weight	21	33	43	21	43	43
Gaining weight	41	53	63	41	63	80

[a] 70% or more normal sperm, physically sound, scrotal circumference over 32 cm, and libido.

WORK TABLE #8
PROBLEM #4

	Bred untested bulls				Bred tested bulls			
	Thin cows		Moderate cows		Thin cows		Moderate cows	
Calving date	Losing	Gaining	Losing	Gaining	Losing	Gaining	Losing	Gaining
February 14 - March 6	20	20	20	20	20	20	20	20
April 18 - May 8	20	20	20	20	20	20	20	20

WORK SHEET #1

Estimated birthday	Weaning date	Days to grow	ADG	Gain, birth to weaning	Birth weight	Weaning weight
March 3 (62)	October 16 (289)	227	2.0	454	80	534
March 23 (82)	October 16 (289)					
April 12 (102)	October 16 (289)					
May 2 (122)	October 16 (289)					

WORK SHEET #2

	FIRST 20 DAYS									SECOND 20 DAYS										THIRD 20 DAYS										
	Julian Date				Expected					Julian Date			Expected							Julian Date			Expected							
						In heat								In Heat										In Heat						
Group no.	Calving period	Average no. calving	At the end of 20 days	Days P-P	No. of cows	%	No.	Conc. rate	Preg-nant	Average calving date	At the end of 40 days	Days P-P	No. of cows	%	No.	Prev. preg.	Heat this per.	Conc. rate	Preg-nant	Average calving date	At the end of 60 days	Days P-P	No. of cows	%	No.	Prev. preg.	Heat this per.	Conc. rate	Preg-nant	Total pregnant
	2-24	55	155	100	20	100	20	60	12	55	175	120	20	100	20	12	8	60	5	55	195	140	20	100	20	17	3	60	2	19
	3-16	75						60										60										60		
	4-5							60										60										60		
	4-25							60										60										60		

WORK SHEET #3
POUNDS OF CALF AND DOLLARS FROM COWS CALVING AT DIFFERENT TIMES

	Cows pregnant					
	First 20 days (average date 3.3)	Second 20 days (average date 3.23)	Third 20 days (average date 4.12)	Lb	Total at 65	$ per cow
Group 1	12	5	2			
No. of calves	534	494	454			
Average weight	6,408	2,470	908	9,786	6,361	318
Total weight						
Group 2						
No. of calves						
Average weight						
Total weight						
Group 3						
No. of calves						
Average weight						
Total weight						
Group 4						
No. of calves						
Average weight						
Total weight						

WORK SHEET #4

FIRST 20 DAYS									SECOND 20 DAYS										THIRD 20 DAYS										
	Julian Date			Expected					Julian Date			Expected							Julian Date			Expected							
					In heat								In Heat										In Heat						
Calving period	Average no. calving	At the end of 20 days	Days P-P	No. of cows	%	No.	Conc. rate	Preg-nant	Average calving date	At the end of 40 days	Days P-P	No. of cows	%	No.	Prev. preg.	Heat this per.	Conc. rate	Preg-nant	Average calving date	At the end of 60 days	Days P-P	No. of cows	%	No.	Prev. preg.	Heat this per.	Conc. rate	Preg-nant	Total pregnant
2-24	55	155	100						55	175	120																		
	T			20			60		T			20					60												
	M			20			60		M			20					60												
	G			20			60		G			20					60												
3-16	75	155	80						75	175	100																		
	T			20			60		T			20					60												
	M			20			60		M			20					60												
	G			20			60		G			20					60												
4-5	95	155	60						95	175	80																		
	T			20			60		T			20					60												
	M			20			60		M			20					60												
	G			20			60		G			20					60												
4-25	115	155	40						115	175	60																		
	T			20			60		T			20					60												
	M			20			60		M			20					60												
	G			20			60		G			20					60												

WORK SHEET #5

FIRST 20 DAYS									SECOND 20 DAYS										THIRD 20 DAYS										
Julian Date				Expected					Julian Date			Expected							Julian Date			Expected							
Calving period	Average no. calving	At the end of 20 days	Days P-P	No. of cows	In heat %	In heat No.	Conc. rate	Preg-nant	Average calving date	At the end of 40 days	Days P-P	No. of cows	In Heat %	In Heat No.	Prev. preg.	Heat this per.	Conc. rate	Preg-nant	Average calving date	At the end of 60 days	Days P-P	No. of cows	In Heat %	In Heat No.	Prev. preg.	Heat this per.	Conc. rate	Preg-nant	Total pregnant
2-24	55	155	100																										
	T			20			60																						
	M			20			60																						
	G			20			60																						
3-16	75	155	80																										
	T			20			60																						
	M			20			60																						
	G			20			60																						
4-5	95	155	60																										
	T			20			60																						
	M			20			60																						
	G			20			60																						
4-25	115	155	40																										
	T			20			60																						
	M			20			60																						
	G			20			60																						

WORK SHEET #6

FIRST 20 DAYS										SECOND 20 DAYS										THIRD 20 DAYS										
Julian Date					Expected					Julian Date			Expected							Julian Date			Expected							
Group no.	Calving period	Average no. calving	At the end of 20 days	Days P-P	No. of cows	In heat %	In heat No.	Conc. rate	Preg-nant	Average calving date	At the end of 40 days	Days P-P	No. of cows	In Heat %	In Heat No.	Prev. preg.	Heat this per.	Conc. rate	Preg-nant	Average calving date	At the end of 60 days	Days P-P	No. of cows	In Heat %	In Heat No.	Prev. preg.	Heat this per.	Conc. rate	Preg-nant	Total pregnant
	2-24	55	155	100																										
		T	L	U	20																									
		T	G	U	20																									
		T	L	T	20																									
		T	G	T	20																									
		M	L	U	20																									
		M	G	U	20																									
		M	L	T	20																									
		M	G	T	20																									
	4-28	115	155	40																										
		T	L	U	20																									
		T	G	U	20																									
		T	L	T	20																									
		T	G	T	20																									
		M	L	U	20																									
		M	G	U	20																									
		M	L	T	20																									
		M	G	T	20																									

L = losing weight; G = gaining weight; U = untested; and T = tested for fertility.

19

HIGH TECHNOLOGY APPLICATION ON MY RANCH

George A. McAlister

In this day and age, technological change is just as inexorable as the tides, and this applies to the cattle industry just as it does to every other industry in America and in the world. If evidence of these technical changes in our industry is required, one needs only to visit any of the dozens of genetic centers in America and observe a demi or quad of a 7- or 8-day old embryo, to observe the placement of each segment of the original embryo into different recipient cows, and to view, ultimately, the birth of exact twins or quadruplets, as the case might be. Just a few short years ago, many of you listening to me today would have deemed this an impossibility.

I think it is fair to say that problems and technology go together like ice cream and apple pie; if a problem exists in any industry, there is always someone seeking a technological answer. Last July at the National Cattlemen's Association in Denver, two major recommendations were made: 1) Do not increase your herds at this time, and 2) use all available technology in managing your herd if you want to be profitable.

Now I'll briefly describe my background. I was born on a farm in east Texas and have been around cattle all my life. However, I studied mathematics and spent some time working with nonagricultural technology before becoming a serious breeder in the mid-1960s. For many years I was content, as are most ranchers, to let the bull take care of breeding the cow when the time was right. But in the early 1980s, my son Kyle and I became very interested in cattle breeding via embryo transfer. We realized we could rapidly expand our herd of Red Brangus cattle by using embryos flushed from the best Red Angus and(or) Red Brahman cows. We superovulated these cows and artificially inseminated them, using semen from some of the outstanding bulls in those breeds. We commissioned Dr. Bill Turner, Chairman of Animal Sciences at L.S.U., to assist us in finding these brood cows and champion sires.

Early on in this venture, I discovered two problems: Our pregnancy rate in embryo transplants was directly proportional to the exactness of heat synchrony of the donor and the recipient cows, and even though we used synchronizing drugs, our pregnancy rate went up when we knew the natural heat cycle of the recipients. Remember that our program began before the advent of Synchro-Mate B. But even now, it is highly desirable in many instances to know the natural heat cycle of the recipients, particularly if you choose not to use heat-inducing drugs.

Many genetic centers with a large recipient herd do not use these drugs. In many cases a genetic center will have enough recipients and exact information on each recipient's natural heat cycle that they can pick those that are in synchrony with the donor to be flushed. This kind of program, however, depends largely on how frequently personnel observe heat and the accuracy of the records they keep. Because the bovine animal is nocturnal in its sexual habits (approximately 65% come into heat after 6:00 p.m. and before 6:00 a.m.), picking the exact moment of peak heat for each cow is a very troublesome chore. It requires that the herd be observed in all types of weather and checked many times during the day and night for cows showing heat. I was challenged by this problem and the possibility of using electronic animal identification.

Most cattle are either firebranded on the hip or tattooed in the ear and(or) have a tag in the ear in addition to the increasingly used brucellosis tag. Regardless of whether you are transferring embryos or merely administering routine shots and medical treatments, it is important to know (in most cases) the identity of the animal in the chute. Standard procedure after the cow's head is in the head gate and the sides of the chute are squeezed to hold her motionless is to have someone read the ear tattoo and(or) the brucellosis tag, if it is still attached to the ear, and(or) the brand if it is readable; and then amid the bawling and squawling of calves and cows and the cursing of men, to call out these numbers to someone with a pencil and pad. Obviously, in all of this din and confusion, errors occur; nines are heard and written as fives, and vice versa. Not to mention the fact that occasionally someone looking for the brucellosis number or ear tattoo will get his teeth knocked out when the cow shakes its head and applies a hard horn to the nearest lip.

My son and I run approximately 1,000 head of cattle, about half of which are recipients. Because we are in the embryo transfer business, both workmen and cattle spend a lot of time in and around the squeeze chute. The thought occurred to me one day that it would be very nice to merely wave a wand over the rump of an animal and immediately see her brand and(or) brucellosis number appear on a CRT screen in the work chute area. Thus began a quest that took me to both coasts and to various companies to see what was available in electronic technology for identification. I found that there was plenty available for electronic human identification, but very little available in electronic animal identification. The work that had been done in animal identification was pointed mostly toward the dairy industry, and various patents had been issued both here and abroad covering devices for animal identification. Most of it fell into the "Rube Goldberg" category.

The best I could find in human identification was a card about the size of an American Express card. When a person carrying this card wanted access to a restricted area and had to go through a locked door, he could merely hold the card near the edge of the locked door in the general area of an embedded sensor. The sensor would read the number on the card and send this number to a computer in the building that would match the card number to the identity of the person presenting it. If this person was indeed on the approved list, a signal came back by wire to unlock the door. Simultaneously, it printed the person's name, the door location code, and the time of entry. It didn't take a great deal of "smarts" to see that a modification of this art would work to identify cattle.

My search took me to the "Silicone Valley," which seemingly is the heartland of all computer technology. There I found a company that could make a card that I could implant under the hide of an animal. I would have to create the computer software to correlate this card number with the actual cow number.

I had already become aware of the need of computerizing my herd records. Therefore, I was on a simultaneous search for computer software that would satisfy my needs for herd management including embryo transfer, health records, accounting records, pedigree records, registration records, physical characteristics of calves, and on and on. I looked at several systems, which all fell short of my requirements. Therefore, I hired my own programmer, Mr. Rick Hancock, and began to create a total computer program for herd management, including electronic identification of animals and all of the aforementioned subjects.

After modification of the card and sensor, I began experiments that continue today to electronically identify my cattle. It works like this. I implant a plastic card under the hide of a cow, in the area between the pin and hook bones. The operation is quite simple. I clip an area about 8 in. in diameter and disinfect it with a Betadine solution. I then use a sterile marker to draw a line in the center of this area and parallel to the backbone. Novacaine is administered. I then make a cut just through the hide about 2 1/2 in. long. Using my gloved finger, I start making a pocket between the hide and the fatty layer underneath. When I can insert all three fingers upward toward the backbone, I withdraw and insert a sterilized card. The wound is then closed with only two stitches. After 7 to 10 days, the stitches are removed; after 6 wk, the hair has grown back -- if its location were not known, it would be hard to find.

Now back to the subject of heat detection. I am not the first person to be challenged by this problem of heat detection. A search of the Patent Office records will show that several devices have been patented to detect the onset of estrus. Some of them are vaginal probes that measure temperature changes in the female's body, and others are as simple as the very popular K-Mar patch that changes color when enough pressure is put on the patch for a sufficient amount of time. Obviously, the K-Mar patch is a mechanical device and does not fall in the category of electronic detection.

It has been determined by various research centers that a cow in heat will stand to be mounted from 20 to 60 or more times during the 12 hr to 18 hr of estrus. Furthermore, if the length of each mount is accurately determined and plotted on a graph, a typical estrous cycle would resemble a bell curve. Even though some cows have a very low intensity linear-type heat, these cows are in a minority. The majority follow the bell-curve pattern.

My idea for heat detection is an extension of the animal I.D. system, with the card implanted in the rump area of each female and each card having a discreet code. A teaser (or gomer) bull or a testosterone-treated female would carry a sensing device mounted where the ordinary chinball marker is worn. As the mounting animal uses its head and neck to mount a female implanted with the card, the sensing device would slide over the card and read it. Also, with the aid of a real-time-clock chip, the exact time when the card was read can be recorded. As the teaser bull dismounts, there would be another reading and the exact time recorded. Obviously, the difference between these two recordings would give the length of the ride. Ideally, the signals from the sending device would be broadcast for approximately one-half mile by a battery-pack-powered transmitter carried by the gomer bull, and the signal would be recorded on magnetic tape at a selected receiving station. This information would be placed into the computer and a graphic curve could be generated reflecting the estrous period for a particular cow. The computer that decodes this information can be programmed to select the longest ride and, therefore, the moment of peak standing heat. We are experimenting with this method of heat detection.

What have others done to solve this same problem? Five or six years ago, the Los Alamos National Laboratory developed a prototype electronic identification and subdermal body-temperature sensing unit. The identification of the animal and the body-temperature sensing are accomplished through the use of a transponder, which is the size and shape of a small cigar (about 5 1/2 in. long). It is implanted beneath the animal's skin just behind the shoulder and parallel with the backbone. This works in conjunction with an electronic sensing device called an "interrogator." The "interrogator" antenna directs a beam of microwave energy to the implanted transponder. Circuit elements on the transponder antenna use the energy to power an encoded answering signal that contains the animal's unique identification number and her subdermal body temperature. In order to use this device and to identify a cow, you must be near the animal -- no more than 3 ft to 5 ft away. There are several of these devices in animals today, and experiments are being conducted by Mr. Dean M. Anderson at the Jornada Experimental Range, Las Cruces, New Mexico. But this device is very expensive -- its cost ranges up to about $400 per animal. Therefore, the commercial application of it is not acceptable. This compares with a cost on my card of approximately $10 per animal, plus the implantation process is very simple and can be done by the rancher himself. In fact, we have done all of our implants without the assistance of a veterinarian.

Another device to electronically identify animals is covered by Patent No. 4,262,632, issued April 21, 1981. This device comprises a capsule or pill containing an electronic transmitter. This device is uniquely coded to broadcast a series of electronic pulses from each animal in which it is implanted. It also has to have an interrogator/receiving unit that is located outside the animal and may be used for sensing and using the electronic pulses to identify the animal. Usually the capsule is implanted subdermally, but it is particularly adapted to be orally implanted in ruminating animals such as cattle, sheep, and goats to take advantage of the peculiar stomach configuration of the ruminant. If it is implanted this way, it may be made to permanently reside in the second stomach or reticulum by weighting it to produce the proper specific gravity. The power supply for this is in the form of a long-life battery. However, the inventor says a passive storage element is desirable into which power may be induced from the interrogating unit. This renders the capsule useful for the life of the animal and makes possible its recovery and reuse after the animal is slaughtered. I know of no herds using this particular system, nor have I been able to contact the inventor to get more specifics.

As late as October 25, 1983, Patent No. 4,411,274 was issued. It consists of a radio-transmitting device mounted on a resilient pad, which in turn is mounted in the rump area of a cow. Underneath this resilient pad is positioned a permanent magnet. When another cow or gomer bull mounts the first cow, then the weight of the breastbone of the gomer bull or mounting cow will apply pressure to the transmitter, which causes the magnetic switch in the transmitter to be activated. Since there is a time-delay-switch circuit, a delay of approximately 15 seconds takes place between the switch application and the signal transmission. When transmission does occur, it is in a short burst of three consecutive signals, and then the transmission automatically shuts off to conserve the battery energy. This device will transmit about one-half mile to the receiving station, which includes a receiver and decoder. It also has a real-time-clock device in it so that the receiver can print out the actual time that each mount occurred; but it does not give the length of the ride. Each transmitter is coded so that the receiving station knows which cow is being mounted. Once again, the cost of this device is very significant, and I know of no herd that is using it at the present.

There have been various other electronic systems, either described in articles or actually invented, but so far none of these have really caught on in the marketplace. Most of them are too expensive for the cow/calf producer. The Intel Company manufactures a device that is gaining popularity in the dairy industry. This device is strapped to the animal's neck with a belt. When the dairy cow goes into a stall, she is identified; her milk production rate is identified, and she is fed accordingly. However, this system costs approximately $80 per cow, which is not cost effective for the typical rancher. Since this device is carried externally, it is subject to loss.

Currently we are implanting our own card device in recipient cattle after they have been confirmed as 90 days pregnant. We do it at that time because the wound requires approximately 2 wk to 3 wk to completely heal, and we do not want the animal ridden during this time. As I mentioned before, the implantation of the card is a very simple operation. Out of approximately 100 insertions, we have had one cow reject the card. This is strictly an experimental project and so we did not implant any of our expensive donors. Also, we are working to reduce the size of the card to approximately 2 in. by 2in. This should be possible in the next few months.

I personally feel that implanted cards will continue to be reduced in size until they are no larger than a quarter, or perhaps even a dime, and can be inserted easily under the skin much like RALGRO or other ear implants. I feel very strongly that by the end of this decade such implants in domestic animals will be commonplace and that this movement will probably be led by a thoroughbred horse association. In fact, I have already been contacted by one such association looking for an implant that would be unobtrusive and cost no more than $10. They feel that my current-sized card is too large. They state that the card they select would be furnished by the association along with the registration papers. Their implant would be mandatory for the owner to place in a specified area and would always remain in the animal.

If all stock could have identification implants, it would make rustling of cattle and theft of horses very difficult. For example, if all cows had a simple ear implant with a positive discreet identification, and if all sales and stockyard centers had a sensor, and if all cattle had to pass through this sensor, than rustled or "hot" cattle could be easily identified. However, the major benefit will be for the rancher in the work chute area where he can positively identify the animal and administer the medical treatments that have been predetermined by the computer.

he can positively identify the animals and administer the medical treatments that have been predetermined by the computer.

Part 3

GROWING AND FINISHING CATTLE

20

WHY DIFFERENT ANIMALS PRODUCE DIFFERENT TYPES OF BEEF

Rex M. Butterfield

Beef cattle breeders, in their decision to pitch in their lot with a particular breed of cattle or in their hesitance to do so, are influenced by a great number of factors. One of the factors is the knowledge that breed selection commits them to the production of beef that will fall into a certain category of "quality" in terms of their particular market. They weigh this against all the other important factors when their cattle reach the point of sale. At that time they hold out their hand for an amount that may vary greatly according to their selection of the breed or mix. There are other factors that influence the all-important price, but we are going to address just one factor, the composition of beef. (It surely is easier being a scientist than a cattle breeder.)

In a previous paper, we talked about the early- and late-maturing breeds. The nature of the beef produced by these two wide classifications will determine our choice of breeding stock. However, in the real world many commercial cattle breeders will obtain some of their better characteristics by using one or more breeds in their females and one (usually) or more in their males.

After selecting a particular breed or combination of breeds, we are interested in knowing what type of meat we will produce if we fiddle around and change the growth process of cattle within our herd.

It may be that because of market changes we need to produce fatter carcasses to meet the new requirements. We remember, however, that the ideal carcass in any market has the maximum amount of muscle, the minimum amount of bone, and whatever amount of fat that the current market wants. That definition of an ideal carcass will fit any market. Therefore, the most likely changes will be either the proportion of fat or in increased muscularity (i.e., muscle: bone ratio).

Let us consider a hypothetical problem -- a market that needs fatter carcasses. We would need earlier-maturing carcasses and so would need to select for early maturity. To please the market and the bank, we would push the fattening curve onto younger (and smaller) animals so that, should they be killed at the same weight, they will be fatter. Of course, we could let them get heavier and fatter but that might or might not suit our market because the carcass might be too big. The alternative is, of course, to feed much more heavily or to feed heifers rather than steers. All of these procedures will produce fatter carcasses.

A demand for fatter beef at present is an unlikely change. In most markets the trend is, and has been for some time, towards less fat. How do we get animals with less fat? And how do we achieve this economically in a country that equates "quality" with intramuscular fat ("marbling")?

If we already have a selected breed or breed mix, how do we achieve leaner beef?

The first step is to select for later-maturing animals, and this simply means selecting animals that fatten at greater weights. The slow way of achieving this is to seek out those animals in your herd or breed that carry least fat at market weights and to concentrate selection on that type of animal. In general, this is possible and gradually the type of slaughter cattle produced will be changed. The changes that have been brought about in the last decade in some of the traditional British breeds in this country are dramatic demonstrations of this procedure.

The same effect can be produced by simply selecting for bigger cattle that will be less mature and, therefore, have less fat at preferred slaughter weights. There are some dangers in doing this too quickly, which we will talk about in the next paper. It is possible, of course, to vary the slaughter weight downwards so that animals are killed at an earlier stage of the fattening process, but this, of course, means that you are producing less beef per breeding unit and hence depression of slaughter weight is not a popular option unless corresponding increases in price are obtained.

The simplest and quickest means of reducing fat on carcasses is to omit the process of castration. In this country, and mine, castration is enshrined in tradition and carried out as though a religious ceremony; in other countries, such as some European countries, it is regarded as an act of ignorance. Under conditions of good nutrition there is no doubt that bulls outgrow steers; and there is no doubt that bulls produce leaner carcasses at the same slaughter weights. Acknowledging that the management problems are not insurmountable, we should consider the production of young bulls for slaughter as perhaps the quickest and least disruptive of the traditional cattle breeding procedures if we wish to seriously take up the challenge of leaner beef (figure 1).

The other method of changing the proportions of muscle, bone, and fat within carcasses is to increase the muscle: bone ratio. In other words, select for more muscular cattle. There are, of course, fairly clear-cut divisions between the breeds so far as this characteristic is concerned. The dairy breeds have the lowest muscle to bone ratio, followed by the British beef breeds, and with the Continental cattle leading the way. There is little doubt that a cattleman with the single objective of more muscle (if only it were that simple!) would start with a Continental breed and would find plenty of opportunity to select up to the ultimate in muscularity. Examples of high muscularity are found in the Piedmont, Belgian Blue, and maybe in other breeds. It is here that the search for what is ideal in the eyes of some consumers may well lead the producer beyond the brink into functionally inefficient cattle. (Discussed in the next paper.)

I must refer to the question of the relative proportions of the various cuts of meat that, of course, could have a marked effect on total price earned by a carcass. Despite our readiness to look at live animals and praise or criticize their relative proportions, there is really very little difference in the proportions of the tissue that really matters, i.e., muscle. Work done initially in Australia, and repeated in many other places with other breeds, has shown that in normal steers the differences in the distribution of muscle weight, and hence in the proportions of red meat in various "cuts," is extremely small and of little economic significance (figure 2).

There is, however, in double-muscled animals a slightly higher proportion of the muscle weight concentrated in the thicker muscles, mainly in the upper regions of the limbs and around the spine. These cattle must be regarded as functionally abnormal animals. Although some specialized areas of beef production are finding them to be economically viable, it is unlikely that they will make much of an impression on the mainstream of beef production.

Why do different animals produce different types of beef? Because their genetic potential to grow muscle, bone, and fat is different; because of the influence of sex on this process; and because cattlemen slaughter at different weights.

Figure 1. Failure to castrate is the quickest way to more red meat.

Figure 2. There is no such thing as a normal steer with well-developed muscle in one part of its body and poorly developed muscles elsewhere.

21

CATTLE ARE NOT CREATED EQUAL

Lee D. Bowen

Why, all of a sudden, are communication and production data so important to purebred, cow-calf, stocker, feeder, and feedlot operators? Why is interchange of information and documentation within these livestock segments so vital in today's agribusiness and economic environment? Their tremendous importance is linked to the estimate that less than 5% of the people in the above segments of the cattle industry have made any <u>real</u> profit in the last 10 yr. <u>Real</u> cattle profit in today's economy can only be calculated when land inflation (land appreciation) is deleted from profit-and-loss statements, along with "extra profits" -- the increase in per capita beef consumption in past years and outside income from minerals, etc.

Let's stop kidding ourselves and take a realistic look at our beef business by taking into account 1) lower land values, 2) level or declining per capita beef consumption, 3) higher interest rates, 4) lower beef prices, and 5) higher taxes, fuel, utility bills, etc. Efficiency in beef production must involve the whole equation -- from conception to consumption. The industry has made great strides in some areas such as, seed stock genetics (although "fads" have gone from one extreme to another), vaccines, medicines, growth hormones, feed additives, and ionophores. Unfortunately, in today's and tommorow's beef business, these improvements aren't enough to survive and make a profit.

Texas Cattle Feeders Association recently ran a survey asking the question, "What type carcass do you want?" They sent copies to five different packers and they got five different answers. I would guess that if we lined up five different feedyard operators and asked, "What type steer do you want to feed?" we would get five different answers -- or maybe 10 different answers if we asked the same five operators the same question the next day. The same kind of question-and-answer response would hold true all the way down the line -- feeder, stocker, cow-calf operators, and seed stock producer.

Ladies and gentlemen, it boils down to one fact -- we don't know what we're doing! I don't know what I am doing and you don't know what you're doing. For instance, let's imagine that we take some bulls and breed them to 100 commercial cows and hopefully get 90 calves. If we have any performance data on the bulls, it is probably "registry" information, not performance information. We have no documented performance data on the individual commercial cows, much less any records. We may feed the calves together in a pen (probably not); but, if we do, all of that information would be only pen averages on gain and feed conversion. We have no correlated weather information, no correlated pen condition information, no net-energy feed information, no comparison information, no individual information. Perhaps, we check to see what the penmates do in the packing house. Even if we do all this (and very few of us do), we still have only AVERAGES -- NOTHING BUT AVERAGES.

Let's fantasize for a minute. While you are in the packing plant, a voice from above whispers in your ear, "This low-choice carcass is from the most efficient steer that ever ate grass on the face of the earth. His birth-to-carcass retail value is the best in history." What could you do to benefit from this inside information? The answer is simple -- absolutely nothing! The only way to trace a combination is by identifying cattle individually and performance testing; then go on from there. In other words, to good business management. A cow is a factory. If a large corporation owns 100 factories, or production plants, they generate 100 profit-and-loss statements and then combine them in a consolidated statement. As businessmen, we must do the same to survive. Each cow must have an individual profit-and-loss statement.

In the U.S. today, a grass steer (slaughtered off of grass) will yield 52% and produce a carcass of lower value than a steer slaughtered out of the feedyard that will yield 64% and have a higher carcass value. Also, carcass merit is, at most, influenced by 50% at the feedyard level, but that 50% of the carcass weight is produced in 25% or less of the total lifetime of the steer. So, the feedyard has a right to know "what the steer will do," because all cattle are not created equal in ability to perform in the feedyard.

Predictability of performance is difficult to achieve because many ingredients go into the makeup of cattle. Unfortunately for you and me, neither the cattle industry, colleges, nor livestock specialists can predict performance for us. The stockman has to make predictions alone by using information from various sources, as well as by using his own records. The main reason the registered breeder places his bulls in the bull test is so that their performance predictability is documented and his bulls will bring higher prices. Likewise, commercial herds in the present and future beef business must have documentation if they are to be sold at a higher price, to both the feeder and the packer. My opinion is that the spread between predictable cattle and nonpredictable cattle will increase tremendously. The cattleman who is in the cattle business for true profit will not survive without predictable, genetically superior, efficient cattle.

The U.S. commercial cow-calf segment is divided into two fairly equal groups in terms of total cow numbers. The first group is made up of cow herds of 50 cows or less; the other has herds of 50 head or more. I am not as concerned for the group having herds of 50 cows or fewer, because the vast majority of people in the smaller-herd category rely on outside income to survive in the cow business. But, please don't take me wrong. It is important for these people to document their cattle, but not as critical as it is for the group with cow herds of 50 head or more. Owners of the larger herds cannot survive without predictable cattle, which requires documentation.

I would like to assume for a few minutes that you agree with me that cattle are worth more when they are predictable or documented. How does one go about documentation? It takes cooperation between the owners of registered, commercial cow-calf, and stocker (if not bypassed) herds, the feeder operations, and the packer.

When ownership changes less often during the cycle, communication and cooperation are simplified. I will touch lightly upon the steps in such a cycle: 1) Cows must be identified (number branded, tagged, tatooed, or a combination), and 2) bulls must be traced as to the cows that they have bred (but on some ranches this can be virtually impossible). Artificial insemination makes this step easier. If only the first steps can be accomplished, that is 50% better than doing nothing. Subsequent steps involve documentation of: birthdates, birth weights, 205-day weaning weights, yearling weights, feedyard weights, conversions, death loss information, and carcass data. The carcass information should include maturity, marbling, quality grade, packer's warm-carcass weight, adjusted fat thickness, ribeye area, kidney-pelvic-and-heart fat, and yield grade.

This gives you a start, especially if you have sire evaluation information. Through this cycle, the more information you have, the better your comparisons become.

This sounds pretty simple. It is easy for me to stand up here and tell you how to do it, but the mechanics of such documentation (especially if you and your help are not dedicated to the project) can be horrendous. In fact, I believe that if the dedication is not there, it probably is impossible. Obviously, the applications are different from ranch to ranch. This is why I once again emphasize that the cowman must be the one to do the documentation; no one else can do it for him.

Once all of the steps are in place and results are in, rapid improvement can be made. If we are to have profitable business operations, these steps must be taken to ensure predictable, documented cattle. The larger rancher who doesn't produce predictable cattle will go out of business, unfortunately, unless our economy returns to rampant inflation.

Since I own a commercial cattle-feeding business called Beaver River Land & Cattle Company, where we custom feed cattle, I am measured in the eyes of customers by my ability to predict cattle performance before cattle are bought for feeders or brought to the feedyard by ranchers. It is vitally important that I have the ability to predict the performance of cattle as closely as possible. To achieve such performance predictability, I need, and the feeding industry in the future will require, complete documentation.

I firmly believe our common goal must be the production of a desirable carcass by the least-cost method.

(Editor's Note: Slides followed the presentation and illustrated the author's work at Beaver River Ranch.)

22

COMPENSATORY GROWTH IN BEEF CATTLE

Iain A. Wright

INTRODUCTION

Compensatory growth has been known to occur in beef cattle for nearly 70 yr. Cattle that have been subjected to a period of feed restriction grow faster when the period of restriction is over than they would have done if no restriction had been imposed. This phenomenon is obviously very useful to the cattle feeder if he can save on the feed bill during periods when feed is scarce or expensive. However, compensatory growth is subject to certain constraints. The aim of this paper is to examine possible explanations for the way in which compensatory growth works.

THE EFFECT OF WINTER FEED LEVEL ON PERFORMANCE OF WEANED SUCKLED CALVES DURING THE FOLLOWING SUMMER

In the United Kingdom, it is normal practice to house weaned suckled calves at the start of the winter, feed them indoors, and then turn them out to pasture the following spring. Two experiments at the Hill Farming Research Organisation in Scotland examined the effects of the feeding level imposed during the winter on the performance during the following grazing season. In both experiments the cattle were Charolais crosses from either Hereford x Friesian or Blue - Grey (Whitebred Shorthorn x Galloway) cows. There were two age groups. The calves in the older group were born in November - December and so were an average of 11 mo old at the start of the winter. The calves in the younger group were born in March - April and so were about 7 mo old at the start of the winter. Heifers and steers were used in both experiments. During the winter (November to mid-May) the cattle were fed 16 kg of grass silage plus a variable quantity of barley so that three different growth rates were achieved. The results for the winter period are shown in table 1. The three levels of feeding created three distinct levels of live weight gain so that by turn-out in the middle of May the cattle varied considerably in live weight. At turn-out the three groups were different not only in live weight, but also in degree of fitness, as shown by both their body condition score and backfat measured ultrasonically between the last and second to the last ribs. (Editor's note: Body condition scores are explained in "The Relationships Between Body Condition, Nutrition, and Performance of Beef Cows.")

TABLE 1. WINTER PERFORMANCE OF WEANED SUCKLED CALVES FED 16 KG SILAGE AND VARYING QUANTITIES OF BARLEY

	Winter feed level		
	Low	Medium	High
Experiment 1			
Initial live wt, kg	279	279	286
Live wt at turn-out, kg	325	367	406
Daily live wt gain, kg	.31	.58	.79
Body condition score at turn-out[a]	2.57	2.78	2.90
Ultrasonic back fat area at turn-out, cm^2	2.55	4.03	4.27
Experiment 2			
Initial live wt, kg	265	281	262
Live wt at turn-out, kg	348	411	421
Daily live wt gain, kg	.44	.69	.84
Body condition score at turn-out[a]	2.65	2.84	3.04

[a]On a 0 to 5 scale.

At turn-out the cattle were divided into two groups and each group grazed perennial ryegrass *(Lolium perenne)* swards maintained at one of two levels. The sward height and herbage masses for the two experiments are given in table 2.

TABLE 2. MEAN SWARD HEIGHT AND HERBAGE MASSES GRAZED BY CATTLE IN EXPERIMENTS 1 and 2

	Short sward		Tall sward	
	Height, cm	Mass, kg dry matter/ha	Height, cm	Mass, kg dry matter/ha
Experiment 1	4.2	1800	7.1	2900
Experiment 2	4.3	1900	6.5	3600

It is normal for cattle to lose live weight when turned out in the spring. Table 3 shows the loss in weight in the animals from Experiment 1 from the day of turn-out until 9 days later. Both the level of feeding before turn-out and the height of the sward after turn-out had an effect on this weight loss. The higher levels of feeding before turn-out resulted in greater weight loss, and the shorter of the two swards also led to more weight being lost. The exact timing of this loss in live weight was examined more closely in Experiment 2 where the cattle on the taller sward were weighed twice weekly for the two weeks following turn-out. Figure 1 shows that although the minimum weight occurred 10 days after turn-out, most of the weight was lost in the first 3 days. Because it occurred so rapidly, it seems that the loss was due to changes in the weight of the contents of the gut and did not involve any real change in the weight of the empty body. It is worth noting, however, that cattle take between 15 and 50 days after turn-out to attain their preturn-out live weight again.

TABLE 3. WEIGHT CHANGE (KILOGRAMS) OF CATTLE ON EXPERIMENT 1 FROM THE DAY OF TURN-OUT TO 9 DAYS LATER

	Winter feed level		
	Low	Medium	High
Summer sward height			
Short	-11.7	-23.8	-29.9
Tall	-2.5	-16.2	-25.2

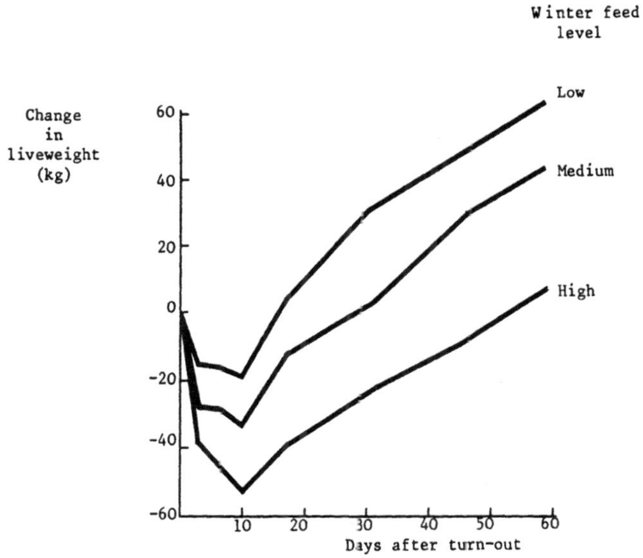

Figure 1. Live weight change following turn-out of cattle from three levels of winter feeding (Experiment 2).

The performance of grazing cattle is given in table 4. Live weight gain on pasture was inversely related to the rate of weight gain during winter. The cattle on the lowest level of winter feeding gained the most weight during summer and those from the highest winter treatment gained least in summer. It is particularly interesting to note that this relationship is evident at both sward heights. Herbage intake figures for Experiment 2 are not yet available, and in Experiment 1 intake was measured only in those cattle grazing the tall sward. Nevertheless these figures show clearly that in relation to their live weight, cattle from the low level of feeding in winter had higher herbage intakes on pasture. On the short sward it would be expected that voluntary intake would be limited by the amount of herbage available, and the difference in mean live weight gain between the animals on the two swards suggests that this was the case. It is therefore interesting that the effect of winter nutrition was still evident on the shorter sward.

As well as affecting live weight gains, the various treatments also affected the length of time taken for the cattle to become fit for slaughter. Throughout Experiment 2 the cattle were slaughtered when they were judged to have reached a particular level of carcass fatness (E.E.C. fat class 4L). Table 4 also gives the mean ages and weights at slaughter of the cattle from the three winter feed levels. Those cattle from the lower levels of winter feeding took longer to reach slaughter condition but were heavier at slaughter. Obviously, the timing of marketing of cattle will be influenced greatly by both the winter feeding level as well as the way in which the animals are managed during the grazing season.

RESTRICTION DURING EARLY LIFE

It has been demonstrated that when cattle are restricted in feed intake between 7 mo and 17 mo of age, they can compensate to some extent when the restriction is removed. But what if the level of nutrition is restricted earlier in life, during the suckling phase? An experiment with calves reared as twins gave an opportunity to examine this. Twin pairs were created by cross-fostering Charolais cross calves from Hereford x Friesian and Blue-Grey cows. Some calves were reared as singles and others as twins, but no calf was reared by its own natural mother. The calves were all born in March - April; their performance during the following summer is shown in table 5. The calves reared as twins were subjected to lower levels of nutrition and by weaning in mid-September were only 82% of the weight of singles, with a weight difference of 34 kg. From early November until mid-May the following year these cattle were housed and fed grass silage ad libitum along with a restricted quantity of barley. There was no evidence of any compensation, as shown by the weights and weight gains in table 5. Both singles and twins gained weight at the same rate. It appears that calves that are restricted during the first 6 mo of their life cannot overcome their weaning-weight disadvantage.

TABLE 4. THE EFFECT OF WINTER FEEDING LEVEL AND HERBAGE HEIGHT ON PERFORMANCE OF WEANED SUCKLED CALVES ON PASTURE

	Winter feed level			
	Low	Medium	High	Mean
Experiment 1				
Daily live wt gain, kg				
Short sward	1.10	1.02	.87	1.00
Tall sward	1.35	1.23	1.19	1.26
Mean	1.23	1.12	1.03	
Herbage organic matter intake, g/kg live wt/ day (tall sward only)	17.7	16.2	15.9	
Final live wt, kg	420	440	464	
Final condition score[a]	2.76	2.91	2.92	
Final ultrasonic back fat area, cm^2	3.20	4.52	4.21	
Experiment 2				
Daily live wt gain, kg				
Short sward	.86	.66	.51	.67
Tall sward	1.25	1.18	.91	1.11
Mean	1.05	.92	.71	
Age at slaughter, days	667	577	548	
Live wt at slaughter, kg	508	484	455	
Carcass wt, kg	296	285	255	

[a]On a 0 to 5 scale.

TABLE 5. THE PERFORMANCE OF SINGLE AND TWIN SUCKLED CALVES FROM TURN-OUT TO WEANING

	Rearing type	
	Singles	Twins
Live wt at turn-out in May, kg	76	68
Weaning wt in September, kg	189	155
Daily live wt gain, kg	.93	.71

TABLE 6. THE PERFORMANCE OF SINGLE AND TWIN REARED CALVES DURING THE POSTWEANING WINTER

	Rearing type	
	Singles	Twins
Live wt at start, kg	239	184
Weaning wt at end, kg	375	327
Daily live wt gain, kg	.76	.75

COMPENSATORY GROWTH AND BODY COMPOSITION

It has been shown that feed restriction can lead to enhanced live weight gains in cattle once the restriction is lifted, provided that the cattle are not too young and that this enhanced performance is associated with higher intakes. It has also been demonstrated in some experiments that the composition of the tissue gained can be affected by compensatory growth.

Baker et al. (1982) fed Friesian steers at two different levels during winter and then turned them out to graze. They slaughtered steers at different stages to measure body composition changes. On pasture the steers from the lower level of winter feeding gained more weight than those that had been on the higher level. This resulted from higher herbage intakes and differences in the composition of the tissue gained. The cattle exhibiting compensatory growth deposited more protein and water and less fat. Since each kilogram of protein gain needs less energy than each kilogram of fat gain, the differences in tissue gained help explain some of the compensation.

CONCLUSIONS

Compensatory growth in cattle seems to result from at least two phenomena. First, cattle subjected to nutritional restrictions have greater feed intakes when the restriction is removed, and this results in faster growth. Second, compensating cattle may deposit more protein and water and less fat in their bodies, resulting in greater efficiency of energy utilization. Body composition may also be involved in the increased intakes during compensation. When animals are restricted, not only do they grow more slowly, but they deposit less fat. Thus lower levels of feeding result in leaner animals. It may be that leaner cattle have a greater appetite. Certainly the figures in tables 1 and 2 support this idea. The cattle from the lower levels of winter feeding were thinner at turn-out (as shown by body-condition score and back fat) and, as a result, had higher herbage intake on pasture.

If body composition plays a central role in compensatory growth, it is clear that we need to examine the effects it might have on the carcass.

REFERENCE

Baker, R. D., N. E. Young, and J. A. Laws. 1982. Changes in the body composition of cattle exhibiting compensatory growth. Anim. Prod. 34:375.

23

NUTRITION OF FEEDLOT CATTLE

Rodney L. Preston

FEED INTAKE

Animals consume feed to meet their energy needs. Assuming the feed has a reasonable balance of nutrients, the major factor governing the amount of feed consumed is the energy need of the animal. With cattle, however, two processes act to govern the amount of feed consumed. Both are related to the digestibility of the energy in the diet (Conrad et al., 1964).

With feeds of lower digestibility (less than 67% digestible dry matter [DM]), the undigestible DM in the feed limits the feed intake because of the undigestible "fill" in the rumen. Thus, with increasing digestibility up to 67%, feed intake will increase as the animal attempts to meet its energy need while at the same time the undigestible "fill" is decreasing. When the diet has more than 67% digestible DM, undigestible fill is no longer a regulating factor. The animal is able to satisfy its energy need and, therefore, as DM digestibility increases above 67%, feed intake will tend to decrease, energy intake and rate of gain will tend to plateau, and feed efficiency will continue to improve since less feed is being consumed. This is diagrammatically shown in figure 1.

A diet with a DM digestibility of 65% to 67% is about a 50% roughage, 50% concentrate diet. Straw, stover, grass hays, and even good quality alfalfa hay when full-fed to growing-finishing cattle will not permit maximum energy intake nor, as a consequence, maximum rate of gain because cattle cannot meet their energy needs due to undigestible "fill" that limits feed intake. Corn silage made from corn that will yield at least 100 bushels of corn grain per acre will have a digestibility of 65% to 67% indicating that this silage compares favorably with diets high in concentrate. Diets with more than 50% concentrate that are typical feedlot diets will have DM digestibilities greater than 67%.

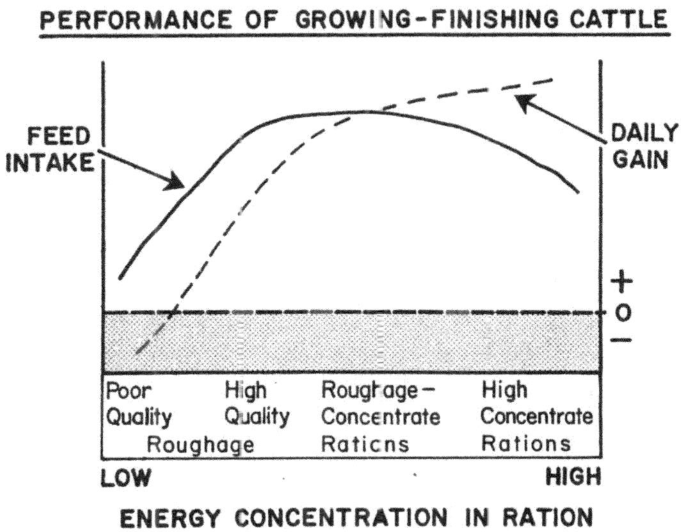

Figure 1. Relationship between energy concentration in the diet, feed intake, and daily gain in cattle.

The major problem in predicting the performance of feedlot cattle is our inability to predict feed intake. Once feed intake is known, predicted and actual performance are nearly the same. Minnesota workers have recently reviewed the feed-intake response of cattle used in their experiments over the last 15 to 20 yr (Goodrich and Plegge, 1984). They found that DM intake was maximum when the metabolizable energy (ME) content of the diet was 1.26 kcal/lb DM. This is equivalent to a diet with about 40% roughage and 60% concentrate. They also found, however, that ME intake was maximum when the diet contained 1.41 kcal ME/lb DM, which is equivalent to a diet with about 10% roughage and 90% concentrate. These results with feedlot cattle indicate that DM intake will be maximum with feedlot diets containing about 40% roughage and 60% concentrate, but ME intake will be maximum, and will result in maximum rate of gain when the diet contains 10% roughage and 90% concentrate.

ENERGY SYSTEMS

The question is often raised as to why so much concentrate (grain) is fed to feedlot cattle. Since about 73% of the cost of feed can be assigned to the energy needs of the cattle (Preston, 1972), the major factor that determines the relative value of feeds is their energy cost. At the time of this writing, with milo grain selling at $100/ton, the following roughages should cost no more than the amount shown to be competitive on a nutrient basis:

Alfalfa hay	\$78/ton
Orchard grass	\$74/ton
Corn silage	\$27/ton (30% DM)
Cottonseed hulls	\$53/ton

Since all of these feeds, except perhaps corn silage in certain areas, sell for considerably more than the prices listed, their use in feedlot diets should be minimized in favor of milo or some other competitive grain to minimize the cost of gain. One should not assume that grains will always supply energy cheaper than roughages. Two years ago, cottonseed hulls sold for \$25 to \$30/ton. With this situation, the cost of gain could be reduced by feeding a higher amount of cottonseed hulls in the diet. The relative cost of grains and roughages should be monitored and roughage: concentrate ratios adjusted to make the maximum use of the cheapest source of energy over the range of 10% to 40% roughage or 60% to 90% concentrate. It is well to point out, however, that the use of home-grown roughages may present a different consideration since returns per acre farmed may be more important than the cost of energy from roughage and concentrate in the feedlot operation. In reviewing several studies comparing returns on corn grain and corn silage (Preston, 1972), feedlot returns are generally maximized with high levels of corn grain and low levels of corn silage; however, returns per acre were greater when maximum levels of corn silage were fed. Thus the question becomes, "Is the operation a farming or feeding operation?"

Energy has been mentioned in relation to feed intake and performance and, therefore, it is appropriate now to discuss energy in the formulation of diets for feedlot cattle. Figure 2 shows how energy is derived by the animal to meet its energy need. The gross energy (GE) of feeds fed to cattle is rather similar even for feeds as different as wheat straw and corn grain. Digestibility loss is the major loss in energy when feeds are consumed. Thus the digestible energy (DE) value of wheat straw is only 50% of its GE whereas the DE value of corn grain is about 90% of its GE value. In terms of DE, corn grain has twice the energy value of wheat straw.

In studying figure 2, it is obvious that as one progresses down the chart, each succeeding energy value better describes the true energy value of the feed to meet the energy need of the animal. Energy in feeds is used either for maintenance or production, and the energy terms used to describe these values are net energy for maintenance (NE_M) and net energy for production (NEp). As noted above, the relative energy value of corn grain compared to wheat straw remains 2 to 1 for DE, ME, and NE_M, and total digestible nutrients (TDN). Notice, however, that the ratio for NEp is about 10 to 1. In other words, corn grain is 10 times more valuable as an energy source than wheat straw.

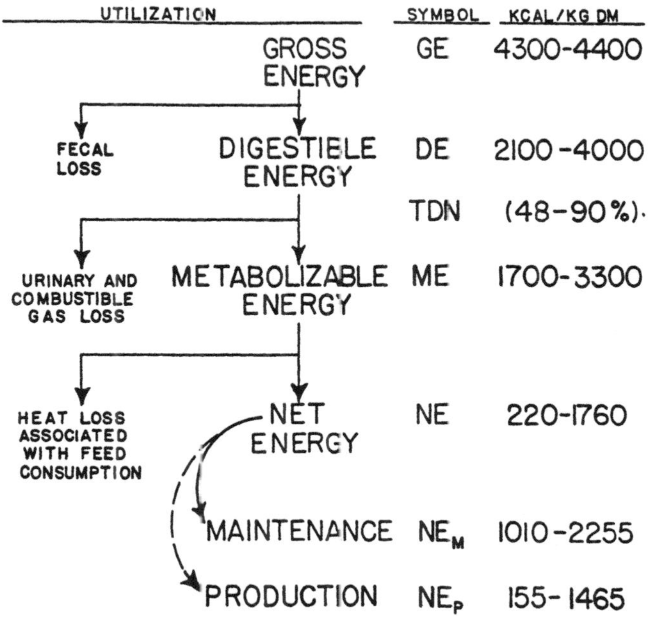

Figure 2. Energy utilization in animals.

Thus, in evaluating the relative cost of energy from different feeds, one should use their net energy value. The problem in doing this is that energy is first used for maintenance and then for production. Normally feedlot cattle eat about 2.5 times their maintenance needs. Therefore, an appropriate split would be 40% of the NE_M and 60% of the NE_P value to arrive at the energy value of a feed to be compared with another feed.

In practice, the other systems of energy, namely DE, TDN, or ME, can be used to formulate feedlot diets with approximate equal results as the net energy (NE) system. In reality, diets are formulated to certain energy levels through roughage to concentrate ratios, selecting the feeds on the basis of those that supply energy and other nutrients the most economically.

The main advantage of the NE system is its ability to predict performance of cattle and therefore close-out time, weight, feed efficiency, and cost of gain. Generally this system will give predictions that are within ± 5% of actual performance.

If actual performance is less than 95% of that predicted, it means that some other management factor is not right. The most common problem is bad weather and(or) muddy lots. Disease, diet imbalances, poor quality feed ingredients, improper feed processing, or poor bunk management can also be causes of less than predicted performance.

Thus, the NE system becomes a good management tool to determine if cattle are performing according to expectations.

It is beyond the scope of this paper to present the equations and method of calculating predicted performance based on the NE system. For those who are interested in this procedure, please refer to the National Research Council (NRC) 1984 publication on the nutrient requirements of beef cattle.

PROTEIN REQUIREMENTS

A discussion of the protein requirements of beef cattle is presented in another paper in this publication. The reader is referred to this paper for more details regarding the following.

The crude protein requirements of cattle are shown in table 1. Thus, for feedlot cattle expected to gain 3.0 lb/day and fed a 20% roughage and 80% concentrate diet, the diet should contain 12% CP on a DM basis. Other values for different rates of gain and roughage to concentrate ratios can be interpolated from this table. For instance, cattle gaining 2.5 lb/day and fed a 5% roughage and 95% concentrate diet would require about 12.2% crude protein in their diet.

TABLE 1. CRUDE PROTEIN REQUIREMENTS[a] FOR CATTLE DIETS

Composition of diet[a]		Daily rate of gain, lb				
% roughage	% concentrate[b]	0	1.0	1.5	2.0	3.0
100	0	6.2	8.1	8.7	9.0	9.4
80	20	6.6	8.7	9.2	9.6	10.1
60	40	6.9	9.1	9.8	10.2	10.8
50	50	7.1	9.4	10.0	10.4	11.1
40	60	7.2	9.7	10.3	10.8	11.4
30	70	7.4	9.9	10.6	11.1	11.8
20	80	7.6	10.1	10.9	11.3	12.0
10	90	7.8	10.4	11.1	11.7	12.3
0	100	7.9	10.7	11.4	12.0	12.7

[a]Dry matter basis.
[b]Grain plus supplement.

As discussed in my other paper, protein needs decrease after the cattle are on grain and their weight is about 75% of their expected slaughter weight. Protein levels can be reduced at this time to 8.5% to 9% of the DM.

Other topics related to protein needs are discussed in another paper presented here. These include the use of nonprotein nitrogen (urea), bypass protein, and protein requirements of feedlot bulls. Please refer to "Protein Requirements of Cattle" for a discussion of these topics.

MINERALS

General requirements for various minerals that may be deficient in feedlot diets are given in table 2. Minerals collectively constitute no more than 6% of all feed costs for cattle. Therefore, it is safe to conclude that supplying adequate mineral levels to cattle does not constitute a major cost factor in cattle feeding.

TABLE 2. MINERAL REQUIREMENTS OF FEEDLOT CATTLE

Mineral	Requirement[a] %	Requirement[a] ppm
Salt	.2	
Phosphorus	.25	
Calcium	.45	
Potassium	.7	
Sulfur	.12	
Zinc		30
Cobalt		1.0
Selenium		.1
Iodine		.1

[a]Concentration required in the diet DM.

Salt is a common requirement of all cattle and is generally added to livestock diets at the rate of .5% of the diet DM. This certainly is adequate, and since salt is cheap, no one pays too much attention to whether this is more than the animal requires. In those situations where volume and saline content of cattle waste is a problem, the level of salt can be cut in half, namely to .25%, without affecting the performance of cattle. Basic research indicated that the minimum salt requirement is probably near .16% of the diet DM. Therefore, a minor savings in feed cost could be obtained by reducing the salt level, but a much more important aspect is the decrease in animal waste and salinity as a result of feeding salt at a level that more nearly equals the animal's actual requirement.

Phosphorus is probably the most expensive mineral required by cattle. About one-third of the total mineral cost in feeding cattle comes from supplying phosphorus alone. Phosphorus requirement figures (from the NRC) for growing-finishing beef cattle are in the range of .17% to .32% phosphorus in diet DM. There has been some tendency to increase supplementary phosphorus levels in beef cattle diets to between .3% to .4% phosphorus, until a shortage of supplementary phosphorus occurred. At that time, the phosphorus requirements of various domestic animals were reappraised by a subcommittee of the NRC and phosphorus requirements were lowered in the case of many animals.

In my opinion, there is no need to exceed .5% phosphorus in the DM of diets designed for growing or finishing cattle. High phosphorus levels, unaccompanied by higher calcium levels, can lead to calculi or water belly problems in steers. Therefore, I think that the phosphorus level in cattle diets should be near .5% of the diet DM.

Calcium needs of feedlot cattle have been questioned recently and sources of calcium have been studied. A summary of recent research (Owens, 1983) indicates that feedlot cattle require .43% calcium in the diet DM.

Diets fed to finishing cattle are high in concentrate, and they are "acid" diets that alter the pattern of phosphorus excretion. High concentrate diets are acid for two reasons: 1) in contrast to roughage type diets, the mineral composition results in an acid urine; 2) the acid type fermentation in the rumen results in the production of lactic acid, some of which cannot be metabolized by the animal and is therefore excreted in the urine. Maintaining the calcium level of the diet so that it is 1.5 to 2 times greater than the phosphorus level will eliminate this problem. Therefore, if .5% phosphorus is the level of phosphorus in a given diet, the calcium level should be .38% to .50% in the diet DM. If the phosphorus level in the diet is higher, the level of calcium should be correspondingly higher. Remember, many calcium sources also contain phosphorus. If calcium alone is needed, limestone or calcium carbonate is the best source of calcium without phosphorus.

Potassium is a forgotten mineral since most diets that contain very much hay or silage will contain a sufficient amount of potassium. A second reason this mineral is sometimes overlooked is that the potassium requirement for ruminants is approximately double that of other animals. This requirement is near .7% of the diet dry matter. Hays and silages usually contain between 1% and 1.5%; however, grains can contain less than .4% potassium. Therefore, any time diets contain more than 50% to 60% grain, they can be deficient in potassium, especially if urea is used instead of oil meals as a source of supplemental protein.

It may be well to review the symptoms of potassium deficiency since these are not commonly known: poor growth, decreased feed intake, and stiffness in the legs and back similar to the early stages of founder. Because of this similarity, I feel that many of the symptoms ascribed to founder in commercial feedlots may actually be early signs of potassium deficiency. Also, potassium deficiency may be the reason for the lack of success in using high concentrate diets in many commercial circumstances. Potassium supplementation also becomes important during the receiving period and supplemental protein withdrawal as discussed in my paper on protein requirements.

Sulfur presents a real enigma in cattle feeding. Sulfur is required and can be supplied in the form of inorganic sulfur, even though the basic requirement of the animal is for sulfur containing amino acids. These amino acids, however, can be synthesized by the rumen microorganisms if the necessary inorganic sulfur and nitrogen are present in the diet. The requirement for sulfur in cattle diets is probably no greater than .12% of the diet DM. Sulfur supplementation becomes an important matter in cattle diets only when nonprotein nitrogen (urea) is being used as a major source of supplementary protein. Sulfur may also be important during periods of supplementary protein withdrawal.

Trace minerals are required by cattle; however, only a few require supplementation in commercial cattle feeding. A discussion of these is difficult since supplementation depends to a large extent on the feeds that are used and the location in which they are grown. Basically, I would start out by saying that supplementary trace minerals are not required with a few exceptions.

These exceptions include **iodine** which can be easily supplemented in the form of iodized salt. **Cobalt** tends to be deficient in feed raised in the northcentral part of the United States and in Florida. In these areas, cobalt should be added to cattle-finishing diets. **Selenium** is deficient in the Pacific Northwest and north of the Ohio and east of the Mississippi rivers. While selenium should be added to the diets of reproducing cattle in these areas, it is not clear that it is deficient in feedlot cattle diets. **Zinc** may require supplementation under certain circumstances.

There is also danger in overfortifying cattle-finishing diets with certain trace minerals. An excess of **copper** can be detrimental and even toxic to cattle. One should not include more than one source of trace minerals in any cattle feeding diet.

VITAMINS

The only vitamin of practical importance in feedlot diets is vitamin A. Green forages and yellow corn grain contain the precursor of vitamin A, carotene, that is converted to vitamin A when absorbed by the digestive tract. Because of uncertain storage conditions and times, these feeds should not be relied on to supply the vitamin A requirements. Dehydrated alfalfa hay with a stated vitamin A level, protected with an antioxidant, is the only feed I would rely on for vitamin A value.

Therefore, vitamin A should be added to all feedlot diets. The minimum level to use is 2,000 International Unit (IU) per 100 lb of body weight. To provide a safety factor for loss of activity in storage or feed processing, I would increase this to 3,000 IU per 100 lb of body weight. Thus, an 800 lb steer should have 24,000 IU of vitamin A daily.

Other vitamins are mentioned from time to time for feedlot diets. Vitamin E levels may be low in the blood of feedlot cattle, but there does not seem to be a performance response when vitamin E is added to diets containing adequate levels of selenium. The B vitamins are generally synthesized in adequate amounts by the rumen microorganisms. Once in a while, cattle will show symptoms of thiamine (vitamin B-1) deficiency, which may be accentuated by lactic acidosis. In these circumstances, thiamine may be required as a therapeutic agent. Niacin has also been studied recently, but the inconsistency of response in feedlot cattle does not warrant its addition to all feedlot diets.

ADDITIVES

It is beyond the scope of this paper to discuss in detail the use of all additives that can be used in feedlot cattle.

Growth stimulating implants (Synovex, Ralgro, Compudose) are still the most beneficial additives to be used in the feedlot. These increase rate of gain (8% to 12%), improve feed efficiency (5% to 10%), and increase the leanness of the carcass.

Melengestrol Acetate (MGA) is useful to eliminate estrus behavior of feedlot heifers and thereby give some improvement in performance. Ionophores (Rumensin and Bovatec) are added to improve feed efficiency. Antibiotics and sulfa drugs are beneficial for newly received feeder cattle. Certain antibiotics will also reduce the incidence of liver abscesses in feedlot cattle.

In deciding on an additive program for feedlot cattle, FDA regulations must be followed to assure that the cattle will be deemed safe for human consumption at the time of slaughter.

REFERENCES

Conrad, H. R., A. D. Pratt, and J. W. Hibbs. 1964. Regulation of feed intake in dairy cows. I. Change in importance of physical and physiological factors with increasing digestibility. J. Dairy Sci. 47:54.

Goodrich, R. D. and S. D. Plegge. 1984. Prediction of dry matter intake by feedlot cattle. Invited Paper, 76th Annual Meeting, American Society of Animal Science.

NAS-NRC. 1984. Nutrient Requirements of Beef Cattle. 6th revised edition. Natl. Academy Press, Washington, D.C.

Owens, F. N., A. L. Goetsch, D. C. Weakley, and R. A. Zinn. 1983. Buffers and neutralizers in beef cattle. In: Buffers, Neutralizers, and Electrolytes Symposium. Natl. Feed Ingredients Assoc.

Preston, R. L. 1972. Nutritional implications in economy of gain in feedlot cattle. J. Anim. Sci. 35:153.

24
PROTEIN REQUIREMENTS OF CATTLE

Rodney L. Preston

INTRODUCTION

The need for protein in animal diets has been well-known for a long time. Yet, protein need and satisfying this need is a major topic among scientists and producers at nutrition meetings.

Protein is determined in feeds by using the Kjeldahl procedure that actually measures the nitrogen content of feed. Since protein contains 16% nitrogen on the average, multiplying the nitrogen content by 6.25 (100/16) gives the crude protein (CP) content of the feed. The animal, however, requires true protein or the amino acids that make up true protein. Since the CP procedure does not distinguish between true protein and nonprotein nitrogen (NPN), a CP analysis of a feed is only a start in describing the ability of a feed to meet the protein requirements of an animal.

PROTEIN DIGESTIBILITY

The first step in measuring actual utilization of CP is to determine the digestibility of the protein contained in the diet. Digestion trials are an important part of ruminant research, but in many cases they yield data that have been misinterpreted. Protein digestibility is a major example. Apparent digestible protein (DP) has been determined on many feeds. An overriding influence on the meaning of these coefficients, however, is the effect of metabolic fecal nitrogen.

Figure 1 compares results obtained by several workers relating CP to DP (Preston, 1980). It can be seen that all five estimates are nearly the same. Table 1 summarizes these five equations. While nearly the same, there is some difference between these estimates over the crude protein range found in ruminant diets.

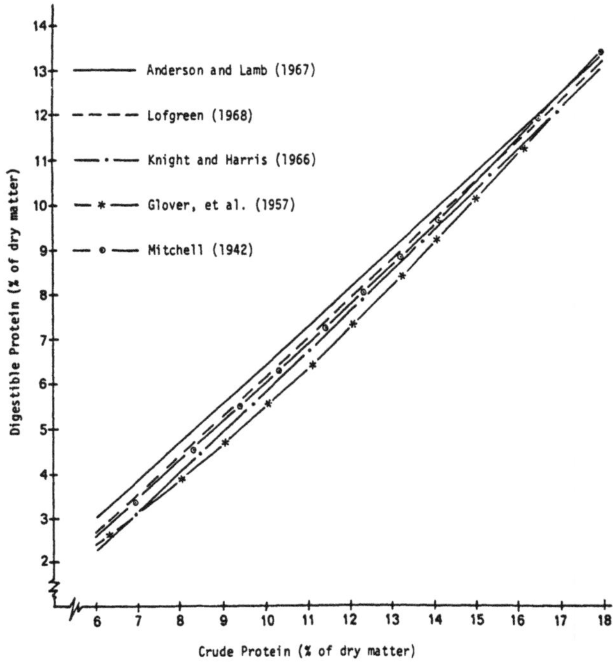

Figure 1. Relationship between crude protein and digestible protein in ruminant diets.

TABLE 1. APPARENT DIGESTIBLE PROTEIN PERCENTAGE[a] OF RUMINANT DIETS AS INFLUENCED BY CRUDE PROTEIN LEVEL

Source[b]	Crude protein (% of dry matter) 8	10	12	14
(1)	3.9	5.5	7.3	9.1
(2)	4.0	5.8	7.6	9.4
(3)	4.3	6.0	7.8	9.6
(4)	4.4	6.1	7.9	9.6
(5)	4.7	6.4	8.1	9.8

[a]Dry matter basis.
[b]Source:
(1) Glover et al. (1957).
(2) Knight and Harris (1966).
(3) Mitchell (1942).
(4) Lofgreen (1968).
(5) Anderson and Lamb (1967).

Use of DP values in feed composition tables can be very misleading as shown in table 2. When four CP levels were fed, digestibility coefficients were very much lower than those using DP values shown in the National Research Council feed composition tables for the feeds included in this experiment (Preston et al., 1965). Calculated digestibility coefficients using the equation of Glover et al. (1957) are very close to those actually determined in this experiment. Therefore, the use of DP levels for feeds to formulate diets is very misleading and is not recommended.

TABLE 2. INFLUENCE OF CRUDE PROTEIN LEVEL UPON THE APPARENT PROTEIN DIGESTIBILITY COEFFICIENT[a]

Source	Crude protein (DM basis), %			
	6.2	8.0	11.7	13.5
Actual	36.0	47.7	56.6	65.9
Calculated[b]	40.5	48.2	59.8	64.1
Calculated[c]	42.4	55.4	68.7	72.9

[a]Preston et al. (1965).
[b]Glover et al. (1957).
[c]National Research Council.

If one wants to estimate the DP content of a diet, any of the equations shown in table 1 could be used. An easy equation that will work in most cases is %DP = 0.9(%CP) - 3, where DP and CP are on a dry matter basis.

NONPROTEIN NITROGEN (NPN)

Ruminants have a distinct advantage over monogastric animals, such as pigs and poultry, in that they can utilize NPN for protein purposes. As previously mentioned, feeds can contain NPN. Rapidly growing forages can have a large portion (more than 30%) of their CP in the form of NPN. A common NPN source used to supplement cattle diets is urea. Ammonia- or urea-treated silages and roughages are another form of NPN. When cattle consume NPN, as well as true protein, microbial metabolism in the rumen (paunch) converts NPN into microbial protein, which is subsequently digested by the animal and is used to meet its protein need.

Cattle have been experimentally fed all of their protein requirement as NPN; however, general rules of thumb developed many years ago are still appropriate for the use of NPN in practical feeding situations:

1. No more than one-third of the total protein in the ration should come from NPN.
2. Protein supplements should not contain more than 3% urea.
3. Concentrates should not contain more than 1% urea.

Certain feeds and diets are better suited for NPN use. Grains are particularly good for allowing maximum use of NPN. The starch in these grains is readily available and more nearly matches the availability of NPN for microbial synthesis in the rumen. Molasses is also good in this regard and often molasses and NPN are mixed together as a supplemental source of CP and energy. With poor quality forages, straws, stover, etc., a small amount of NPN will enhance their digestibility. However, high forage diets are not well suited for maximum NPN utilization since their energy is only slowly available.

BYPASS PROTEIN

Recently, bypass protein has received considerable attention. This is true protein that is not metabolized by the microorganisms in the rumen and therefore is digested further down in the gastrointestinal tract, thus "bypassing" the rumen. Theory states that high-producing cattle with a high protein need cannot obtain sufficient protein from rumen microbial metabolism; therefore they benefit if some of the protein bypasses the rumen without microbial breakdown. A clear and consistent demonstration of the need for bypass protein has yet to be made. A few studies show a response to bypass protein (e.g., newly received calves at the feedlot) but many other studies have failed to show a response.

A few bypass-protein values are shown in table 3. Many of these are estimates based on indirect observations. Until the real need for bypass protein is more clearly defined, the following is recommended. If cattle require more than 10% CP in their diet dry matter, the additional protein should be in the form of bypass protein. For instance, if cattle require 12% CP, then 2% units or 17% (2/12) of their total protein should be bypass protein.

TABLE 3. TYPICAL CRUDE PROTEIN AND RUMEN BYPASS-PROTEIN VALUES FOR CATTLE FEEDS[a]

Feedstuff	% CP[b]	Bypass, % of CP
Alfalfa hay	17	25
Alfalfa, dehydrated	19	60
Barley, grain	12	30
Brewers grains, dried	28	65
Corn silage, mature	8	40
Corn, grain	10	50
Corn gluten meal	45	65
Cottonseed meal	46	40
Distillers grains	30	65
Orchard grass hay	11	30
Sorghum grain (milo)	11	60
Soybean meal	50	35
Timothy hay	8	35
Urea	288	0
Wheat, grain	13	25

[a]Preston (1984).
[b]Dry matter basis.

PROTEIN REQUIREMENTS FOR MAINTENANCE AND GROWTH

After reviewing a considerable body of information (Preston, 1980), the following equation best describes the DP requirements of cattle:

$$DP/W^{0.75} = 1.6 + 5.2\ G$$

W [live weight of the animal] and G [daily gain] are both in kilograms.

The value $W^{0.75}$ is an estimate of the animal's metabolic body weight. The DP requirement calculated in this equation is in grams per day.

As will be noted later, there are exceptions to this general equation, but this is a good general recommendation for meeting the DP requirements of cattle.

PROTEIN: ENERGY RELATIONSHIPS

When the above DP equation is divided by the equation describing digestible energy (DE) requirements (Garrett et al., 1959), optimum DP/DE diets required for varying rates of gain can be calculated as shown in table 4. Also shown are the relative energy intakes required to achieve these rates of gain. It can be seen that at maintenance the ratio is nearly 12 g DP/Mcal DE, and as rate of gain increases, the ratio increases.

TABLE 4. RELATIVE ENERGY INTAKE AND PROTEIN TO ENERGY RATIOS REQUIRED BY GROWING-FINISHING CATTLE

Rate of gain, lb/day	Energy intake[a]	Optimum DP/DE ratio[b]
Maintenance	1.0	11.6
1.0	1.6	18.2
1.5	1.9	20.0
2.0	2.2	21.3
3.0	2.8	23.0

[a]Energy intake relative to maintenance energy requirements.
[b]Grams DP/Mcal DE.

These ratios can be converted to percentage DP requirements by multiplying the ratio and the DE content of the diet, expressed as Mcal/100 g DM. Since roughages and concentrates contain about 1.15 Mcal and 1.72 Mcal DE/lb of DM, respectively, percentage DP requirements can be calculated for diets containing varying proportions of roughage and concentrate fed to give varying rates of gain as shown in table 5. Using one of the conversion equations for estimating CP from DP (table 1), these DP requirements can be converted to CP percentages. These are shown in table 6. This table should prove very useful in formulating cattle diets. Body weight, rate of gain, and energy concentration of the diet have been considered in their derivation.

TABLE 5. APPARENT DIGESTIBLE PROTEIN REQUIREMENTS[a] FOR CATTLE DIETS

Composition of diet[a]		Daily rate of gain, lb				
Roughate, %	Concentrate, %[b]	0	1.0	1.5	2.0	3.0
100	0	2.9	4.6	5.1	5.4	5.8
80	20	3.2	5.1	5.6	5.9	6.4
60	40	3.5	5.5	6.1	6.5	7.0
50	50	3.7	5.8	6.3	6.7	7.3
40	60	3.8	6.0	6.6	7.0	7.6
30	70	4.0	6.2	6.8	7.3	7.9
20	80	4.1	6.4	7.1	7.5	8.1
10	90	4.3	6.7	7.3	7.8	8.4
0	100	4.4	6.9	7.6	8.1	8.7

[a]Dry matter basis.
[b]Grain plus supplement.

TABLE 6. CRUDE PROTEIN REQUIREMENTS[a] FOR CATTLE DIETS

Composition of diet[a]		Daily rate of gain, lb				
Roughate, %	Concentrate, %[b]	0	1.0	1.5	2.0	3.0
100	0	6.2	8.1	8.7	9.0	9.4
80	20	6.6	8.7	9.2	9.6	10.1
60	40	6.9	9.1	9.8	10.2	10.8
50	50	7.1	9.4	10.0	10.4	11.1
40	60	7.2	9.7	10.3	10.8	11.4
30	70	7.4	9.9	10.6	11.1	11.8
20	80	7.6	10.1	10.9	11.3	12.0
10	90	7.8	10.4	11.1	11.7	12.3
0	100	7.9	10.7	11.4	12.0	12.7

[a]Dry matter basis.
[b]Grain plus supplement.

At this point, consider the meaning of the DP equations, ratios, and requirements that have been presented. None of these can be used in a predictive way; that is, rate of gain cannot be predicted from these values alone. Performance of cattle is determined by their energy intake relative to maintenance needs. Requirements for the remaining nutrients are largely a function of relative energy intake. However, for each level of energy intake, and thus for every rate of production, there is an optimum ratio between protein and energy where both are used most efficiently. This is what has been proposed here, namely the optimum amount of protein to be fed in relation to energy intake.

SUPPLEMENTAL PROTEIN WITHDRAWAL

Certain observations have led to the postulation that at times cattle may be able to perform optimally, and therefore more economically, on levels of protein that are very much below the levels shown in tables 5 and 6.

A series of tests were set up to study the need for supplemental protein during the entire growing-finishing period. These experiments have been published (Preston and Cahill, 1972; Preston and Parrett, 1972; Preston and Cahill, 1973; Preston et al., 1973) and recently reviewed (Preston, 1976). Therefore, they will only be summarized here.

Experiments were conducted with steer calves that replicated certain treatments in two different years. The calves were fed 15 lb limestone-treated corn silage, a full-feed of dry crimped corn, supplemental minerals, vitamins, and an antibiotic. The important feature about this work was that when soybean meal was discontinued as the source of supplemental protein, it was replaced with the same amount of a supplement made up with corn but containing the major minerals (Ca, P, K, S) in amounts equal to soybean meal.

Table 7 presents the combined daily gain data by period for these two experiments. Table 8 presents the feed efficiency data. In terms of the overall feeding period, performance was reduced in calves that did not receive any supplemental protein from the start of the experiment. This difference occurred during the first 56 days, however, with performance after this time equal to or even better than that observed in steers fed supplemental protein. Performance of steers fed supplemental protein for the first 56 or 112 days did not differ from that of steers fed supplemental protein continuously for 168 days. It is of interest to note that gain and feed efficiency of steers fed supplemental protein continuously were not as good during the last third of the trial as they were for steers that did not receive supplemental protein during this time. Also of interest is the fact that when supplemental protein was withdrawn, the diet contained only 8.2% to 8.6% CP in the DM. In these two experiments, the weight of the steers at the time when supplemental protein was first removed from the diet (56 days) ranged between 760 lb to 790 lb.

Complete carcass evaluation (table 9) did not reveal any difference between the various supplemental protein feeding times. Rib-eye area was significantly decreased in those cattle that did not receive any supplemental protein.

Research results from many other laboratories have confirmed these findings. Therefore, it seems safe to conclude that steers can be finished during the latter part of the finishing period on levels of CP that are much lower than standard feeding recommendations (e.g., 8.2% to 8.6% of the DM). It is not known whether the point at which supplemental protein can be decreased is determined by time on feed or live weight. It was concluded, however, that con-

TABLE 7. INFLUENCE OF SUPPLEMENTAL PROTEIN WITHDRAWAL ON THE GAINS OF GROWING-FINISHING STEER CALVES

Item	Length of supplemental protein feeding			
	None	56 days	112 days	Continuous
No. of steers	10	30	20	20
Avg. daily gain, lb				
0 to 56 days	2.84[a]	3.24	3.44	3.47
56 to 112 days	2.16	2.25	2.13	2.36
112 to final	1.97	2.04	1.98	1.84
0 to final	2.25[a]	2.52	2.52	2.56

[a]Significantly less ($P<.05$) than other treatments.

TABLE 8. INFLUENCE OF SUPPLEMENTAL PROTEIN WITHDRAWAL ON THE FEED EFFICIENCY OF GROWING-FINISHING STEER CALVES

Item	Length of supplemental protein feeding			
	None	56 days	112 days	Continuous
Dry matter/gain				
0 to 56 days	5.23[a]	4.68	4.60	4.58
56 to 112 days	7.57	7.76	8.21	7.38
112 to final	7.13[a]	8.21	8.20	8.92[a]
0 to final	7.15[b]	6.53	6.55	6.48

[a]Significantly less ($P<.05$) than other treatments.
[b]Significantly greater ($P>.01$) than other treatments.

TABLE 9. INFLUENCE OF SUPPLEMENTAL PROTEIN WITHDRAWAL ON THE CARCASS CHARACTERISTICS OF GROWING-FINISHING STEER CALVES

Item	Length of supplemental protein feeding			
	None	56 days	112 days	Continuous
Dressing, %	63.2	62.7	62.9	63.3
Conformation score	Choice	Choice	Choice	Choice(+)
Marbling score	Small	Small	Small	Small
Final grade	Choice(-)	Choice(-)	Choice(-)	Choice(-)
Fat thickness, in.	0.6	0.6	0.6	0.6
Kidney, pelvic, heart fat, %	3.3	3.3	3.4	3.3
Rib-eye area, cm^2	10.6[a]	11.2	11.1	11.4
Cutability grade	3.7	3.5	3.6	3.5

[a]Significantly less ($P<.05$) than other treatments.

ventional type beef steers require substantially less CP after they reach 760 lb to 790 lb; when fed a corn-corn silage type diet, no supplemental protein is required after this time if the diet contains at least 8.6% CP on a DM basis.

In all experiments where supplemental protein was not included in the diet from the beginning of the feeding period, performance of the cattle was moderately to severely reduced. Thus, CP levels at least equal to those shown in table 6 should be fed when cattle are initially started on finishing diets.

Several questions related to these findings require answers. What is the importance of certain minerals during the time of supplemental protein withdrawal? Potassium appears to be of special importance. Can similar results be expected when cattle are fed higher roughage diets? For instance, can supplemental protein be withdrawn when cattle are being full-fed on roughage? It is postulated that some grain will be required in the diet to achieve successful supplemental protein withdrawal results. Young (1978) concluded that steers weighing at least 710 lb and consuming at least 15 lb of corn grain DM daily did not require supplemental protein if at least 1.9 lb CP is present in the total daily diet. Adequate feed intake at the time of supplemental protein withdrawal is important to assure a minimum protein intake (e.g., 1.8 lb to 2.0 lb/day) after removal of supplemental protein.

At what body weight should supplemental protein be withdrawn from heifers, steers, and bulls? Will it be the same for exotic breeds of cattle and dairy-type steers as well as conventional beef breeds? The answer is that when beef animals are about 75% of the slaugher weight required to yield mainly choice carcasses, their supplemental protein needs diminish.

Two obvious conclusions are apparent from the research reported here:

1. Supplemental protein needs of feedlot cattle do not remain constant during the feeding period. Once the cattle are started on grain, CP levels considerably below those recommended can be fed without sacrificing feedlot performance or carcass desirability.
2. Recommended CP levels for feedlot cattle are as follows:
   - For the first 28 to 56 days, feed a diet containing at least the levels shown in table 6. If the cattle are calves weighing less than 500 lb, slightly higher levels of CP may be desirable (e.g., 12% to 13% of the DM).
   - After this initial feeding period, CP levels can be reduced to the following amounts at the indicated live weights or percentage of live weight required to yield mainly choice carcasses:

Live weight, lb	or	% of weight at choice grade	CP in diet Dm, %
750		75	9.4
850		85	8.9
950		95	8.5
1050		105	8.2

- Because of the critical importance of protein to the performance of feedlot cattle, feedstuffs should be analyzed for CP to assure that the above levels of CP will be contained in the final ration.

If feed intake is below normal at the time of supplemental protein withdrawal, protein should not be decreased if the resulting CP intake will be less than 1.9 lb/head/day.

There is a common belief that bulls require a higher level of protein in their diet compared to steers. Research has shown this to be incorrect, however. Bulls will generally eat slightly more feed than steers and in this way will consume more total protein. Recommended protein levels for bull feeding are shown in table 10. As yet, research has not been conducted with bulls on reducing protein levels to 8.5% of the diet dry matter as was mentioned earlier for steers. Until research defines minimum protein levels for bulls, those shown in table 10 should be followed.

TABLE 10. SUGGESTED PROTEIN LEVELS FOR GROWING-FINISHING BULLS

Days on feed	CP in DM, %
0 to 56 days	12
After 56 days	10.5

SUMMARY

Cattle protein needs have been reviewed and protein recommendations have been made that consider weight, rate of gain, and proportion of roughage and concentrate in the diet. Research on reducing protein levels in feedlot cattle diets after the cattle are on a full-feed of grain and weigh at least 750 lb was reviewed. Protein levels at this time can be substantially reduced. Protein requirements for feedlot bulls appear to be similar to steers.

REFERENCES

Anderson, M. J. and R. C. Lamb. 1967. Predicting digestible protein from crude protein. J. Anim. Sci. 26:912.

Garrett, W. N., J. H. Meyer, and G. P. Lofgreen. 1959. The comparative energy requirements of sheep and cattle for maintenance and gain. J. Anim. Sci. 18:528.

Glover, J., D. W. Duthie, and M. H. French. 1957. The apparent digestibility of crude protein by the ruminant. I. A synthesis of the results of digestibility trials with herbage and mixed feeds. J. Agric. Sci. 48:373.

Knight, A. D. and L. E. Harris. 1966. Digestible protein estimation for NRC feed composition tables. J. Anim. Sci. 25:593.

Lofgreen, G. P. 1968. As published in NRC Nutrient Requirements of Beef Cattle, 5th ed., Pub. 2419. Natl. Acad. of Sci., Washington, D.C.

Mitchell, H. H. 1942. The evaluation of feeds on the basis of digestible and metabolizable nutrients. J. Anim. Sci. 1:159.

Preston, R. L. 1976. Protein withdrawal from feedlot cattle. Proc. 11th Annual Pacific Northwest Animal Nutr. Conf., p 69.

Preston, R. L. 1980. Empirical value of crude protein. Oklahoma State Univ. MP-109, pp 201-217.

Preston, R. L. 1984. Typical composition of feeds for cattle and sheep. Feedstuffs 56(36):32.

Preston, R. L., D. D. Schnakenberg, and W. H. Pfander. 1965. Protein utilization in ruminants. I. Blood urea nitrogen as affected by protein intake. J. Nutr. 86:281.

Preston, R. L. and V. R. Cahill. 1972. Source of supplemental protein and time of supplemental for growing-finishing steer calves. OARDC Res. Summary 63, p 37. Wooster, OH.

Preston, R. L. and N. A. Parrett. 1972. Form of corn grain and protein levels for yearling steers. OARDC Res. Summary 63, p 75. Wooster, OH.

Preston, R. L. and V. R. Cahill. 1973. Withdrawing supplemental soybean meal from steer calves at varying times during the finishing period and the value of supplemental amino acids, with observations on DES implant levels. OARDC Res. Summary 68, p 1. Wooster, OH.

Preston, R. L., S. W. Kock, and V. R. Cahill. 1973. Supplemental protein needs and utilization of dry and crimped corn for yearling steers. OARDC Res. Summary 68, p 9. Wooster, OH.

Young, A. W. 1978. Supplemental protein withdrawal from corn-corn silage rations: effect of weight and corn intake at withdrawal. J. Anim. Sci. 46:505.

25

PREPARATIONS FOR RECEIVING AND PROCESSING STOCKER CATTLE

E. J. Richey

The success of any stocker program revolves around being well prepared to receive, process, and medicate stressed cattle. Proper planning, plus adequate equipment and supplies, are essential for this operation. The following outline -- based upon practical experience -- is offered as a guide to cattlemen.

PREPARATIONS FOR RECEIVING AND PROCESSING STOCKER CATTLE

When you purchase shipped stocker cattle, you must consider that the cattle on arrival are:

- Potentially sick
- Experiencing a diet change
- Hungry
- Thirsty
- Tired (if not completely worn out)

You must assume that they need to be:

- Watered
- Fed
- Vaccinated
- Dewormed
- Deloused and degrubbed
- Implanted
- Castrated
- Dehorned
- Branded
- Given vitamins
- Treated (if they are sick)

You need to know:

- Approximate age of the animals (calves or yearlings) to determine the feed type.
- Average weight of the cattle to 1) calculate dosage of wormers, insecticides, and medication, 2) be sure that the equipment fits the size of the cattle.

- Number of cattle in the shipment -- to determine the number of vaccine doses needed.
- Origin -- to be prepared for a reaction to a grub kill.
- Medication given before shipping.

BE PREPARED -- DON'T WAIT UNTIL THE CATTLE ARRIVE TO GET READY

Nothing can be more frustrating when processing newly purchased cattle than to need something and not have it; to break something and have no replacement; to have equipment that does not fit the animals; or to have a crew that won't tend to business.

Facilities: (Must match the size of the cattle)
Holding pens
Crowding chute
Squeeze chute
Trough space (sufficient to feed all cattle at one time)
Watering tanks

Feeding: (Consider the following)
Age (energy vs protein for growth)
- Calves require more energy.
- Yearlings respond better to a high protein feed.

Roughage content of the ration
- Protein **enhances** fiber digestion.
- Energy **inhibits** fiber digestion.

Low intake (A 400 lb steer requires approximately 1 lb of protein to gain 1 lb of body weight.)
- A 60% concentrate ration containing 13% crude protein results in a protein-deficient diet until consumption reaches 7.7 lb/day.
- A 40% crude protein supplement requires an intake of 2.5 lb to meet the 1 lb protein requirement.

Physical form (pellet size and pellets vs meal)
- 3/16" pellets are readily accepted and crumble easily; 1/2" cubes are hard and cause the animal to drop the cube.
- Fewer animals sick and more weight gain when pellets are fed.

Protein source (soybean meal [SBM] vs cottonseed meal [CSM])
- Total intake of the SBM nearly 2 times that of the CSM base feed during the first few days.
- Potassium level is higher in SMB than CSM. Grass hay preferred to legume hay.

Watering:

Have adequate tank space for drinking water rather than using an automatic water tank.
Some range cattle have never seen a water tank.
Have capability to control water intake.

Equipment and supplies:

Unloading area
- Hot shots (new batteries)
- Pencil and paper for cattle tally
- Pry bar for frozen doors
- Gravel for slick unloading ramps

Holding pens
- Hand whips or poopers to move the cattle
- Hollers

Crowding chute
- Hot shots
- Antibackup devices or pipes
- Grease markers to identify sick animals

Working chute area
- Vaccines (injectables)
  IBR, PI3, BVD
  Leptospirosis
  Clostridia bacterins (4 to 6 way)
  Pasteurella and hemophilus
- Vitamins (injectables)
  A, D, B12
- Epinephrine (injectables)
- Needles, automatic syringes, and plastic syringes
- Wormers
  Injectables - automatic syringe
  Drench - automatic syringe or dose syringe
  Boluses - balling gun
  Paste or gel - paste gun
- Implanting
  Implant pellets
  Implant gun
  Extra needles
  Cotton
  Alcohol
- Branding
  2 branding irons
  Heater
  Torch
  Full propane bottle or a spare
  Matches
  Wood blocks to burn
  Crescent wrench
- Dehorning (tipping)
  2 irons to cauterize the tip
  Supplies as needed in branding
  Bucket for water
  Water to rinse off the blood

    Dehorner to fit size of cattle
    Nose bar or strong halter
- Castration
    3 buckets (1 for testicles, 1 for rinsing your hands, and 1 for disinfecting equipment)
    Water and disinfectant
    Hand towels
    Newberry castrating knife
    Double-crush emasculator for heavy cattle
    Sharp knife or scalpel and extra blades
    Trimming scissors
    Wound dressing (hand pump)
    Long lasting penicillin (syringe and needles)
- Lice and grub control
    Injectable (IVERMECTIN) - syringe and needles
    Pour-on application - should be done when unloading to spread vapors
        Dipper (with ounce measurements)
        Bucket (mark for permanent use)
    Spot application
        Automatic-filling applicator
    Medication compatible with the wormer
    Water to rinse off spills
- Identifying and treating sick animals
    Electronic thermometer
        Fully charged or extra batteries
    Have treatments worked out in advance
        Oxytetracycline and sulfa
        Neomycin and amprolium
    Identification tags
        Numbered ear tags and tag pliers
        Numbered back tags and tag cement
    Record cards for temperature, medication, and remarks
    Pencils for recordkeeping
    Eye medication and wound dressing
- Miscellaneous equipment and supplies
    Trash barrel
    Scoop shovel and broom
    Hammer and nails
    Oil can and oil
    Pliers and screw driver
    Bucket of gravel for slick areas
    Water jug and drinking cups
    Cooler for vaccines and medications

PURCHASING SYRINGES, NEEDLES, VACCINES, BACTERINS, WORMERS, AND OTHER ANIMAL HEALTH EQUIPMENT

Syringes and equipment

Automatic syringes

- Get all one size and brand to be able to exchange parts.

- Use pistol grip with metal or plastic plungers.
- Buy 50 ml -- the preferred size.
- Check the calibration on each syringe (some administer 1 ml more accurately than they do 5 ml).

Plastic reusable syringes
- Get all one size and brand to be able to exchange parts.
- Buy those made of heavy plastic (ARDES).
- Use 30 ml size (can administer the medication using only one hand).

Needles
- Buy all-metal disposable.
- Use one size (1" x 16 gauge).
- Keep 100 needles on hand.
- Replace them often.

Balling gun (bolus administration)
- Use plastic head without the restraining springs or a multiple bolus balling gun.

Dose syringes
- Use 4 oz capacity.
- Have a measuring cup for adequately measuring the dose.

Vaccines and wormers (Note: the shelflife of a reconstituted vial of vaccine or a package of wormer drench powder is very limited.)
- Vaccines -- Purchase 50 dose and 10 dose vials so you will not have to reconstitute another 50-dose vial to vaccinate a few head, unless the 50-dose vial is cheaper than the 10-dose vials needed to vaccinate them.
- Wormer drench powder -- You need to have either worm boluses or an injectable wormer available to finish worming a load rather than open a package of drench powder for a few head.

RULES FOR THE CREW

- Leave the dogs at home.
- Come early to drink coffee and eat donuts.
- Come early to get the stories out of the way.
- Everyone stays with his/her assignment for the day.
- Discourage visitors -- they interrupt the routine.
- No breaks until the shipment is finished.
- Be serious and save $$$$$$.
- This is not a social function -- unnecessary conversation interrupts the routine.

26

MANAGEMENT OF SICK, NEWLY ARRIVED STOCKER CATTLE

E. J. Richey

A successful program for the management of sick cattle must be simple and systematic. Sick animals must be easily identified and the treatment must be routine and require a minimum of judgment decisions by the working crew.

The key elements in this program include:

- Identifying sick cattle as soon as possible
- Keeping adequate records
- Treating sick animals systematically
- Evaluating sick cattle daily
- Changing treatment, if necessary, until an improvement is noted

A cattleman wanting to use this program should consult a veterinarian prior to implementing these procedures.

IDENTIFYING SICK CATTLE AS SOON AS POSSIBLE

On the morning following the arrival of newly purchased cattle, begin the routine processing. At this time identify the animals that are visibly sick and then separate them from the healthy-looking animals.

Visible symptoms of illness include excessive nasal discharge, labored breathing, harsh deep coughing, moderate to severe depression, or bloody diarrhea. Animals exhibiting only loose stools or diarrhea without a show of blood are not kept from the rest.

As soon as the sick-looking animal is restrained in the chute, **take its temperature** using a rectal thermometer.

The processing crew now follows two rules for determining sick cattle:

1. **All cattle with a rectal temperature of 104°F or greater.**
2. **All visibly ill cattle regardless of the body temperature.**

Animals designated as sick are identified with numbered back tags glued to the forehead or numbered ear tags. The sick animals as well as the healthy-looking animals are vaccinated, wormed, and injected with vitamins. Castration and dehorning are too stressful for sick cattle and are delayed if at all possible. These procedures require additional time and it is important that the processing and treatment of the sick cattle be completed well before noon.

RECORDS

Complete and accurate records are a necessity. They tell tomorrow what was observed and done today. The processing crew uses a record card with a different form printed on each side. The first form used is the **examination** portion, which is only filled out on the first day the animal is pulled as sick.

The assigned tag number, body temperature, and estimated body weight are recorded at the top of the card. Visible symptoms are recorded by checking appropriate spaces. If a diagnosis is made, it is recorded. The severity of illness is rated as slightly ill (S), moderately ill (M), or very ill (V). Space is also provided for any additional remarks that pertain to the sick animal. (See examples 1-4.)

Remember, all this is to **jog the memory tomorrow**, or in someone's absence to inform a substitute why the animal was pulled as sick. The reverse side of the card is the **treatment** portion of the record. This portion will be discussed in subsequent sections.

SYSTEMATIC TREATMENT

When cattle are sick on arrival or shortly thereafter, there is usually insufficient time to diagnose or identify the causative organisms before treatment begins.

The bovine respiratory diseases (BRD) and diseases that result in diarrhea are the cause of 99% of the health problems during the first 3 wk after arrival. The BRD are caused by a combination of respiratory virus infections and stress that are compounded by bacterial infections. The major stresses have already occurred and the viruses are essentially untreatable. Therefore the bacterial infections must be controlled by the use of antibacterial drugs (antibiotics and sulfas). The sick cattle are medicated for bacterial infections in general rather than for specific diseases, with the exception of bloody diarrhea. In all cases cattle designated as sick are medicated while in the squeeze chute for routine processing. The **sick** animals are medicated in **one** of **two** ways, the deciding factor being the presence or absence of **bloody diarrhea.**

1. Sick animals **without** bloody diarrhea but pulled because of visual signs or a high body temperature are treated with:

   Injectable oxytetracycline
   and
   sulfamethazine boluses

EXAMPLE 1. USING BODY TEMPERATURE TO MONITOR IMPROVEMENT WHERE ANIMAL RESPONDED QUICKLY

Symptoms when 1st pulled as sick: Body temp. 106.0 Weight 500 lb Tag No. 8

Nose	Dry ___	Crusted ___	Discharge MP	Clear ___
Eyes	Clear ___	Cloudy ___	Ulcers ___	Watery X
Lungs	Heavy Breather ___	Labored ___	Rapid ___	Cough X
Diarrhea	Bloody ___	Watery ___	Black ___	
Digestive	Bloat ___	Drawn ___	Full X	
Foot Rot	Yes ___			
Nervous System	Staggering ___	Convulsions ___	Muscle Twitch ___	
Depression	Slight ___	Moderate X	Severe ___	

Other ___

Diagnosis: Upper Respiratory

Severity of Illness: M

Remarks: MP: Refers to Muco-purulent discharge

Tag 8

Date	Temperature	Severity of Illness	Oxytet. 100	Sulfa bol.	Erythromycin	Tylosin	Penicillin	Amprolium	Neomycin +	Remarks
1-13	106°	M	25	3 1/2						
1-14	102.1°	S	25	2 1/2						Eating
1-15	102.5°	S	25	2 1/2						Eating
1-16	$\bar{1}$01.8°	±	25	2 1/2						?
1-17	101.9°	0	25	2 1/2						Released

The 1st treatment with oxytet. and sulfa boluses reduced the temperature from 106°F to 102.1°F. Because the temperature dropped more than 2°F or to below 104°F within 24 hr, the animal was designated as improved. The treatment was repeated until the animal was nearly normal for 2 consecutive days -- then released.

EXAMPLE 2. USING BODY TEMPERATURE TO MONITOR IMPROVEMENT WHERE ANIMAL RESPONDED SLOWLY

Symptoms when 1st pulled as sick: Body temp. 106.1 Weight 400 lb Tag No. 1

Nose	Dry ___	Crusted ___	Discharge ___	Clear ___
Eyes	Clear ___	Cloudy ___	Ulcers ___	Watery X
Lungs	Heavy Breather ___	Labored ___	Rapid ___	Cough X
Diarrhea	Bloody ___	Watery ___	Black ___	
Digestive	Bloat ___	Drawn X	Full ___	
Foot Rot	Yes ___			
Nervous System	Staggering ___	Convulsions ___	Muscle Twitch ___	
Depression	Slight ___	Moderate X	Severe ___	

Other ______________________________

Diagnosis: ______________________________

Severity of illness: Very

Remarks:

Tag 1

Date	Temperature	Severity of illness	Oxytet. 100	Sulfa bol.	Erythromycin	Tylosin	Penicillin	Amprolium	Neomycin +	Remarks
2-7	106.1°	V	20	3						
2-8	105.6°	V			20					
2-9	105.7°	V				20				
2-10	103.5°	M				20				
2-11	102.0°	S-M				20				Eating
2-12	101.9°	±				20				Eating
2-13	101.9°	0				20				Eating-released

Oxytet. plus sulfa and erythromycin were apparently ineffective since little reduction in temperature occurred. Medication was changed daily until tylosin resulted in a noticeable improvement. The fever dropped from 105.7°F to 103.5°F within 24 hr following treatment. Treatment with tylosin continued until the animal appeared normal for 2 consecutive days.

EXAMPLE 3. USING OBSERVED SEVERITY OF ILLNESS TO MONITOR IMPROVEMENT

Symptoms when 1st pulled as sick: Body temp. 103.3 Weight 500 lb Tag No. 6

Nose	Dry ___	Crusted ___	Discharge X	Clear ___
Eyes	Clear ___	Cloudy ___	Ulcers ___	Watery X
Lungs	Heavy Breather ___	Labored ___	Rapid ___	Cough X
Diarrhea	Bloody ___	Watery ___	Black ___	
Digestive	Bloat ___	Drawn X	Full ___	
Foot Rot	Yes ___			
Nervous System	Staggering ___	Convulsions ___	Muscle Twitch ___	
Depression	Slight ___	Moderate X	Severe ___	

Other ______________________________

Diagnosis: Pneumonia

Severity of illness: M

Remarks: Sick on processing -- pulled because of visible illness.

Tag 6

Date	Temperature	Severity of Illness	Oxytet. 100	Sulfa bol.	Erythromycin	Tylosin	Penicillin	Amprolium	Neomycin +	Remarks
1-8	103.3°	M	25	3 1/2						
1-9	103.9°	M			25					Not eating
1-10	102.5°	S			25					?
1-11	102.5°	±S			25					Eating
1-12	101.8°	±0			25					Eating
1-13	102.0°	0			25					Released

This animal was pulled on processing due to visible signs (nasal discharge, eyes red and watery, coughing, drawn, and moderately depressed). Since his body temperature was below 104°F, we used "severity of illness" to monitor treatments. Oxytet. and sulfa - no improvement. But erythromycin resulted in improvement. The erythromycin treatment continued until the animal was free of clinical signs and eating for 2 consecutive days.

EXAMPLE 4. ANIMAL WAS PULLED FROM THE PEN FOR WATERY DIARRHEA 5 DAYS AFTER ARRIVAL -- USED BLOODY DIARRHEA TREATMENT SCHEDULE

Symptoms when 1st pulled as sick: Body temp. 105.3 Weight 500 lb Tag No. 15

Nose	Dry ___	Crusted ___	Discharge sl.	Clear ___
Eyes	Clear ___	Cloudy ___	Ulcers ___	Watery X
Lungs	Heavy Breather ___	Labored ___	Rapid ___	Cough X
Diarrhea	Bloody ___	Watery XX	Black ___	
Digestive	Bloat ___	Drawn slight	Full ___	
Foot Rot	Yes ___			
Nervous System	Staggering ___	Convulsions ___	Muscle Twitch ___	
Depression	Slight ___	Moderate X	Severe ___	

Other ______________________________

Diagnosis: Diarrhea - no blood

Severity of illness: S

Remarks: Animal arrived 5 days ago -- severe diarrhea, very watery, no blood in stool.

Tag 15

Date	Temperature	Severity of illness	Oxytet. 100	Sulfa bol.	Erythromycin	Tylosin	Penicillin	Amprolium	Neomycin +	Remarks
1-13	105.3°	S	25	--				150	15	
1-14	104.5°	S			25			150	15	Diarrhea
1-15	100.5°	±S			25			150	15	Stool less loose, eating
1-16	101.0°	0			25			150	--	Eating
1-17	101.8°	0			25			150	--	Stool OK, released

The animal had a body temperature of 105.3°F the morning it was pulled. The fever did not respond to the oxytetracycline injection -- therefore the injectable portion of the treatment was changed to erythromycin, which was continued until the 5 days of oral treatment were completed.

2. Animals exhibiting **bloody diarrhea** with or without a fever (104°F or greater) are treated with:

Injectable erythromycin
and
amprolium - neomycin drench

Nonbloody diarrhea is essentially ignored when the cattle are processed. The loose stools may be caused by a change in diet or a heavy worm load. Putting cattle on a high roughage diet and worming them often alleviates this problem. However, if an animal is also exhibiting visible signs of illness or an elevated body temperature (104°F or greater), it should be systematically medicated with injectable oxytetracycline and sulfamethazine boluses.

The processing crew makes no treatment decisions. The treatments are designated in advance, usually when a crew member is obtaining the body temperature.

It is necessary to record the medication administered to the animal on the treatment portion of the record. The tag number, date of treatment, body temperature, severity of illness (S, M, or V), and the **amount** of **each** medication administered are recorded.

The use of the examination and treatment portions of the record card is demonstrated in examples from several treated sick animals (see examples 1 and 2).

As cattle leave the squeeze chute, sick animals are separated from those that don't appear ill by use of a cutting gate. All animals (sick and healthy) are held near the working area until the processing operation is complete. This permits easy observation for the detection of reactions to vaccinations or medications.

After processing is complete, the healthy-looking cattle are moved to drylots or small pastures, observed twice daily for 14 days, and pulled if they appear sick. They are not run through the chute to have their temperature taken each day. The cattle that appear to be ill in the days following the initial processing are subjected to the same treatment program as the group of sick cattle from the first day.

EVALUATING SICK CATTLE DAILY

It is imperative that sick animals be rechecked and evaluated each day. The rechecking is done as early in the day as possible (well before noon) to obtain useful body temperatures. Sick cattle are observed before they are moved to the working-chute area. If possible, each animal is observed for appetite, respiratory difficulties, consistency of stool, and degree of depression.

When checking the sick animals, the first thing to do after restraining the animal in the chute is to take its temperature! Next, the record card should be checked to see why the animal was pulled the first day. Then the new body temperature should be recorded. At this point determine if the animal appears less sick than the day before based on the change in appetite, respiratory difficulty, fecal consistency, or change in depression.

Next, the severity of illness should be rated (S = slight, M = moderate, or V = very ill) using the visual observations. The record card for this day shows the date, body temperature, and severity of illness, as well as the previous day's information and treatment. It is not necessary to change the examination portion of the sick animal's record card. Changes in the physical signs can be noted in the section for remarks on the treatment card.

When an animal starts on a medication, improvement must occur within 24 hr or it is presumed the medication is not effective. An animal is showing improvement if:

1. It shows a 2° reduction in fever from the first day's body temperature or a temperature less than 104°F within 24 hr following treatment.
2. It has a rectal temperature of less than 104°F on the first day and improves physically within 24 hr following treatment (i.e., from very ill to moderately ill, moderately ill to slightly ill, or slightly ill to less than slightly ill).

If the animal shows improvement, the same medication should be continued. In sick animals **without** bloody diarrhea, treatment no. 1 should be repeated (oxytetracycline and sulfamethazine boluses). In animals pulled with **bloody diarrhea,** the amprolium-neomycin drench is repeated even if a significant firming of the stool is not observed. If the fever has dropped 2°F or to below 104°F, or if a fever was not present the previous day, the injectable oxytetracycline medication should be repeated also. The injectable oxytetracycline and amprolium-neomycin drench should be continued for a total of 3 days. Then the animal should be treated another 2 days with the injectable oxytetracycline and amprolium drench (omitting the neomycin).

If improvement does not occur, medication should be changed to the next drug on the treatment schedule.

CHANGING TREATMENTS

If an animal does not improve, the medication should be changed following a set schedule of treatments. Changes continue at daily intervals until the animal begins to improve. A successful treatment should be continued until fever, depression, lack of appetite, and other clinical signs of illness are absent for 2 consecutive days.

Using a predetermined sequence of treatments eliminates having to determine what drugs to use next if the previous treatment did not work. A change to the next treatment is systematic. The advantage of this procedure is that if a disease organism is resistant to certain drugs, medication is changed frequently until an effective treatment is found.

We are essentially using the sick animal as a laboratory test. If the treatment does not work, the animal will not improve; therefore, we must change treatment.

If an animal does not improve after medicating with treatment no. 1 (injectable oxytetracycline and sulfamethazine boluses), the medication should be changed to treatment no. 2 (injectable erythromycin). If the body temperature does not drop 2°F or to below 104°F in cases of bloody diarrhea with fever, change the treatment to injectable erythromycin rather than injectable oxytetracycline.

Treatment sequences for medicating sick, newly arrived, and stressed cattle, with and without bloody diarrhea, are outlined in tables 1 and 2. Remember, if an animal fails to improve within 24 hr after a treatment is administered, change to the next treatment on the treatment schedule.

If an animal improves on a treatment, that treatment should be continued until clinical symptoms are absent for 2 consecutive days. Then release the animal.

If you change treatments and find a medication that works, use this medication as treatment #1 on subsequent sick cattle pulled from this load. However, remember that what works on cattle from this load may not work on cattle from another load. Search for the drug that works by systematically changing treatments when animals do not respond to medication.

Guidelines for drug use are outlined in table 3.

MISCELLANEOUS

Often situations are encountered that do not follow a pattern. It is virtually impossible to cover all situations that arise. However, several of the more common ones are discussed in this section.

The normal body temperature of newly arrived stressed cattle has not been determined. In this program, any body temperature below 104°F upon arrival is accepted as normal. However, the body temperature of some animals receiving treatment will settle between 103.4°F and 104°F. In these animals, change the treatment if the animal exhibits a poor appetite or any degree of depression. If the appetite is good and the animal is alert, continue the treatment for a total of 5 days, and then release the animal.

A large rise in the body temperature from one day to the next is probably due to a poor thermometer reading caused by air in the animal's rectum when the temperature was taken. In that case the thermometer probe cannot get a true reading of the body temperature. Be alert for this situation. It is like placing the probe in a balloon. If a sudden increase is noted in the body temperature one day after a sudden drop, change to the next treatment.

If diarrhea continues for more than 2 days after processing in an animal that is not being treated, or if diarrhea suddenly occurs in an animal, pull the animal from

TABLE 1. TREATMENT FOR SICK ANIMALS NOT EXHIBITING BLOODY DIARRHEA

---

* If animal fails to improve within 24 hr, change to next treatment.

* If animal improves on a treatment, continue that treatment until clinical symptoms are absent 2 consecutive days.

TREATMENT NO. 1:	OXYTETRACYCLINE (under the skin): 5 cc per 100 lb (100 mg/ml) PLUS SULFAMETHAZINE BOLUSES (oral): 1 bolus per 150 lb.
TREATMENT NO. 2:	ERYTHROMYCIN (GALLIMYCIN) (deep in the muscle): 5 cc per 100 lb.
TREATMENT NO. 3:	TYLOSIN "TYLAN 200" (in the muscle): 5 cc per 100 lb.
TREATMENT NO. 4:	PROCAINE PENICILLIN G (under the skin): 10 to 20 cc per 100 lb.
TREATMENT NO. 5:	
TREATMENT NO. 6:	
TREATMENT NO. 7:	Treat for 3 days with oxytetracycline. If no improvement, treat for 3 days with sulfamethazine. If no improvement, treat for 3 days with penicillin. Subsequent treatments, if required, should consist of 3 days of sulfamethazine <u>or</u> penicillin.

It is advisable to consult your veterinarian pr or to implementing this procedure. His experience in your area may result in the selection of a different sequence of treatments. Also, treatments no. 5 and no. 6 are left open so that you or your veterinarian can add prescription medications to this sequence. If you choose not to add to the sequence, move no. 7 into the no. 5 slot.

The route of administration and the dosages reported may be different than the manufacturer's label guidelines. However, the charges are implemented to provide adequate serum concentrations for approximately 24 hr, thus susceptible bacteria are controlled by treating once a day. The withdrawal periods have been substantially lengthened to compensate for the higher dosages. (Hjerpe and Routen: Treatment of bacterial pneumonia in feedlot cattle, with special reference to antimicrobic therapy. Proc. 9th Ann. Am. Assoc. Bov. Pract., Dec. 1976.)

TABLE 2. TREATMENT FOR ANIMALS EXHIBITING BLOODY DIARRHEA

1st DAY: Drench with 1 oz of diluted AMPROLIUM **and** 3 ml of NEOMYCIN SULFATE for each 100 lb body weight.

PLUS

Inject with OXYTETRACYCLINE (100 mg/ml) at a rate of 5 ml for each 100 lb body weight.

2nd DAY: Repeat the above treatment unless a fever, if present, has not dropped 2°F or to below 104°F. In that case, drench as described above and replace the oxytetracycline with injectable ERYTHROMYCIN as prescribed in table 3.

The injectable treatment will require changing to another injectable antibiotic if the fever persists. Use the same criteria for improvement as previously described. Once improvement is observed, continue using that antibiotic for the injectable portion of the treatment.

3rd DAY: Repeat the drench used days 1 and 2, but change the injectable treatment to the next antibiotic listed on table 3 if the required improvement in body temperature is not noted.

4th DAY: Drench with 1 oz of diluted AMPROLIUM per 100 lb body weight. But, OMIT the neomycin. **Continue** the appropriate injectable antibiotic medication.

5th DAY: Drench with 1 oz of diluted AMPROLIUM per 100 lb body weight.

AND

Treat with the appropriate injectable antibiotic.

OTHERS: If necessary, continue the injectable antibiotic portion of the treatment past the 5th day to control a fever. The successful injectable treatment needs to continue until the fever is absent for 2 consecutive days before the animal is released.

The route of administration and the dosages reported may be different than the manufacturer's label guidelines. However, the changes were implemented to provide adequate serum concentrations for approximately 24 hr, thus susceptible bacteria are controlled by treating once a day. The withdrawal periods have been substantially lengthened to compensate for the higher dosages. (Hjerpe and Routen: Treatment of bacterial pneumonia in feedlot cattle, with special reference to antimicrobic therapy. Proc. 9th Ann. Conv. Am. Assoc. Bov. Pract., Dec. 1976.)

TABLE 3. SUGGESTED GUIDELINES ON DRUG USE

I. OXYTETRACYCLINE, 100 mg/ml
For subcutaneous use in cattle with respiratory disease. Use 5 cc/100 lb (5 mg/lb). Inject no more than 10 cc per site. **Do not sell or slaughter for 20 days after the last treatment with oxytetracycline.**

II. SULFAMETHAZINE
For oral use (15 gram boluses)
- Use 1 bolus for every 150 lb for initial treatment (1 1/2 grain/lb).
- Repeat once daily with 1 bolus for every 225 lb (1 grain/lb).

The following precautions should be observed in using sulfamethazine:
- Do not overdose. Sulfas may be injurious to the kidneys. Closely follow recommended doses.
- Do not treat with sulfamethazine for longer than a 5-day period.
- Avoid use of sulfamethazine in severely dehydrated cattle or in cattle that are not drinking.
- **Do not sell or slaughter for 10 days after the last treatment with sulfamethazine.**

III. ERYTHROMYCIN (GALLIMYCIN 200)
Use 5 cc/100 lb (10 mg/lb) in treating respiratory disease in new cattle. Inject deep into the muscle of the rump or thigh. Use no more than 10 cc per injection site. **Do not sell or slaughter for 30 days after the last treatment with erythromycin.**

IV. TYLOSIN (TYLAN 200)
Use 5 cc/100 lb (10 mg/lb) injected into the muscle of the neck. Use no more than 10 cc per injection site. **Do not sell or slaughter for 20 days after the last treatment with Tylosin.**

V. PROCAINE PENICILLIN G
Use 10 cc/100 lb (30,000 units/lb) or 20 cc/100 lb (60,000 units/lb) injected subcutaneously. There is no limit on the volume used per injection site. **Do not sell or slaughter for 20 days after the last treatment with penicillin.**

VI. AMPROLIUM (CORID 9.6% SOLUTION)
To be diluted and used **orally** as a **drench** for coccidiosis. Pour 6 oz of corid in a quart container and then fill with water. Use 1 oz (30 ml) for each 100 lb of body weight daily for 5 days.

VII. NEOMYCIN (NEOMYCIN SULFATE)
To be used **orally** as a **drench** for bacterial gut infections. Use 5 cc/150 lb body weight. Dilute the dose needed in at least 3 times as much water and administer orally.

The route of administration and the dosages reported may be different than the manufacturer's label guidelines. However, the changes were implemented to provide adequate serum concentrations for approximately 24 hr, thus susceptible bacteria are controlled by treating once a day. The withdrawal periods have been substantially lengthened to compensate for the higher dosages. (Hjerpe and Routen: Treatment of bacterial pneumonia in feedlot cattle, with special reference to antimicrobic therapy. Proc. 9th Ann. Conv. Am. Assoc. Bov. Pract., Dec. 1976.)

the healthy group and begin the **bloody diarrhea treatment schedule.** In cases of severe diarrhea in which the animal is rapidly losing condition (gaunt, caved in), it is necessary to replenish amino acids, electrolytes, B vitamins, and water. In addition to the designated treatment for bloody diarrhea, administer 1 gallon of an oral nutrient-electrolyte solution via a stomach tube twice a day, inject B-complex vitamins, and administer 500 ml of an amino-acid-electrolyte-vitamin solution intravenously each day.

The daily checkup, evaluation, and treatment of sick animals should be conducted well before noon, preferably completed by 10:00 a.m. The elevated body temperatures exhibited by newly stressed cattle in the afternoon and evening can result in erroneous evaluations of treatment.

For cattle that show first signs of illness in the afternoon or evening, set rules must be followed:

1. Pull because of visible signs only.
2. Start medication with treatment #1. Tag and put the animal in the sick pen.
3. If the body temperature is above 104°F, change the treatment the next morning.
4. If the body temperature is below 104°F, continue the treatment.

After an animal has received 3 to 4 treatments of erythromycin and the temperature has dropped, a 1°F to 1.5°F rise in body temperature may occur. If this happens, evaluate the animal's appetite and other physical signs. After multiple treatments, the animal may appear sore. This soreness or lack of movement may be interpreted erroneously as "depression." In this case force the animal to the feed trough.

ELECTRONIC THERMOMETER/BODY TEMPERATURE

The electronic thermometer is very useful for detecting animals with an elevated body temperature. What is not widely known is that cattle do not maintain a body temperature within a very narrow range such as humans exhibit. Cattle temperatures fluctuate many degrees rather than holding constant. Under stress conditions, cattle temperatures may range from about 100°F to 108°F and follow a diurnal pattern (low temperatures in the morning and high temperatures in the afternoon, not dropping until 4 to 6 a.m.). The magnitude of the fluctuation is affected by the environmental temperature and humidity as well as the stress of transportation.

The magnitude of the fluctuation may be as great as 6°F to 7°F immediately after arrival of the cattle and then begins to settle into a normal diurnal variation of 2°F to 3°F after the cattle have become adjusted to the new environment, usually about 7 to 14 days.

Another fact is that the normal animal's body temperature rises with movement or excitement of the animal.

When body temperatures are at their diurnal lows and not elevated by the stresses of movement, animals with fevers from infections may be determined by using the electronic thermometer, with the following limitations:

1. Newly arrived cattle should be divided into groups of not over 25 head and allowed to rest overnight with free access to hay and water.
2. Processing and temperature taking should start at dawn and be completed within 3 hr.
3. No animal should be out of its pen or be waiting for processing for more than 30 min.
4. Cattle should be moved through processing with a minimum of excitement or stress; body temperatures should be taken when the animal first enters the chute.

Ignoring the above conditions for using the thermometer could give misleading readings. The sick animal does not elevate its body temperature above the high diurnal variation in a consistent manner. There may be a little more latitude in using the thermometer to monitor treatment responses in the sick pen where temperatures and changes are recorded daily. However, it is still advisable that the body temperature be taken during the early morning hours.

An elevated body temperature in the afternoon is seldom meaningful and should not be used as a criteria for sickness. The diurnal variation occurs regardless of the seasonal environmental changes. However, the highest low-diurnal body temperatures have been observed on the coldest days while the lowest were seen on the hottest days of the year.

INFLUENCE OF INFECTIOUS DISEASES ON BODY TEMPERATURES

First, a brief explanation of our philosophy about the influence that infectious diseases have on the body temperature and clinical symptoms (figure 1).

In untreated infected animals, the body temperature begins to elevate after the incubation period. Some animals (as shown by curve A) will recover without exhibiting clinical symptoms, and in others (as shown by curve B) the body temperature will continue to elevate and clinical symptoms of illness appear. Gradually the animal's defense systems overcome the infection, and as the animal begins to recover, the body temperature drops and clinical symptoms begin to disappear. Finally, the body temperature returns to normal and the animal is said to be in a convalescent state -- on the way to recovery.

However, in some animals (as shown by curve C) the body defenses fail to overcome the infectious process and the animal begins to succumb to the disease. The clinical symptoms continue to worsen and eventually the body temperature begins to fall. If the animal cannot overcome the infec-

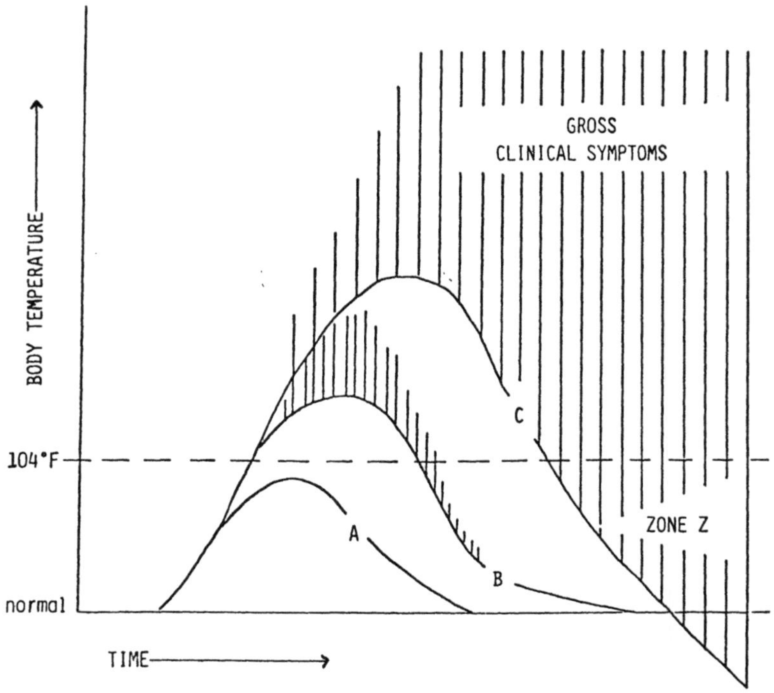

Figure 1. Body temperatures and gross clinical symptoms.

tion, the body temperature will drop well below normal and death usually occurs.

Animals that **are infected** yet have less than a 104°F body temperature and still show no clinical signs of illness can be missed. If the animal's natural defenses do not respond, or if our loving care is not sufficient, those animals will begin to exhibit visible clinical symptoms a day or two later.

During processing if we pull those animals that "have a rectal temperature of 104°F or greater," we will pull most of the animals that are exhibiting clinical symptoms as well as those that will develop clinical symptoms and illness if not treated.

Animals that "exhibit visible signs of illness regardless of body temperature," should also be pulled. These animals can be seen in the chute by anyone having just minimal knowledge of cattle; they have severe clinical symptoms and body temperatures below 104°F (zone Z). Without the correct treatment, these cattle are on the way out!

27

ANAPLASMOSIS IN BEEF CATTLE

E. J. Richey

Anaplasmosis is an infectious disease of cattle caused by a minute parasite, *Anaplasma marginale*. The *Anaplasma* organism invades the red blood cells. In an attempt to destroy the parasite, the animal's defense system destroys the infected red blood cells, which subsequently results in anemia. The disease is transmitted from infected animals to healthy animals by insects or by surgical instruments contaminated with infected blood.

All ages of cattle may become infected with anaplasmosis. The severity of illness and the percentage of deaths increase with age. Calves under 6 mo of age seldom show sufficient symptoms to detect that they are infected. Cattle 6 mo to 3 yr of age become increasingly ill and more deaths occur with advancing age. After 3 yr of age, 30% to 50% of the cattle with severe clinical anaplasmosis die.

Anaplasmosis outbreaks are related to the lack of a control program, the percentage of the herd that are anaplasmosis carriers, the amount of vector transmission, and the presence of susceptible animals.

[No controls + carriers + vectors + susceptible animals = outbreaks]

The management of anaplasmosis must include the development of control programs to prevent outbreaks, the elimination of the carrier state, the reduction of vector transmission, the treatment of ill animals, and the protection of susceptible animals during an outbreak.

CONTROL PROGRAMS FOR ANAPLASMOSIS

Various control programs for anaplasmosis are listed below:

Test the herd and separate carriers from susceptible animals.

This program necessitates bleeding, testing, and identifying each animal in the herd. It requires keeping carriers separate from susceptible animals during the vector season, or disposing of one group of the animals (carriers or susceptible animals).

Test the herd and clear up the carriers with tetracycline antibiotics. (See section on "Clearing of Carrier State")

Administer anaplasmosis vaccine (ANAPLAZ from Fort Dodge Laboratories) for a control program.

For an effective vaccination program, these recommendations should be followed:

- The initial vaccination (1st year) consists of 2 doses given 4 wk apart, scheduled so that the second dose is given at least 2 wk before the vector season begins.
- The first booster should be administered 2 wk or more before the next vector season.
- After the 1st booster, additional boosters should be administered at least every other year to provide adequate protection.
- If at all possible, breeding females should not be vaccinated while pregnant. The vaccine does not prevent infection but aids in prevention of clinical symptoms of bovine anaplasmosis. A vaccinated animal can become a carrier if infected with *A. marginale*.

Administer a continuous <u>oxytetracycline</u> medication during the vector season.

An injection of oxytetracycline is administered every 28 days, beginning with the start of the vector season and ending 1 mo to 2 mo after the vector season ends. The repeated oxytetracycline treatments will prevent clinical anaplasmosis from occurring but will allow most exposed animals to become carriers. If the oxytetracycline treatments are not continued past the end of the vector season, clinical anaplasmosis may occur in the more recently exposed animals. The recommended dose is 3 to 5 mg/lb of body weight (BW) when using the 50 to 100 mg/ml oxytetracycline or 9 mg/lb BW when using LA-200 (LIQUAMYCIN LA-200 [200 mg/ml]: Pfizer Inc.).

Administer a continuous <u>chlortetracycline</u> medication during the vector season.

Chlortetracycline administered daily at the rate of .5 mg/lb BW will prevent clinical anaplasmosis and prevent the development of the carrier state.

The common vehicles used for the daily administration of chlortetracycline are:

- Medicated feeds.
- Medicated salt-mineral mixes offered free choice.
- Medicated feed blocks. Consumption data should be available from the feed block or salt-mineral manufacturer.

Administer a continuous chlortetracycline medication the year around.

Chlortetracycline administered continuously throughout the vector season at the rate of .1 to .25 mg/lb BW per day can prevent clinical anaplasmosis from occurring in susceptible animals. This low dose of chlortetracycline, while preventing clinical disease, will allow the carrier state to establish in exposed susceptible animals. If the medication is withdrawn too soon following an exposure to *A. marginale*, the newly infected animal may become clinically ill. It is advisable to administer the low level of chlortetracycline continuously throughout the year. The medicated supplement is normally formulated by adding 1500 g of chlortetracycline to 1 t of a 35% to 50% NaCl mineral mix. The medicated mix is offered to the cattle free choice throughout the year.

NOTE: When administering chlortetracycline in this manner, it is essential that cattle receive an adequate uptake of the medicated mixes and blocks. This requires placing the medicated mix or blocks near water holes, providing sufficient protection from the sun and rain, and replenishing the mix at frequent intervals. Since cattle often prefer salt deposits (associated with salt water spills around oil wells) more than medicated mixes, it is advisable to routinely check to ensure that the cattle are consuming the medicated mix.

Bulls frequently do not consume adequate chlortetracycline and require additional protection, such as vaccination.

REDUCING VECTOR TRANSMISSION

Anaplasmosis is spread by the transfer of *A. marginale*-infected red blood cells from a diseased animal to a susceptible one. Primarily, the transmission is mechanical; that is, it is transmitted by biting insects or by blood-contaminated instruments used by man. However, there are biological vectors such as certain species of ticks.

Man commonly transmits anaplasmosis organisms from one animal to another on dehorning saws, castrating knives, vaccinating and bleeding needles, tattoo instruments, and ear notchers. When processing a suspected anaplasmosis-infected herd, it is a good practice to change vaccinating needles between animals. A quick rinse of the contaminated surgical instruments in clean water or disinfectant between each animal will usually prevent transmission. When this type of transmission occurs, a large number of cattle in the herd show signs of anaplasmosis at nearly the same time, without earlier cases having appeared.

Three biting insects (horseflies, stableflies, and mosquitoes) are known to mechanically transmit anaplasmosis from diseased cattle to healthy cattle.

Ticks carry *A. marginale* differently than other insects. The organism can live in ticks and may be passed through several generations of ticks. Ticks may transmit the disease months, and perhaps years, after biting an infected animal.

Control of biting insects quite often can be frustrating and generally is not considered to be a practical, reliable method of totally preventing transmission of anaplasmosis. However, applications of insecticides that reduce the biting-insect population will substantially reduce the number of clinical anaplasmosis cases occurring in a herd. Periodic spraying and dipping, as well as forced use of dust bags and back rubbers, are the common methods of insecticide application.

CLEARING THE CARRIER STATE

Anaplasmosis carrier cattle may be cured of the infection by treatment with certain tetracycline antibiotics. Carrier-state elimination programs must include postmedication serologic testing. The animal may test positive for several months following treatment but the positive reactor's blood may not be infective. All animals that test positive 6 mo after treatment ceases should be considered as "treatment failures." Failures should be retreated or separated from the rest of the herd. Animals cleared of the carrier state are susceptible to reinfection but exhibit resistance to clinical anaplasmosis for as long as 30 mo.

Programs for the elimination of the carrier state should be conducted after the vector season has ended. Oral administration of chlortetracycline permits treatment on a herd basis as well as the use of economical antibiotic premixes.

Oxytetracycline (50-100 mg/ml): 5-day treatment.
Administer 10 mg/lb body weight daily for 5 days. Intramuscular: Administer not over 10 ml per injection site. Intravenous: Oxtetracycline should be diluted with physiological saline and administered by a veterinarian.

Oxytetracycline (50-100 mg/ml): 10-day treatment.
Administer 5 mg/lb body weight daily for 10 days. Intramuscular: Administer no more than 10 ml per injection site. Intravenous: Oxytetracycline should be diluted with physiological saline and administered by a veterinarian.

Oxytetracycline (LA-200): 4 treatments at 3-day intervals.
Each animal receives 4 treatments of LA-200 at 3-day intervals at a dosage of 9 mg/lb BW. The medication for each treatment should be divided between two injection sites and given by deep intramuscular injection.

Chlortetracycline: 60-day treatment.
It is recommended that chlortetracycline be fed at a level of 5 mg/lb BW daily for 60 days. This level of chlortetracycline will aid in the elimination of the carrier state of anaplasmosis in beef cattle.

Chlortetracycline: 120-day treatment.

Chlortetracycline fed at the rate of .5 mg/lb BW per day for 120 days will eliminate the carrier state of anaplasmosis. The medication must be administered each day. An attempt to eliminate the carrier state by administering twice this dose of chlortetracycline (1 mg/lb BW) every other day for 60 feedings resulted in only 75% of the treated animals being cleared of anaplasmosis.

METHODS OF HANDLING AN ANAPLASMOSIS OUTBREAK

The proper handling of an outbreak should include the treatment of clinically ill animals and provide adequate protection for the remainder of the herd. The clinically ill animal may only be the first of many that will become ill or exposed to anaplasmosis.

Treatment of Sick Animal

By the time one sees an animal with clinical anaplasmosis, it is almost over the acute infection and is suffering from anemia. Any excitement or exertion could cause the animal to collapse, resulting in death. A veterinarian should be notified immediately for the confirmation of anaplasmosis and subsequent treatment of the affected animals.

If treatment is initiated, it is recommended that a single treatment with LA-200 (200 mg/ml oxytetracycline) at the rate of 9 mg/lb BW be administered rather than repeated treatments with a lower concentration of oxytetracycline. Blood transfusions may be indicated and should be administered by and on the advice of the veterinarian.

Temporary and Prolonged Protection for the Remainder of the Herd

In addition to treating the clinically ill animals, the remainder of the herd must be adequately protected. The unexposed animals and the animals in the incubation stage must be provided with temporary protection until prolonged protection can be established.

Temporary protection is accomplished by administering parenteral injections of oxytetracycline. A single injection of oxytetracycline, administered to animals incubating the disease, will delay clinical symptoms of anaplasmosis for approximately 28 days. This delay will allow sufficient time to begin a procedure to establish prolonged protection.

Prolonged protection can be accomplished by using repeated oxytetracycline treatments, vaccination, or feeding chlortetracycline.

Use one of the following methods:

Injectable oxytetracycline.

At the first indication of anaplasmosis, gather all susceptible animals and administer intramuscular injections of oxytetracycline, injecting no more than 10 ml per injection site. This treatment must be continued at 28-day intervals throughout the vector season. After withdrawal of the medication, close observation should continue for symptoms of anaplasmosis that may have been only delayed, not aborted, in some cattle. (Oxytetracycline dosages: 3 to 5 mg/lb BW when using a 50 to 100 mg/ml concentration or 9 mg/lb BW when using LA-200.)

Vaccine and oxytetracycline combined.

At the first indication of anaplasmosis, gather all susceptible animals.

1. Give each animal the 1st dose of ANAPLAZ vaccine and injectable oxytetracycline.
2. Four weeks later, give the 2nd dose of ANAPLAZ vaccine and another dose of oxytetracycline.

If anaplasmosis occurs because an ANAPLAZ booster injection was skipped, administer a booster vaccination and a single treatment of oxytetracycline to all previously vaccinated animals. The previously nonvaccinated animals must be given a second dose of vaccine and another treatment of oxytetracycline in 4 wk. (Oxytetracycline dosages: 3 to 5 mg/lb BW when using a 50 to 100 mg/ml concentration or 9 mg/lb BW when using LA-200.)

Injectable oxytetracycline and oral chlortetracycline.

At the first indication of anaplasmosis, gather all susceptible animals.

1. Administer a single dose of oxytetracycline.
2. Immediately offer chlortetracycline (CTC) free choice in a medicated salt-mineral mix or feed block (.5 mg CTC/lb BW).

Chlortetracycline medicated mixes or blocks should be offered for at least 60 days. Check animals for adequate consumption of the medicated mixes or the feed blocks. (Oxytetracycline dosages: 3 to 5 mg/lb BW when using a 50 to 100 mg/ml concentration or 9 mg/lb BW when using LA-200.)

28

USING PASTURE EFFECTIVELY FOR STOCKER-FEEDER OPERATIONS IN THE SOUTH

H. Allan Nation

The American beef industry suffered two major shocks during the 1970s: first with the export grain market hike in grain prices at the beginning of the decade and the huge cow-herd reduction in the mid-70s; and secondly, with the tremendous rise of energy prices in the mid- to late-70s, which caused tremendous inflation in transportation, fertilizer, and other input costs, including credit.

These two shocks, coupled with the inability to pass along the higher costs of grain and other beef-production costs to the consumer, have brought producers and analysts to the conclusion that beef must be produced more cheaply, if the industry is to survive. This soul-searching must start with the question, "What business are we really in?"

Many producers are realizing now that they are not in the cattle business but are in the grass business. Cattle are merely a way of harvesting and marketing an otherwise worthless grass crop. Once our producers come to this conclusion, the way to raise more profitable beef is clear -- we must maximize the amount of grass in our cattle and minimize the grain and fossil fuels in our operations.

Cow/calf operations have always been a grass/roughage-based business, but these operations usually only produce animals in the 300 lb to 500 lb weight category. Attempts to increase weaning weights dramatically through the use of higher-quality grasses and heavier cows often increase costs even more dramatically, especially in the lower-quality grass regions of the South.

Producers and animal scientists now realize that the most cost-efficient calf comes from a cow that is managed to maximize her scavenging ability on low-quality, inexpensive roughages; her weaned calf (where all of the added gain is sold) is fed the higher-quality and higher-priced forages. However, grass management for a mature cow and an immature weanling are dramatically different.

The large rumen of a mature cow allows her to get the most out of relatively low-quality grasses and forages, while the small rumen of a calf needs the rapid passage of high-quality forages to produce economically acceptable gains. High-quality grass for a weanling calf must consist primarily of young, tender green leaves with an absolute minimum of stems and dead leaves. For years, animal scientists have pointed out that southern perennial grasses lacked the "strength" of western grasses and did not produce acceptable gains in weanling calves. This was not the fault of the grass or the cattle but of the management of the grass. The high rainfall of the South produced a bountiful, rapidly growing grass that quickly deteriorated into stems and dead leaves after its peak growth.

Research by Dr. Bill Oliver at the Hill Farm Experiment Station in northwestern Louisiana in the mid-1970s showed that as stocking rates per acre were increased, individual animal gain also increased on highly nitrated Coastal Bermudagrass. Subsequent research found that an initial stocking rate of approximately 2,000 lb/acre of yearling weight produced the best per acre gain. Stocking rates of less than 2,000 lb/acre reduced gains because of an increase in mature forage; rates of more than 2,000 lb/acre reduced gains because of inadequate forage. The best gains were achieved when the grass was kept at a height of 1 in. to 3 in. This is a very difficult management chore because of the high variability of summer rainfall in the South. Too much rain produced a stemmy, rapidly growing pasture and too little rain provided an inadequate supply.

The southern grass problem has largely been solved by using "Tumblewheel" power fences that allow a pastureman to easily enlarge or contract the calves' grazing area in response to rainfall. The excess pasture can be cut for hay or grazed by mature cows. Pasturemen that follow the above management recommendations and use 1 lb to 1 1/2 lb/acre of nitrogen per day of grazing (applied every 28 days), have been able to produce from 800 lb to 1,000 lb of gain per acre during a mid-April to mid-September grazing season.

This type of management regime is recommended for warm-season perennials such as bermudagrass, Bahia, and spring fescue pastures. It is not recommended for cool-season annuals such as small grains and ryegrass.

With cool-season annuals we have another set of problems and opportunities. Most stocker-feeder operations in the South have been based upon the use of cool-season annuals. These grasses and grains are slow maturing and have high leaf-to-stem content for most of the grazing season. As a result, stocker cattle do very well on such pastures, with average daily gains often exceeding 2 lb/day. The best performance comes with a grass height of 3 in. to 8 in. Shorter grass restricts the leaf area for sunlight photosynthesis and severely slows the regrowth. Taller grass is susceptible to freeze damage, rust, and an early maturity.

The grass farmer's dilemma with cool-season annuals has been the widely varied growing conditions. These range from a hot, dry fall, to a cold low-or-no-growth period, to a very rapid-growing period in the spring. Our traditional management program has been to split the difference between the seasons by using lightweight calves during the spring-growth season and supplementing the cattle during the midwinter period. This management system works relatively well but is an inefficient way to harvest grass. Unfortunately, grazing calves are unconcerned about profit and will step on and sleep on more grass than they eat. They also concentrate on the young and tender regrowth, ignore ungrazed grasses, and allow the pasture to deteriorate into islands of stemmy, undergrazed old grasses surrounded by overgrazed young grasses. These grasses are extremely slow to cover because of the reduced leaf area for photosynthesis.

Ideally we need a system that would allow our calves to graze 6 in.- to 8 in.-high grasses down to 3 in. but no more; this ideal system would restrict the calves' natural tendency to explore their environment and mash a mouthful of grass into the ground with each step. On cool-season annuals, we need to control the calves' grazing patterns.

New Zealand has met this challenge by subdividing its pastures into as many as 90 or 100 grazing paddocks and concentrating all of its calves into one of the paddocks until they have grazed the grass down to the proper 3 in. level. Once they have done this, they are moved to a fresh paddock and the process is repeated. This not only gives the grass farmer control over the height of the grass but minimizes grazing selectivity and trampling.

Such grass management systems are slowly but surely coming to the southern U.S. However, there are two reasons why these systems are meeting resistance: 1) some cattlemen are not willing to commit themselves to the daily management of opening and closing a gate and 2) the traditional "turn-out" method is still marginally profitable. We still have no farming system that offers more profit from less management.

The biggest problem for farmers who try to practice controlled grazing is that of keeping calves from over-grazing fall growth. If the grass is grazed too closely, it will not have time to regrow before the onset of cold weather. Thus, the cattleman often finds himself short of grass in midwinter. By subdividing pasture into 15 or 20 grazing paddocks and practicing a high level of daily grass management (i.e., watching the grass consumption closely), the normal stocking rate can be doubled for continuous grazing on winter annuals. As in any winter grazing program, backup feed sources should be available for severe weather periods. Most of this feed backup can be hay or grass silage from the excess growth in the spring. The same controlled grazing system used for winter annuals can also be used on such summer grasses and legumes as Johnsongrass, crabgrass, bluestem, and alfalfa.

Successful grass farmers need to understand some principles of animal science and nutrition. For example, young growing calves have a higher need for protein and a lower need for energy. However, as cattle get older and heavier, the need for energy greatly increases and the need for protein decreases, thus they need grasses with higher and higher levels of total digestible nutrient (TDN) content.

A year-round grazing system should have lightweight calves starting on high-protein, low-energy grasses and continually moving to higher and higher TDN forages. In other words, calves that have been grazing on Coastal bermudagrass do extremely well when moved to ryegrass pastures, but calves that have been grazing on ryegrass do not do well when moved to a Coastal bermudagrass pasture.

It is important to remember that gains of yearling steers are greater than those of yearling heifers, but weaned steers and weaned heifers show equal gains. To maximize their compensatory ability, cattle backgrounded on grass but that will return to grass, should not be fed to gain in excess of .7 lb/day. The closer cattle are to a 50% Brahman cross, the better they gain on cool-season grasses and make dramatic gains on warm-season grasses.

Because most calves in the U.S. are produced from small cow herds, most postweaning grazing will be done by producers who purchase calves.

Purchased calves may often have health problems and are severe market risks. For grass farmers who can handle these two problems effectively, an intensive stocker-feeder grazing operation can yield profits of more than $100 to $200/acre. Much of the production profit in stocker-feeder operations is lost in freight and marketing charges that can be eliminated with an on-farm finishing system.

Legume/grass pastures are becoming more important because they 1) need less nitrogen, 2) improve the quality of the pasture, and 3) are easier to maintain in a controlled grazing system. Because the grasses compete less for sunlight, they allow companion grasses to regrow before being grazed again. This helps to keep the legume/grass mix in balance and avoids bloat problems.

Intensive grass farming with weaned calves is still in its struggling, formative, "wonder" stage and is considered by many as a "speciality crop." However, for the cattleman who is willing to get off his horse and look at what his calves are eating, intensive grass farming offers a rich and exciting future.

29

IMPLANTING CATTLE ON GRASS

Roger D. Wyatt

INTRODUCTION

Profitablity of beef production systems is largely dependent on the growth rate of animals and the efficiency of feed utilization. Many factors affect the growth rate and efficiency of feed utilization in cattle. The most important of these include: genetic composition, disease status, absence of internal and external parasites, plane of nutrition, and sexual status of the animal. The beef producer can exert a measure of influence on each of these factors through thoughtfully planned management programs.

Beginning in the early 1900s, researchers became interested in the concept of increasing animal productivity via manipulation of the endocrine system. Most of the early interest was focused on the use of gonadal steroids (estrogens and androgens) or synthetic compounds producing effects similar to those of gonadal steroids. More recently, proteins (growth hormones), peptides (growth hormone-releasing hormones), and amino acid derivatives (thyroxine) have attracted considerable interest as growth-stimulating chemical mediators.

The efficacy of implanting cattle with anabolic compounds to increase growth rate and feed efficiency is well established. Implanting is probably the most cost-effective management practice available to the cattle producer.

In spite of the compelling evidence to support implanting as a management practice, many producers have failed to incorporate it into their programs. Recent surveys indicate that fewer than 50% of the cow-calf producers implant their calves. Larger commercial producers are more apt to implant as a routine management practice, but the level of understanding regarding the proper use of various available products is still low.

The purpose of this discussion is to focus attention on several important aspects of implant management as it relates to managing cattle on forage-grazing production systems.

PERFORMANCE CRITERIA USED TO EVALUATE IMPLANT RESPONSE

Several performance criteria are commonly used to measure implant response. The most important of these include: 1) feed efficiency, 2) rate of gain, 3) carcass parameters, and 4) incidence of side effects.

Each of these criteria is measurable and economically important. However, under forage-grazing production situations, a direct measure of feed efficiency is not practical; thus, efficiency measures, such as pounds of beef produced per acre, are more widely used. Carcass parameters are important but are highly influenced by management practices imposed during the finishing phases of production; thus, they do not receive critical attention during forage-grazing production situations in the U.S. In countries where forage-finishing is practiced, producers using forage-grazing production systems for finishing cattle are more aware of the influence of their management practices on carcass parameters. For practical purposes, producers using forage-grazing production systems must evaluate implant responses on the basis of: 1) additional pounds of beef produced during the production cycle as a result of implanting, 2) the incidence of adverse side effects, and 3) ease of application.

IMPLANT MANAGEMENT FOR OPTIMUM RESPONSE

The response to anabolic implants tends to be correlated to the growth rate of the animal. When animal performance is high, responses to anabolic implants tend to be higher. Conversely, when animal performance is low, responses to anabolic implants tend to be low. This relationship should be kept in mind when planning the implant program for a production system. The implant management program should be considered an integral part of the total cattle management program. Implants should be administered at timely intervals when cattle are being handled for parasite control, branding, vaccination, castration, pasture rotation, or other routine handling chores. There is no single, "best" implanting system that will apply to all production situations. Under ideal conditions, single implant responses as high as 40 lb to 50 lb have been observed in 100 day to 120 day grazing tests. Under adverse conditions, such as drought or poor winter feeding conditions, little or no response to implanting may be observed. A good rule of thumb is that a properly designed implanting program can easily result in a 10% improvement in growth rate. This can be roughly applied across a wide range of levels of performance.

Implant Products Approved for Use in Suckling Calves (U.S.)

At the present time three products are approved for use in suckling calves in the U.S. These include: RALGRO (International Minerals & Chemical Corporation) and SYNOVEX-C (Syntex Agribusiness, Inc.) implants, which are approved for use in both steers and heifers, and COMPUDOSE (Eli Lilly and Company) implants, which are approved for use in steers only. A summary of several tests involving the comparison of RALGRO, SYNOVEX-C, and COMPUDOSE implants on suckling calf performance is shown in table 1.

The data provide compelling evidence that implanting pays back in pounds and ultimately in dollars to the producer. Economic analyses of implant test results indicate a return of $10 to $20 for each dollar invested in an implant program. No other available management alternative has as favorable a cost-benefit ratio as an implant program.

Implant Products Approved for Use in Yearling-Stocker Cattle (U.S.)

At the present time, five implants are approved for use in yearling-stocker cattle in the U.S. These include: RALGRO for use in both steers and heifers; SYNOVEX-S for use in steers and SYNOVEX-H for use in heifers; STEERoid (Anchor Laboratories, Inc.) and COMPUDOSE for use in steers only. A summary of several tests involving the comparison of RALGRO, SYNOVEX, and COMPUDOSE implants on yearling-stocker cattle performance is shown in table 2.

ECONOMICS OF IMPLANT PROGRAMS

As previously indicated, the ratio of cost to benefit for implanting is extremely favorable. Consider the typical cow-calf production unit in the U.S. Annual cow costs average about $400/cow unit across the U.S. If the average weaned calf weight/cow were 450 lb, this would result in a break-even calf selling price requirement of $88.89/cwt (table 3).

A realistic expectation of benefit from administration of a single implant to suckling calves is about 25 lb of additional weight at weaning. This additional 25 lb would have the effect of reducing the break-even selling price requirement to $84.42/cwt. Thus, an advantage of $4.47/cwt is realized.

Likewise, a realistic expectation of benefit from administration of two implants (reimplanting) during the suckling period is about 40 lb. An additional 40 lb would have the effect of reducing the break-even selling price to $82.04/cwt. Thus, an advantage of $6.85/cwt is realized for the reimplant system. Considering the current production-cost dilemma of cow-calf producers, one cannot afford

TABLE 1. COMPARISON OF RALGRO, SYNOVEX-C, AND COMPUDOSE IMPLANTS ON SUCKLING CALVES[a]

Location	No. of animals	Length of trial, days	Control	RALGRO	SYNOVEX-C	COMPUDOSE	RALGRO	SYNOVEX-C
Kentucky	25	214				1.60	1.72	
Colorado	78	154	2.33			2.37	2.44	
Colorado	125	131	1.98			2.14	2.20	
Kansas	34	187		1.91		1.71		
Kansas	124	175	2.11	2.17		2.14	2.30	
Kansas	45	160	1.99	2.11		2.12		
Washington	97	179	2.27	2.46		2.46	2.53	
Tennessee	295	195	1.50			1.61	1.71	
Kansas	232	170	2.06	2.12		2.14	2.19	
Kansas	N/A	120	2.03	2.18	2.18			
Kansas	N/A	120	1.81	1.94	1.87			
Kansas	N/A	120	1.41	1.52	1.56			
Colorado	116	157	1.96	2.15	2.12			
Oklahoma	366	105	1.56	1.66	1.66			
Colorado	78	131	1.98		2.10	2.14	2.20	2.11
Oklahoma	168	175	1.36	1.47	1.47			
Oklahoma	188	175	1.54	1.67	1.58	1.60		

[a] Average daily gain, pounds.

TABLE 2. COMPARISON OF RALGRO, SYNOVEX, AND COMPUDOSE IMPLANTS IN YEARLING-STOCKER CATTLE[a]

Location	No. of animals	Sex	Length of trial, days	Treatments[b] Control	RALGRO	SYNOVEX	COMPUDOSE
Colorado	85	S	191	617	667		671
Colorado	68	S	196	699	785	762	753
Colorado	64	H	160	853	939	907	
Kansas	94	S	152		499	499	
Canada	64	S	84	472	522	513	
Montana	176	S	54	1,061	1,139	1,134	
Georgia	78	S	98	712	857	794	848
Texas	150	S	111	857	943	916	
Kansas	110	S	171	767	857	826	
California	163	H	72	717	853	821	
S. Dakota	108	S	93	948	1,080	1,080	839
Kansas	124	S	113		544		490
Arkansas	106	S	104	789	939	880	
Arkansas	99	S	96	867	1,080	1,066	
Texas	106	S	77		821	776	
Canada	224	H	92	640	789	744	
Texas	111	S	124	590	866	649	
Texas	72	S	113	572	640	662	
Kansas	38	S	190		567		576
Kansas	135	H	138	494	562	585	
Kansas	80	S	202	798	866		889
Texas	34	H	70		803	771	
Texas	35	S	70		925	912	
Idaho	2,319	S	134		590	522	376
Kansas	318	S	112		1,080		1,025
Pennsylvania	103	S	112	794	875	884	

[a] Average daily gain, grams
[b] All implanted cattle received a single implant on day 1 of these trials.

to overlook a management practice with this payback potential.

TABLE 3. BREAK-EVEN SELLING PRICE REQUIREMENT FOR THE COW-CALF PRODUCER WITH A $400 ANNUAL COW COST

	No implant	Single implant[a]	Reimplant[a]
Weaned calf weight, lb	450	475	490
Break-even selling price, $/cwt	88.89	84.42	82.04
Advantage, $/cwt		-4.47	-6.85

[a] Assume that implants cost $1 each.

A similar example can be constructed for yearling-stocker cattle production systems. During a 130 day to 170 day grazing season, yearling cattle can be expected to respond as well as suckling calves to implant administration. Table 4 shows the cost dilution achieved from an implant program with production costs of $120/head and 200 lb of gain per head during the season.

TABLE 4. COST DILUTION FOR A YEARLING GRAZING PROGRAM PROJECTED FROM THE USE OF IMPLANTS BASED ON A $120/HEAD PRODUCTION COST

	No implant	Single implant[a]	Reimplant[a]
Gain during grazing season, lb	200	225	240
Production cost, $/cwt	60.00	53.33	50.00
Advantage, $/cwt		-6.67	-10.00

[a] Assume that implants cost $1 each.

An additional 25 lb realized from a single implant administered at the beginning of the grazing season would reduce per hundredweight production cost from $60 to $53.33 ($6.67/cwt advantage). An additional 40 lb produced as the result of a reimplant system would reduce per hundredweight production cost to $50 ($10/cwt advantage). Again, this represents a substantial production cost dilution that can be critical to the profitability of a yearling grazing program.

THE IMPORTANCE OF PROPER IMPLANTING TECHNIQUE

Considerable confusion exists in the field regarding the proper implant site for the available products. Early implants marketed (DES and SYNOVEX) recommended implanting in the middle third of the ear. This implant location became established practice before later products, such as RALGRO, were marketed. While SYNOVEX, STEERoid, and COMPUDOSE are properly located in the middle third of the ear, RALGRO should be implanted under the pocket of loose skin in the lower part of the back side of the ear near its base (figure 1). Proper implanting technique is critical to deriving the optimum biological response.

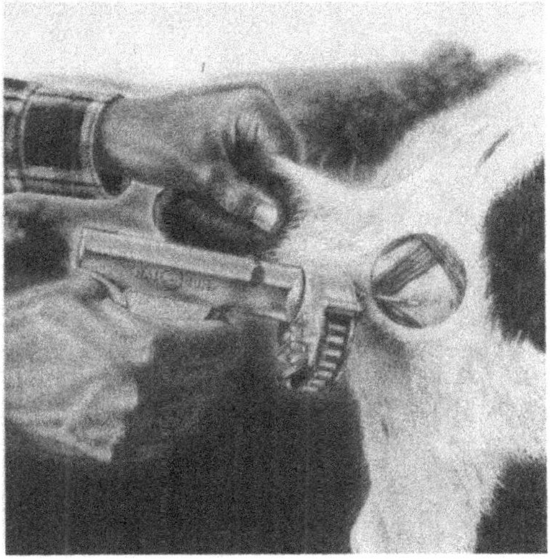

Figure 1. Proper RALGRO implant site.

SUMMARY

The efficiency of implanting cattle with anabolic compounds to increase growth rate and feed efficiency is well established. The implant management program should be considered as an integral part of the total cattle management program. There is no single "best" implanting system that will apply to all production situations. A properly designed implanting program can easily result in a 10% improvement in growth rate and feed efficiency. Proper implanting technique is critical to deriving the optimum biological response.

Part 4

FORAGE, RANGE, AND PASTURE

30

PRINCIPLES OF RANGE MANAGEMENT WITH UNIVERSAL APPLICATION

R. Dennis Child

INTRODUCTION

Throughout modern history, people have learned new management strategies for farm and ranch units by observing the strategies of others and adapting them to their own units. In the past, these strategies were often seen on neighboring ranches with similar ecological and socio-economic situations. Today, through communication systems, land managers are learning about other management strategies from remote parts of the world, and many of them seem to offer tremendous potential for a variety of farm and ranch situations.

Webster defines a principle as "a fundamental truth, law, doctrine, or motivating force upon which others are based; the law of nature by which things operate." This definition implies that a principle is an absolute truth, something that is inflexible and does not change or adapt to the environment it operates within.

Even though the rangelands of the world are extremely varied, they have some common characteristics that still classify them as rangeland. Despite their differences, are there any principles of range management that do have universal applications?

SELECTED ECOLOGICAL PRINCIPLES THAT RELATE TO RANGE MANAGEMENT

Rangeland as an ecosystem. Rangeland fits any modern conceptualization of an ecosystem: an arbitrarily delineated area having interacting physical and biological components to which social, cultural, economic, institutional, and political components must be added. Units of rangeland can be 1) units of landscape where human activities take place, 2) units of production, 3) habitat for a great variety of wildlife, and 4) ecosystems connected to many other ecosystems. The relationships between components of the rangeland ecosystem are commonly illustrated with a net-like diagram (figure 1). Some linkages that connect each component with all other components are direct and easily observed. However, other linkages are often somewhat indirect and more difficult to observe.

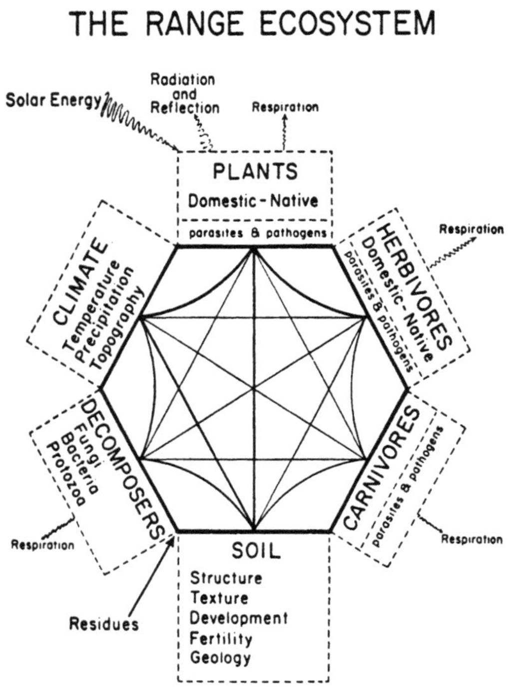

Figure 1. A flow chart showing structural characteristics of the ecosystem in six compartmental areas and connecting interaction linkage among compartments as a function process (Cook, J. Range Manage. 23:387).

EXAMPLE: Direct relationships between livestock (primary consumers) and coyotes (predators) are generally easy to observe. But indirect interactions resulting from ranchers' (humans) direct management of livestock (primary consumers) on a quail population (another primary consumer) may not be as obvious. Many ranchers have altered their resource base in an effort to increase the numbers of livestock and afterwards have wondered what happened to their once-numerous quail population.

Management of units as a total system. Ranchers and other resource managers must visualize and manage their unit as a total system and should not become preoccupied with one component.

EXAMPLE: Managers are often confronted by salespersons and promoters trying to persuade them to use a specific product, breed, or management strategy. The decision to select a larger breed of cattle can impact on many components of an operation; larger calves might result in calving problems and higher related costs; feed requirements during critical periods might require different grazing and feeding strategies that could change land requirements; physical handling facilities might need to be altered, etc.

Uniqueness of each individual and each ecosystem. Each ecosystem, delineated area, ranch, or pasture is unique and will respond to management in a different way from any other. This uniqueness not only adds beauty and variety to this earth, but it also adds resilience and flexibility to management systems as they face a variety of stresses and demands.

EXAMPLE: The variability within a species has provided the opportunity for genetic selection of varieties of plants and animals that will better meet the needs of a particular production system. Genetic selection and genetic engineering in the field of agriculture will continue to provide improved plant and animal resources to meet the additional food needs of the future.

EXAMPLE: Management practices used on one pasture, ranch, or in a region may not be appropriate for any other pasture or ranch within the same region. Ranch managers have often observed pastures that appear to be identical but still respond differently to a particular management strategy (e.g., plant response to a fertilizer treatment or animal response to a grazing method/system).

Energy flow in rangeland ecosystems. The sun is the source of energy by which all human food is produced. In the process of photosynthesis, plants capture the sun's energy for man to use. However, energy is lost during each step in the food chain, and so chains that have fewer steps are more efficient than those going through additional steps. This loss is illustrated by an energy-flow diagram of a community in figure 2. The direct consumption of plant products by man is generally more efficient than passing the energy through an animal.

However, much of the world's land resources do not favor the production of plant foods for direct consumption by man. The most efficient way for man to utilize these lands is to use animals for harvesting the vegetation and coverting it into meat. The efficiency of this conversion is affected by many factors.

EXAMPLE: The overstory/understory relationship is a factor in areas where shrubs and trees in significant number are an important component of the ecosystem. The growth rate of forage under a tree canopy in various types of plantations is reduced as the canopy closes over. The competition for light and nutrients affects the production of forage plants on millions of acres throughout the world.

EXAMPLE: Different photosynthetic pathways vary in their conversion efficiency of the sun's energy into plant biomass (carbohydrates). Warm season grasses (C4 pathway) are more efficient than cool season grasses (C3 pathway) in this conversion process (Sosebee, 1977). This difference in efficiency accounts in part for the variety of management strategies using combinations of coastal Bermudagrass (warm season) and Kentucky fescue or annual ryegrass (cool season) in the southeastern U.S. When temperatures and light are favorable for plant growth, warm season species that are growing with cold season species will out produce them through more efficient conversion of energy. In contrast, the cool season grasses do well during the spring and fall when the growth of warm season grasses is reduced.

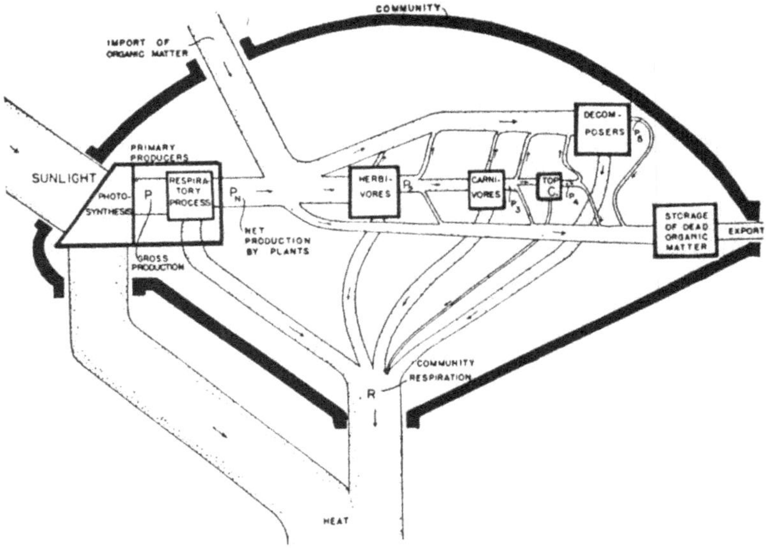

Figure 2. Energy-flow diagram of a community with a large import and a smaller export of organic matter, showing successive fixation and transfer by components and the large respiratory losses at each transfer. P = gross primary production, P_n = net primary production, and P_2, P_4, P_5 = secondary production at the indicated levels. (Redrawn from H.T. Odum, 1956.)

Nutrient cycles within the ecosystem. Nutrient cycles within an ecosystem can be illustrated by the nitrogen and phosphorus cycles in figure 3. Harvesting the products of agricultural land and rangeland will deplete the soil of some essential and potentially limiting nutrients. The rate of this depletion can be altered by using different management strategies. It is also possible, though expensive, to recharge the supply of nutrients in a particular rangeland ecosystem.

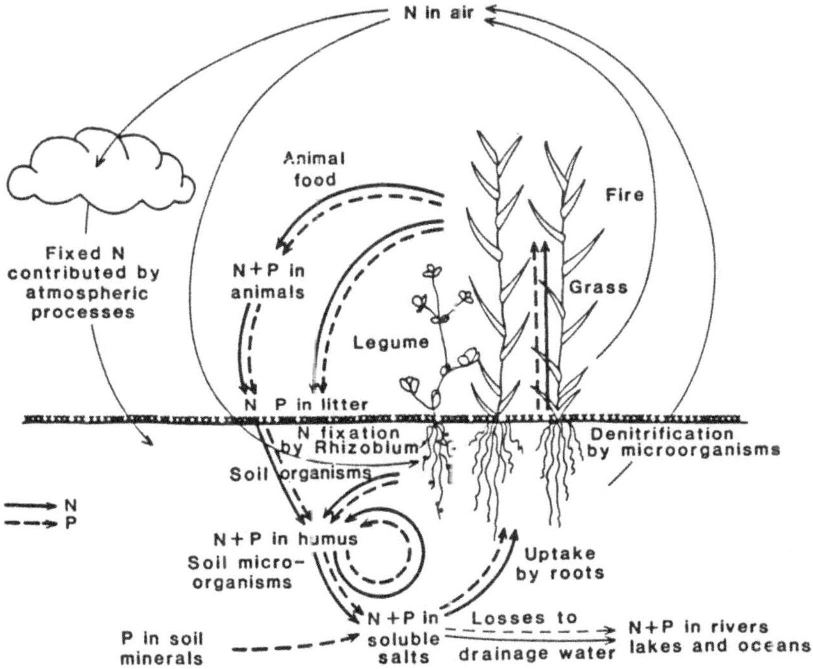

Figure 3. The nitrogen and phosphorus cycles and the soil. [Source: Bradshaw and Chadwick (1980)]

EXAMPLE: Forage grazing results in a lower net loss of nutrients than does harvesting the forage and feeding it as hay or silage to animals. The nutrients lost from grazing are only those that remain in the tissues of animals removed from the management unit. The vast majority of the nutrients consumed are excreted in the feces and urine and only about 4% of those consumed are actually removed via the animal. The return of the excreted nutrients into the plant growth portion of the cycle is actually faster when passing through an animal than when the plants die and decompose.

EXAMPLE: The replenishment of nutrients into the active production cycle is expensive. Fertilization is practiced on some improved pastures but many managers have found that fertilizing rangeland is not economical. Because of the link between fossil fuels and commercial fertilizer, the cost of fertilizer is not likely to decrease enough to be used economically or extensively on marginal lands.

Other soil treatments have been proposed that may accelerate the growth of soil organisms and the development of humus (Walters and Fenzau, 1979). The treatments being tested speed the process that releases nutrients in the soil and makes them available for plant growth, thus allowing them to become active in the nutrient cycle. If effective, they may reduce the amount of nutrients required to replace those removed through agricultural harvesting.

Successional development and stability. More mature successional stages are more stable, though less productive, than developmental stages. Many of the agricultural systems used today are still destroying the diverse native vegetation from the land and planting a monoculture. This procedure is used to maximize the gross production of exportable products from the land base. Monocultures (developmental successional stages) have proved to be less efficient in the conservation of nutrients and more susceptible to natural catastrophies. This balance between higher production (lower stability and higher risk) and lower production (greater stability and lower risk) must be considered in the development of long-term management strategies.

EXAMPLE: Agroforestry and mixed crop/livestock farming systems are designed to increase the ecological and economic diversity on the management units and should be encouraged.

OTHER PRINCIPLES IMPORTANT TO RESOURCE MANAGEMENT

It is not enough to simply know that a technology or management strategy works. Does it pay? Does it fit within the sociopolitical environment? These questions sound simple but in practice are very difficult to answer. The complexities of national and international agriculture and the attempts to rapidly increase food production to meet the needs of our growing population have spawned some inappropriate management strategies both at home and abroad.

The sustainable use and management of the rangeland resources in developing countries will require that a more holistic approach be taken during project planning, implementation, evaluation, and monitoring (Child et al., 1984). Only in recent years have social scientists been involved in agricultural development projects. In the past the social scientist and the agricultural scientist were often only brought together after the fact. The full involvement of the people in the host country is also critical to the success of development programs.

SUMMARY AND CONCLUSIONS

Are there any range management principles that have universal application? There are probably no basic principles that are unique or peculiar to range management. Rather, there are sets or combinations of basic ecological, social, and economic principles relating to rangeland that have been used to develop management strategies for a particular unit of rangeland. These strategies are often developed to meet short- and long-term goals for a particular management unit. The management strategy may include a grazing method, vegetation management, integration of wildlife harvesting with livestock management, or a livestock management/marketing plan.

There is a tendency to observe the success of a particular management strategy on a unit of rangeland in one area and attempt to transfer the strategy to a similar rangeland unit. This has often occurred without an understanding of the principles involved or the management goals critical to the strategy's success. And in some instances, management strategies have unfortunately been assumed to be principles.

Since every unit of rangeland has its own set of management goals, environmental characteristics, available external resources, and combination of basic ecological principles, the particular management strategy developed for that unit of rangeland will not necessarily be appropriate for any other unit. Good range management must view each management unit as a system having interrelated parts and utilize basic principles from many disciplines including ecology, economics, political science, and sociology.

REFERENCES

Bradshaw, A. D. and M. J. Chadwick. 1980. The Restoration of Land. Univ. of California Press. Berkeley.

Child, R. Dennis, H. F. Heady, W. C. Hickey, R. A. Peterson, and R. D. Pieper. 1984. Ecological Use and Management of Rangeland Resources in Developing Countries. Winrock International (in press).

Cook, C. Wayne. 1970. Ecosystem approach in teaching. J. of Range Manage. 36:6.

Odum, Howard T. 1956. Primary production in flowing waters. Limnol. Oceanogr. 1:102.

Odum, Eugene P. 1971. Fundamentals of Ecology. Third Edition. W. B. Saunders Company. Philadelphia.

Sosebee, Ronald E. (Ed.). 1977. Rangeland Plant Physiology. Range Science Series No. 4. Denver.

Walter, Charles Jr. and C. J. Fenzau. 1979. An Acres U.S.A. Primer. Acres U.S.A. Raytown, MO.

31

PUBLIC EDUCATION IN LAND AND WATER MANAGEMENT: A NEW USE FOR COMPUTERS

John R. Amend

THE PROBLEM

Every year we as citizens must make some tough decisions about the management of our natural resources. Sometimes these decisions are on the ballot, sometimes they must be made by elected groups, and sometimes ordinary citizens have to look at opposing views and figure out who is right. Responsible management of these resources must be based on sound technical principles and requires the support of an informed public.

Building public understanding of principles of land and water management is a difficult task. The learner's attention is often difficult to capture, the time he or she has available is limited, and there is a shortage of people with adequate technical background and time to conduct public workshops.

A SOLUTION

Computer simulation has proven to be an extremely effective way to help youth and adults understand the major problems involved in management of natural resource systems. Forty-five minute to 90 min workshops place participants in management situations involving real problems and alternatives. They are not offered solutions, but instead are offered an opportunity to experiment with different management strategies and to observe the probable consequences of their actions. The unique nature of computer simulation captures attention at fairs, service club presentations, and in school classes. Active participation maintains interest and assures learning. And the fact that the computer model for each natural resource system was developed by a group of experienced engineers, scientists, and resource managers assures the workshop leader or teacher that problems and concepts encountered will be presented honestly and in context.

Over the last 11 yr, Montana State University has developed computer simulations for public education programs on energy, water resource management, electrical power management, grazing land management, and home energy conservation. At this time the energy, water resource, and grazing land programs are used by more than 120 universities, school districts, and agencies in the U.S. and Canada. This article describes the grazing lands and water resource programs.

GRAZING LANDS: A PUBLIC EDUCATION PROGRAM

One-third of the total land area of the United States is grazing land -- land on which forage is routinely harvested by grazing animals. When land that is occasionally grazed is included, the fraction increases to one-half. It is our largest use of land. Public understanding of the value of this resource and of the principles involved in its wise management are essential.

The Grazing Land Education program consists of resource materials, workshop plans, instructor training, and a grazing-land-management simulator. The program is designed for use with community and school groups. Its development was sponsored by the National Cattlemen's Association, the Soil Conservation Service, the Bureau of Indian Affairs, the U.S. Forest Service, the Cooperative Extension Service, and the Bureau of Land Management.

The Grazing Land Simulator (figure 1) is an interactive computer that depicts the ecology of grazing lands. Each year it provides its operators with actual herbage growth data based on precipitation and soil conditions for the region modeled. Livestock and wildlife consume the forage while the human population demands food, fiber, and water from the grazing land. Participants must develop and implement management strategies that will meet the demand for food, fiber, and clean water while maintaining the productivity of the land. As time progresses, the economic and environmental consequences of their management strategies are projected by the simulator.

The simulator poses problems to participants, acts according to their decisions, and forces them to live with the consequences of these decisions. Through a process of successive trial, error, and optimization of variables (figure 2), participants develop an understanding of the operation of the system and of the alternatives available.

Participants also often change their idea of what is "optimum" after being faced with the realities of the system. If the variables and problem are real, the participants will see their decisionmaking process as having consequences far beyond the simulation.

Figure 1. Students give full attention as Dr. John Lacey, MSR Extension Range Specialist, demonstrates the Grazing Land Simulator.

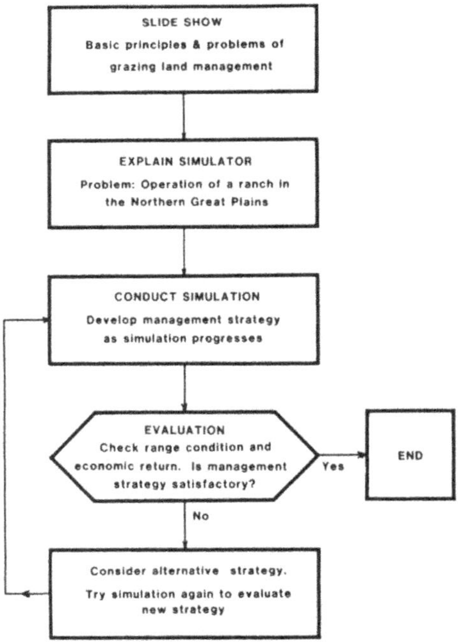

Figure 2. Organization of a group workshop with the Grazing Land Simulator.

This type of simulation differs from standard digital computer simulations in that it operates in "real time" (10 to 20 sec/mo) and presents information concerning all of its variables simultaneously during the run. It is an **interactive** computer; the group conducting the simulation may interact with the model at any time, using simple controls to implement their management decisions. After observing the result of a given simulation, participants may discuss the strong and weak points of their management policy, modify their strategy, press the reset button, and try again.

THE GRAZING LAND SIMULATOR

The front panel of the Grazing Land Simulator (figure 3) depicts a ranch with three pastures -- Arrowhead, Black Butte, and Cottonwood. Size and range site for each pasture is set at the beginning of the simulation. Displays show the number of wildlife and livestock grazing the pasture, percentage use, range trend, and vegetation rating.

A clock in the upper right corner of the panel shows passage of time in months and years, and a display in the upper left shows precipitation. Quantity and quality of run-off water are shown by a downstream display in the lower right, and the balance between food and fiber production and population demand is shown by a balance indicator in the lower center. Colored lamps display animal health and reproductive capability. Management practices implemented by livestock and forage management groups -- hay production and feeding, hunting, seeding, and fencing -- are indicated by displays on the panel. The economic impact of these management decisions -- cost per animal unit month, projected percentage calf crop for the current and coming year, and cumulative profit or loss -- are displayed on indicators in the center and lower right.

Forage growth is programmed from the historical records of 10 representative years for several different range sites in the region modeled. The current model represents the Northern Great Plains; data have been provided by the Ekalaka Experiment Station in Eastern Montana. The simulator may be programmed for other climate and soil conditions by obtaining forage growth records for the region of interest.

THE WATER RESOURCES EDUCATION PROJECT

A mosaic in the rotunda of the Colorado State Capitol concludes with the statement, "This is a land where life is written in water." The combination of land, sunlight, and water is essential to life and to the agricultural community that supports life. The problems of water management, however, are even less understood by the general public than the problems of grazing land management. This is demon-

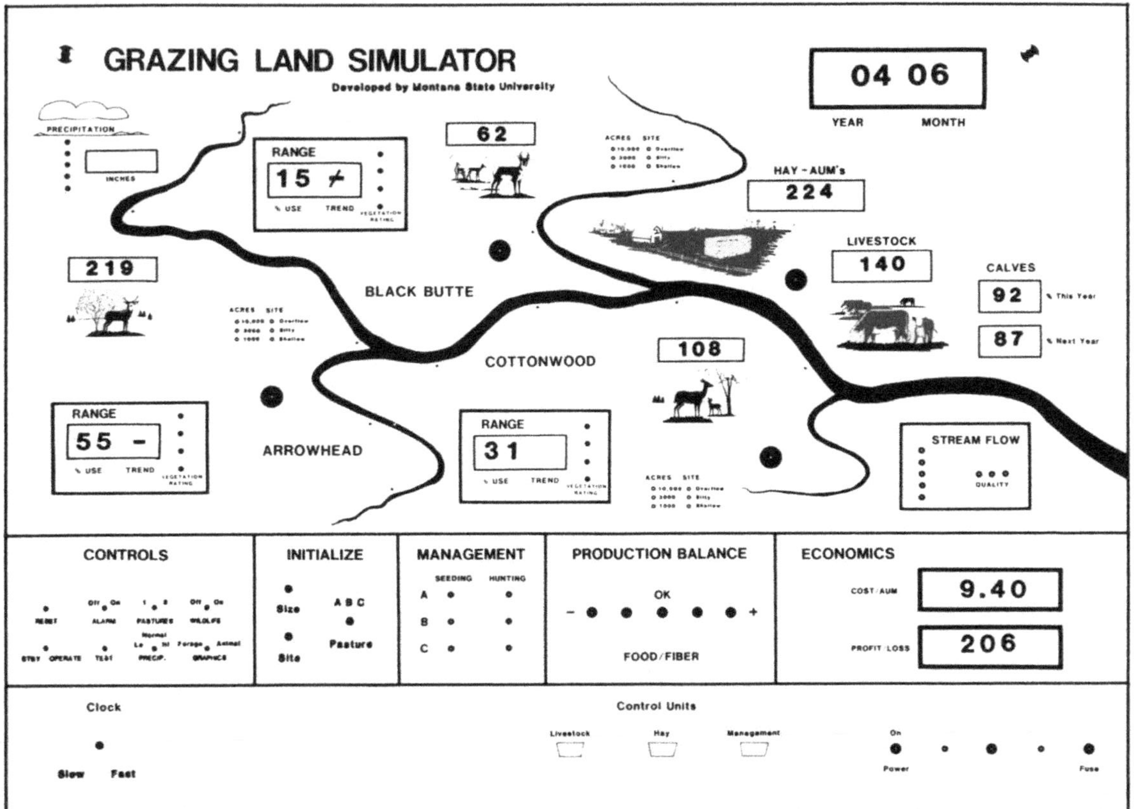

Figure 3. The front panel of the Grazing Land Simulator. The size of this version is approximately 18 in. x 24 in. It is mounted in an aluminum suitcase for easy portability and contains two digital computers using approximately 102 K of memory. Small control consoles placed in the audience (figure 4) are used to implement management decisions.

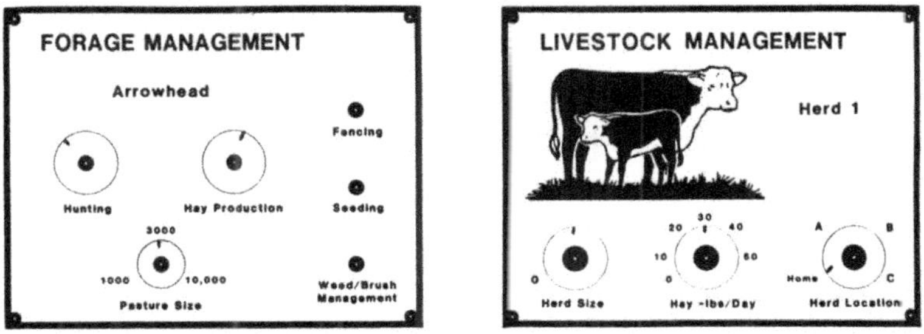

Figure 4. Small control consoles are used by participants to implement their management decisions.

strated by the statement of a workshop participant who, when faced with the fact that the time of maximum stream flow in the western U.S. precedes the peak demand for water by about 2 mo (figure 5) said, "That's easy. Just irrigate early."

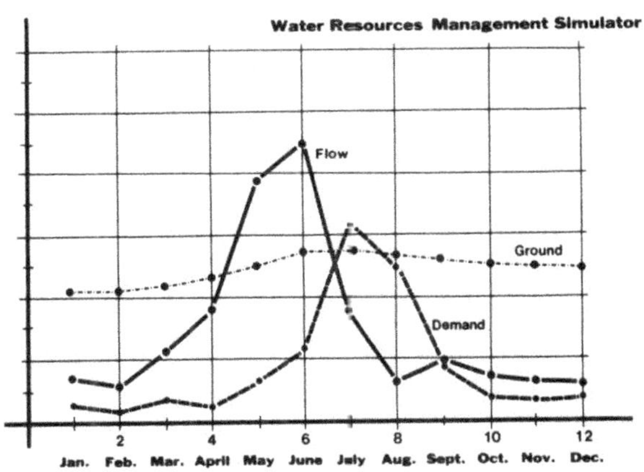

Figure 5. The lack of overlap in time of peak stream flow and water demand for irrigation -- the principal water management problem in the western United States -- is clearly illustrated by this plot reproduced from the TV graphics display of the Water Resources Management Simulator. Infiltration and ground water recharge are also illustrated.

Development of the Water Resources Education Project was a cooperative effort involving the Bureau of Reclamation, state water agencies in several states, the Old West Regional Commission, and several universities and school districts. The results of this effort are instructional materials, a leadership training program, and the Water Resources Management Simulator. This program is again designed for use with community and school groups.

THE WATER RESOURCES MANAGEMENT SIMULATOR

The Water Resources Management Simulator, illustrated in figure 6, is used in our water resource education workshops. Four problem areas are treated in these workshops: 1) sources and quantity of water; 2) uses of water; 3) quality of water; and 4) political management of the water resource. Hydrologic information of a general nature is provided through a short slide talk. Workshop participants are then given a hands-on opportunity to develop and evaluate water management strategies through use of the simulator.

The Water Resources Management Simulator is a digital computer that models a region's water-supply-and-demand situation. Groups from the audience use remote consoles (figure 7) to make water management decisions. Controls on the simulator's back panel permit selection of a number of different stream basins throughout the country. Artwork on the panel may be changed to represent the region of the country in which the simulator is principally used.

The Water Resources Management Simulator asks its operators to make some hard decisions. Each year it provides them with snow-pack and stream-flow data that represent the historical behavior of their region. For some years there will be adequate water. In other years there will be too much or too little. Using the available supply of water, the operators must provide for their region's water needs -- irrigation, energy, livestock, and municipal and industrial uses. They will want to prevent drought or flood from destroying crops or land. They will want to reserve adequate stream flow to support downstream users and fish and wildlife. For each of their water uses, they must decide if the water will come from surface or underground water. They must decide the technology for each of their water uses. And, if they wish, they may flood some of their basin to create a reservoir for storage of surface water.

As the simulator operates, a colored graphic display of stream flow, water demand, and surface or ground water reserve is plotted on a TV monitor. After observing the results of a given simulator run, participants discuss the strong and weak points of their water management policy, modify their strategy, press the reset button, and try again.

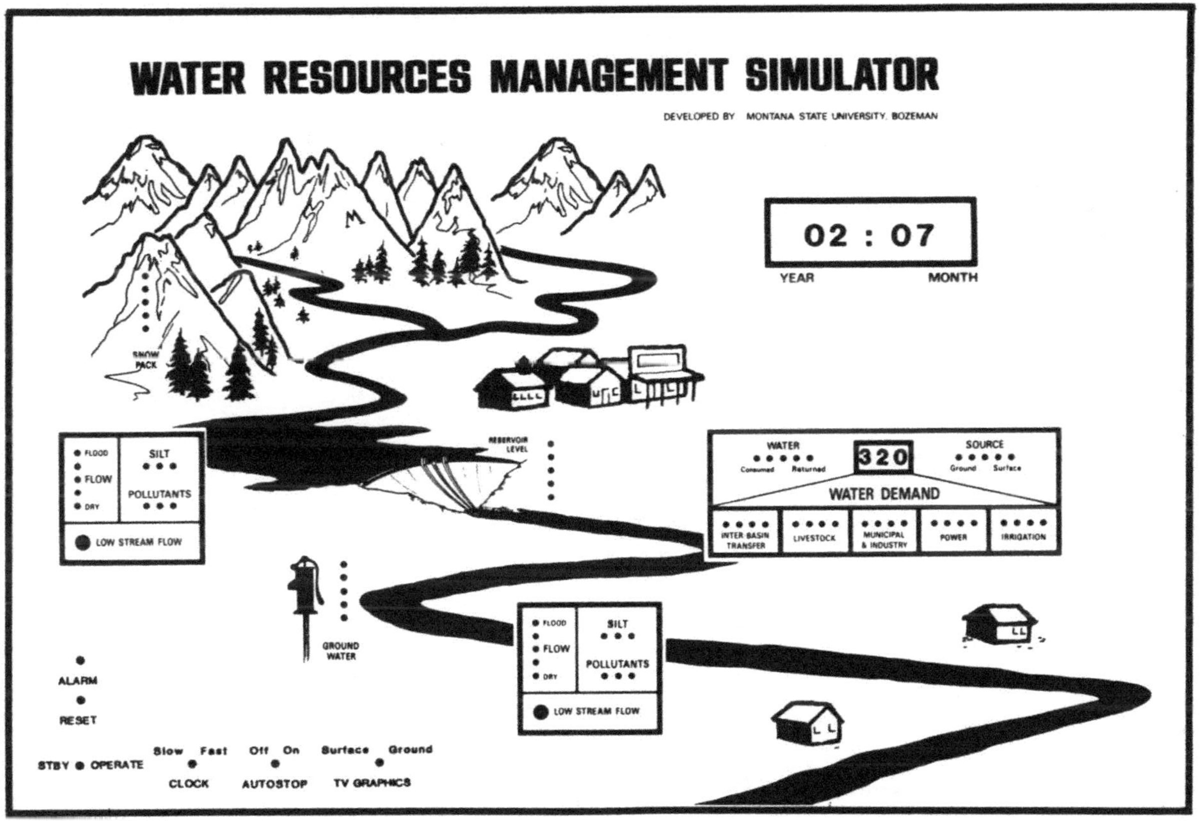

Figure 6. Information concerning stream flow, water use, and water reserve is shown by displays on the panel of the Water Resources Management Simulator. The reservoir overlay is attached to the panel with velcro.

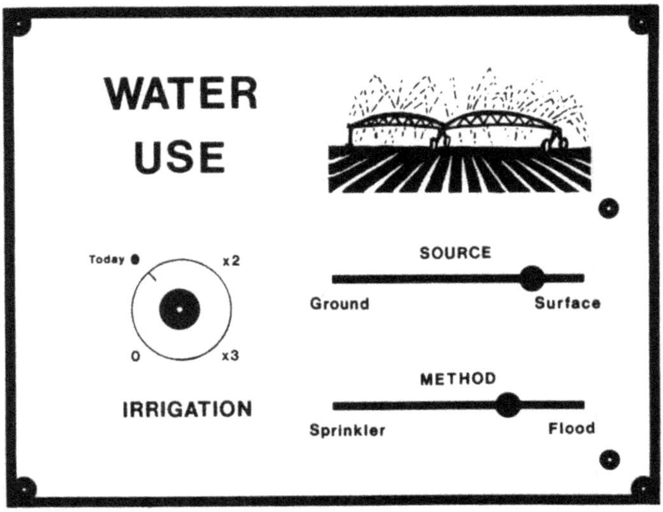

Figure 7. Participants use remote consoles, such as this irrigation control module, to make water management decisions regarding sources of water, water use, and storage of surface water.

SIMULATION AS A TEACHING TOOL

A major advantage of computer simulation lies in the fact that the workshop leader does not present a certain point of view. Participants are presented real problems, an opportunity to experiment with different strategies and policies, and a chance to observe the probable consequences of different courses of action (figure 8). The workshop leader's role is one of moderator and helper rather than lecturer; the integrity of the program is assured by the computer model's honest presentation of problems and its realistic response to participant actions. In effect, the teacher or workshop leader's background is supplemented by the scientific and management background of the simulator's design team.

The importance of active participation in any educational effort cannot be overemphasized. The value of experience has been well understood from the time of the ancient apprentice. One could learn to plow by walking behind the ox and to shoe a horse by working beside the blacksmith, but learning to fly an airplane and to experiment with land or water management policy are far more difficult problems. Computer technology makes it possible to experience these complex problems, and to experiment with alternatives, in a safe, low-cost manner.

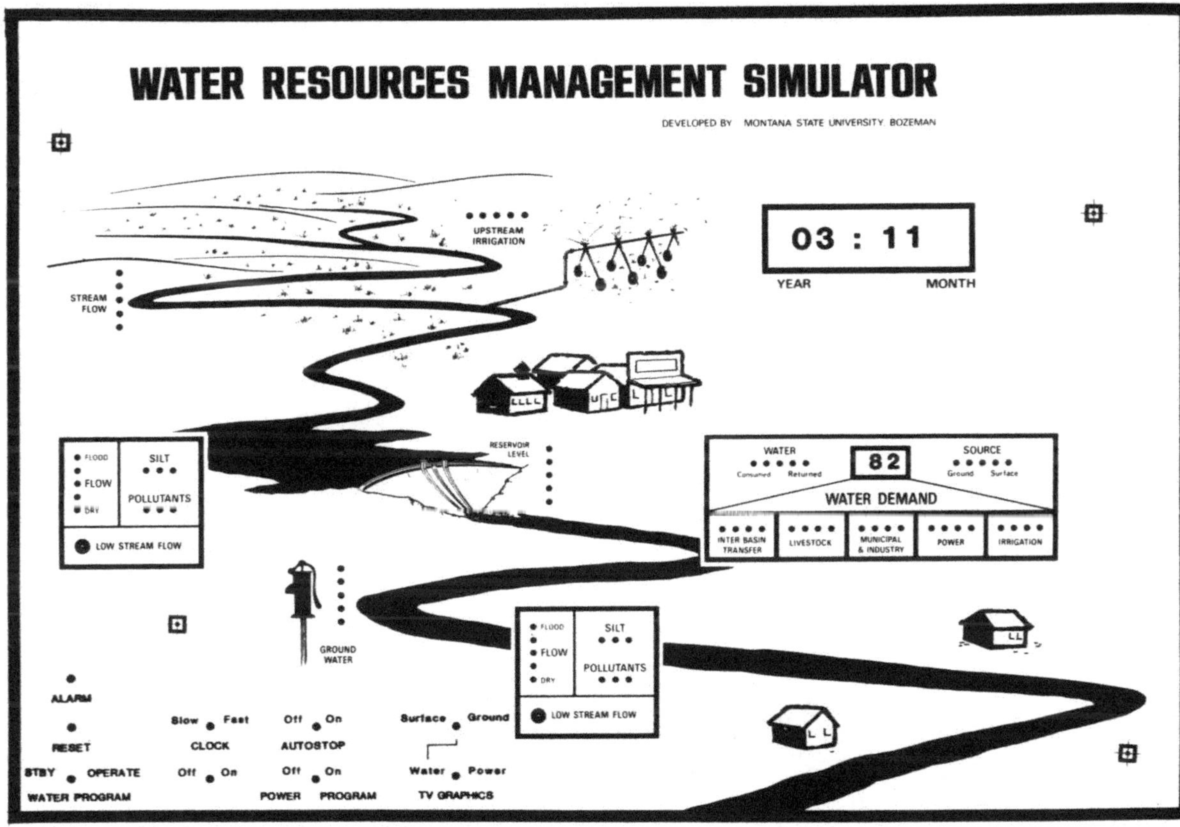

Figure 8. The simulator's front panel artwork may be changed to represent the region of principal use. This panel was designed for the central Great Plains and demonstrates the effect of upstream irrigation on reservoir management.

32
COMPUTERS AND THE LIVESTOCK ENTERPRISE

Gary Conley

Management of a livestock enterprise has always required versatility. The successful manager needs to be knowledgeable of accounting, finance, nutrition, genetics, veterinary service, marketing, personal management, law, tax law, mechanics, and range management (including soils, agronomy, etc.). Most livestock operations are too small to economically employ a person from each of these fields of expertise, but most should hire consultants from several of the areas.

During the past year, the number of firms and individuals providing software (programs) and consulting services to the microcomputer users has more than tripled. Many general programs are available to the livestock producer. If he is knowledgeable in the operation of his computer, they can be applied (with modification) to his operation. But for the most part, the producer should pay a programmer or the supplier to modify a program and train him in the use of the program. Programs are available that can substitute for an expert's assistance, such as a financial analysis of a project, comparison of two or more enterprises, scheduling of a preventive medical program, genetic selection indices for cows and heifers, nutritional least-cost rations using available feedstuffs, and nearly any other task if an expert and a programmer have collaborated in preparation of the program.

Analyses that are too large (very few are) for the micro may be solved by using a network to access a mainframe computer over telephone lines. The availability of these programs, programmers, and communication systems permit the manager of a livestock enterprise instant access to answers to his management problems as well as expert advice in advance planning of future enterprises.

HARDWARE TECHNOLOGICAL PROGRESS

The exponential curve plotting advances of microcomputer technology has continued during 1984. Improvements in computers and their support peripherals have been announced at ever shorter intervals. Modems (telephone communicators) are now available to automatically dial the mainframe computer at the lowest cost time (while the operator sleeps) and retrieve market information or to analyze nutritional, genetic, or financial information. Ink-jet printers can provide colored graphs or high-quality letters. Preprogrammed and reprogrammable memory chips are available so that frequently used programs do not have to be loaded from the disk each time the operator needs to add data or print a current report.

Prices of hardware have continued to decline so that a unit that would have sold for over $12,000 in 1983 now can be purchased for less than $6,000. Even more dramatic is the fact that a computer with this same capacity would have sold for more than $1 million in 1971. Experts are in general agreement that this price decline will continue, at least for memory storage chips.

New technology will continue to develop new products. Many of these products will be useful in livestock enterprises and make it easier for the operator to use his hardware. Some recent advances are 1) voice communication and programming; 2) small portables in which field data can be entered and electronically transferred to the office computer; 3) autodial modems that can retrieve market information at night; 4) data networks that have programs to calculate least-cost rations, genetic indices, and tax planning options; and 5) electronic bar-chart readers to transfer programs directly to the computer without typing or buying a disk.

PURCHASING A SYSTEM

The most important segment of the computer system will be software. Many brands of hardware are equivalent with only small differences, which are advantages only for special applications. Software dealers and programmers can, in many cases, supply the user with a complete package at lower cost than one bought piece by piece from several discount dealers. Manufacturers give volume software suppliers (i.e., programmers) a wholesale price on the purchase of equipment. Thus, by selecting the software first, the buyer may ultimately save money and certainly be assured of a compatible system.

There are several national and local software companies as well as consulting programmers. Select a local outlet that can provide convenient service and training, or a national supplier who offers free consulting time using a modem, which should be adequate for follow-up service and program additions. Insist that a representative personally install and test the system and give an initial training session.

The livestock producer should insist on several basic components and capabilities in his system. First, and most important, each data item should require only a single entry. For example, sale or purchase data entered into the accounts payable or receivable program file should automatically transfer to the 1) tax accounting program, 2) inventory program, and 3) financial statement program, and other categories of your choice. Another example, once cattle weights are filed in the selection program, they should transfer to all other programs requiring these data.

Secondly, the producer of today needs the ability to analyze and evaluate competing enterprises, both current (based on past years' data) and projected (based on industry averages). The analysis should be for the total enterprise, and all pertinent questions should be based on produced input rather than on default assumptions. For example, analysis of a proposed cow herd should take into account the rate of replacement of cows by heifers using projected values for each, along with the tax consequences. Too many of the software packages currently available do not contain this type analysis and so are little better than the producer's best guess (shooting from the hip).

LOCATING SOFTWARE SUPPLIERS

Manufacturers of computers are publishing catalogs of software for their machines. Select several software dealers from these catalogs and call or interview their sales representatives. Also, local dealers can supply the names of local programmers. The state extension services have programs available along with lists of programs from other states; in some cases a producer may hire a university programmer as a consultant to customize programs for specific operations. There are at least three major periodicals dealing with agricultural computing that publish software evaluations as well as advertisements for software suppliers. The American Society of Agricultural Consultants will provide a listing of consultants and their area of specialization. Producers who use the above sources should be able to locate software that will fit their operation at a reasonable cost.

EVALUATION OF THE SOFTWARE

Never buy a computer until the software has been tested in an actual operational test. Enter your own data in the program and check the reports they generate. Ask if entry errors can be corrected if they are not found until after the data have been stored. Know the maximum size numbers the program will handle and how many entries may be made per year or month. There is no reason to accept restricting limits. For example, if you feed 1,500 steers and the program will not accept gross income above $999,999.99 per enterprise, then you are forced to split cattle feeding entries into two enterprises. These are the types of problems that producers have paid for and lived with, needlessly. Buy only when you are sure that the computer, software, and your operation are compatible. If you do not find the software you want, consider hiring a consultant to develop software. The consultant probably has a general program available that can be quickly modified to fit your operation. The consultant may also be in a position to save the buyer money on the total hardware-software package. Make sure the system will be large enough to allow operational growth and for the addition of other enterprises. A check should also be made on the cost of program updates as the supplier makes improvements in the basic software.

When selecting the hardware, make your decision based on service availability and cost, all other items being equal. Insist on a 10-key pad because many entries will be numerical. Also make sure that there are expansion slots available so new chips may be installed at low cost. There should be ports (plug receptacles) for adding disc drives, printers, plotters, digital input devices, etc., at your convenience. Speed is not very important to a single-user situation since even the slowest computers are faster than most printers, typists, and communication modems. The printer is the hardware item requiring the most service. Therefore, buy a high-quality printer to avoid service delays in processing information.

How the software is designed and how well it is explained (documentation) will determine how much a computer is used. The system should interact with the operator so that entry of data is easy and logical. Once data are entered into one data file, they should transfer to other data files without reentry. If calf weaning weights are entered in the cow-production records file, then the program should transfer these weights to the cattle-inventory and financial-statement files. As purchases are made, enter bills in the accounts-payable file. These should be transferred by the program to the check writing, general ledger, enterprise analysis, and other files. Many well-designed programs allow the user to define his own account categories and numbers, which customizes the ledger and enterprise analysis.

A screen (data entry unit) should be available for each division of the program. For example, separate data entry screens should exist for the physical cow herd, pastures and fields, equipment expenses and income, etc. A well-designed data entry screen that is organized in a logical sequence is important. The top left corner should contain the identification number. Sex, birth data, purchase data, etc., should be across the top of the screen. The cursor (a flashing light) should be programmed to automatically move from entry to entry in the logical order of data collection. This same type of organization is necessary for the format of reports, both on screen and hardcopy (printed) reports.

USING THE SYSTEM

Now that you have the equivalent of an accounting firm and several bookkeepers at your disposal, what records should you keep? The answer: all the reports needed to assist in the management operations.

The physical records needed would include the costs for:

- Land acquisition
- Land rental
- Feed
- Equipment purchase
- Equipment lease
- Fuel and repair
- Chemical and fertilizer
- Labor

These records should be coded with account numbers that identify them with the pasture, field, or group of cattle to which they belong.

The cattle records needed would include:

- Identification for each animal and group of animals
- Breeding history of the herd (pedigree information)
- Birth dates
- Weights (birth, weaning, yearling)
- Grazing location and time
- Feed consumed
- Carcass data (yield, fat, loin, etc.)

The reports needed for the computer to generate would include:

- Cattle inventory, value, and location
- Equipment inventory and value
- Feed inventory, value, and location
- Adjusted weights (weaning, etc.)
- Cow productivity (calf weights per unit of time)
- Estimated breeding values or selection indexes
- Sorted lists of calves, cows, and bulls for each trait
- Cash flow analyses
- Budget analysis
- Financial statement
- Tax accounting
- Enterprise analyses

Timeliness is important in all agricultural operations such as planting, cultivating, harvesting, vaccinating, breeding, and feeding. Timeliness is equally important to data entry in the farm computer. The advantages a computer offers are realized only if the data are entered on a regular and prompt basis. When the opportunity arises to sell cows at an advantageous price, if last season's calving data and this year's carcass information are not already entered, it will be difficult to get the data entered quickly enough. The seller will want to print a list of cows in ascending order of their maternal values and another list by their ages. Before hedging or selling, the prudent operator needs to know the cost of production. And so the manager and bookkeeper will need to make a schedule for data entry that is regular and can be followed throughout the year.

Financial reports should be studied monthly but enterprise reports need to be completed at the end of each season (harvest, calving, weaning, etc.). These reports are not printed unless they will be used, but at least once a year all reports are helpful in management decisions. The enterprise analyses tell last year's strong areas. But, price relationships may change next year, so do not drop an enterprise based on just one year's data unless you are sure that management and increasing prices cannot improve its performance.

SUMMARY

The livestock producer can increase his management capabilities by utilization of the modern microcomputer. He must spend the necessary time and money to select a software-hardware package that fits his operation. Selection of good user-friendly programs allows the computer to be utilized by all personnel. This gives the small- and medium-sized livestock producer enterprise evaluation capabilities that used to be available only to large well-financed operations.

REFERENCES

AgriData Resources Exclusive Report. 1983. All about on-farm computing. AgriData Resources, Inc., Milwaukee, Wisconsin.

Fox, Danny G. 1983. Symposium on use of computers in the livestock industry: Use of computers in beef cattle research, extension, and teaching programs. J. Anim. Sci. 57:771.

33

THE NEED FOR OLD-FASHIONED RANGE MANAGEMENT IN LATIN AMERICA

Donald L. Huss

THE SITUATION

About 50% of Latin America produces a kind of vegetation that can only be harvested by animals and converted into products usable by man. These are the region's rangelands, which some would classify as woodland ranges. Ranges provide nearly all of the nutrients consumed by 267 million head of cattle, 17 million sheep, 29 million goats (FAO, 1981), around 4 million alpacas and llamas (Fernandez-Baca, 1975), and an unknown number of wildlife. Cattle are by far the most numerous and about 86% of them are in Argentina, Brazil, Colombia, Mexico, Paraguay, Uruguay, and Venezuela (FAO, 1981).

Rangeland productivity has drastically deteriorated and the deterioration continues. Man-made deserts have been formed (figures 1 and 2). Livestock production is only a fraction of its potential. For example, average calf crops are between 50% and 55%, first calving normally occurs at 3 yr to 4 yr of age with calving intervals exceeding 400 days. Fattening to acceptable slaughter weights is attained at 4 yr to 6 yr of age. Annual extraction rates vary from around 10% in Paraguay to 20% in Argentina (FAO, 1977).

There are examples of cattle ranches that have reached economic reproductive rates of 80% or more and extraction rates of 20% and above (FAO, 1977). Fattening-to-slaughter weights of 880 lb can also be economically reduced to 2 yr or less (Huss et al., 1972). These improvements were achieved through better nutrition and animal husbandry, both of which are within the competency of a qualified range manager in Latin America. Universities, with the exception of some in Mexico, do not offer courses and none offer degrees in range management. There are no agencies with mandates and adequate financial and human resources to foster and advance range management. Information gathered on the limited number of research stations often does not trickle down to the producers.

Figure 1. This is not a natural desert in Mexico, it was made by man's abuse. The original vegetation that it produced is shown in figure 2 (FAO photo by author).

Figure 2. The original vegetation of this range in Mexico has been maintained because of its excessive distance from drinking water. Rangelands close to water have been transformed into man-made deserts as shown in figure 1 (FAO photo by author).

However, let us assume that all of the policy constraints can be removed and that a rangeland development program can be initiated. The question would then be, "What would the program be based upon, old-fashioned or modern range management?"

THE OLD AND THE NEW

Range management is relatively young. It was conceived in the U.S. during the first part of the century by scientists who were concerned about the deterioration of public and private rangelands and decreases in animal production. The study of range management was created by scientists for stockmen, but the latter were skeptical of it. This new discipline advocated many principles and practices that were contrary to traditional ranching methods. The stockmen's adage -- "more livestock, more product" -- was challenged. Therefore, in order for the new discipline to flourish, range management had to be sold to the stockmen who in turn could become partners in its development. The best way to sell something to stockmen is through their wallets, which was what the disciples of the new science set out to do.

The early definitions of range management stated that its purpose was to obtain maximum livestock production from rangelands consistent with conservation of the land resources (Stoddart and Smith, 1955). Range research, management, and extension activities were then geared towards producing forage for livestock and efficiently and economically converting it into produce. Thus, range management and animal husbandry were closely allied. A spinoff was that practices that increased forage production also conserved soil and water. This was old-fashioned range management.

It appears that the objective of obtaining maximum animal production has been or is being forgotten through the sophistication of range management. In 1964, range management was defined by the American Society of Range Management (now Society for Range Management) as "the art and science of planning and directing range use to obtain sustained maximum animal production consistent with the perpetuation of the natural resources" (ASRM, 1964). Ten years later, the Society defined it as "a distinct discipline founded on ecological principles and dealing with the husbandry of rangelands and range resources" (SRM, 1974). Emphasis is now placed on the ecosystem and measurements such as energy inputs, flows, and outputs. There is no objection to this because this approach provides answers to the old questions "how" and "why." However, rangeland development in Latin America is about the same stage that it was 40 yr or 50 yr ago in the U.S. In order for the stockmen to embrace range management, it must be sold to them just as it was in the U.S. Can you imagine how successful one would be trying to sell something to Latin American stockmen and policymakers that is defined as "a distinct discipline founded on ecological principles." There is a need for range management in the region, but it must first be sold in the old-fashioned way -- increased income.

SOME STAND-BY PRINCIPLES AND PRACTICES

The evolution of the discipline since its inception has generated some practical stand-by practices and principles that are essential to the range manager. While space limitations do not permit a thorough description of these, some of the more important ones are briefly highlighted.

Plant-animal Relationships

Range management is related to ecology because it consists of manipulating the environment in which both plants and animals live in such a way as to provide each, as far as practicable, with its more favorable habitat (Stoddart et al., 1975). Therefore, a successful operation requires the simultaneous application of plant and animal production technologies. This is illustrated as follows:

- Poor forage husbandry + poor animal husbandry = very low livestock production.
- Good forage husbandry + poor animal husbandry = low livestock production.
- Poor forage husbandry + good animal husbandry = low livestock production.
- Good forage husbandry + good animal husbandry = high livestock production.

Plant Nutrition

Nutrition is a word that probably conjures up humans or animals. The assumption would be quite right, but there is something else to know. Plants are living organisms and they too have nutritional requirements. A most important difference between plants and animals is their source of food. Plants use energy supplied by the sun to change inorganic matter taken from the earth and air into organic compounds. This process is called photosynthesis. Animals cannot do this. They must live on the organic compounds built by plants.

The plants' role is thus two-fold; first to feed themselves and second to feed man and his animals. It is imperative to successful range management that the plants be properly fed so that they can carry out their dual roles. This means that half of the plant's production should be left for the plant and the other half for the animal. Thus, the range manager's old adage "take half and leave half" is still valid.

Stockmen often confuse production with height. In bunch grasses, for example, about one-third of their height represents about one-half of their production. Perhaps the adage for bunch grasses should be "leave one-third of the height and take the other two-thirds."

Plant Succession

Plant succession or the ability of one community of plants to replace another is an ecological phenomenon of great importance to range management. Primary succession is the original and simultaneous development of plants and soils under prevailing climatic conditions. This continues through a series of changes until a state of equilibrium between climate, soil, and vegetation is reached. This state is called "climax."

In the case of rangelands, climax represents potential forage productivity. If the climax is abused with prolonged overgrazing, highly productive communities will be replaced by poor-producing ones. Soil condition will also deteriorate. This is termed "retrogression." Dyksterhuis (1949) quantified retrogression and placed it in its proper perspective regarding range management. He proposed that range condition be evaluated on the basis of the amount of climax vegetation remaining and introduced the terms "decreasers," "increasers," and "invaders," which have become key words in range management (figure 3). Animals eat first the species they like best, which are the dominants in a climax (figure 3). With continued overgrazing, these will weaken and die and decrease in abundance. Thus, they are called "decreasers." While the decreasers are decreasing, less palatable species increase to a point, but with continued overuse, animals turn to them until these plants also weaken and die. They are known as "increasers" (figure 3). Only unpalatable and evasive species can survive such a system of overgrazing, and eventually these will invade and dominate. They are called "invaders." Rangelands with 0% to 25%, 25% to 50%, 50% to 75%, and over 75% of climax remaining are classified as being in either poor, fair, good, or excellent range condition, respectively.

Fortunately, there is also a secondary succession that is the opposite of retrogression. It is much faster than primary succession, which took thousands of years, because the soil is already developed and some decreaser and increaser plants exist for natural regeneration. Planned succession in all of its aspects provides a tool in which a range situation can be analyzed, synthesized, and eventually integrated into a practical operation of growing and harvesting vegetation for maximum livestock production.

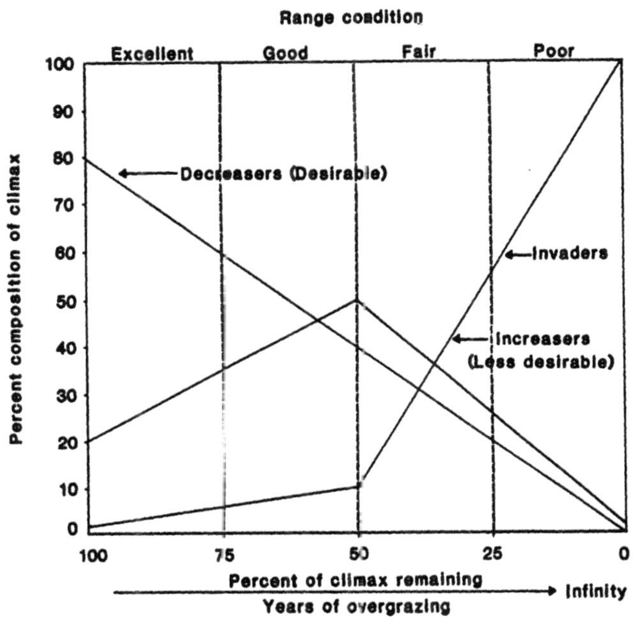

Figure 3. An example of retrogression and range condition related to percentage of climax remaining (Dyksterhuis, 1949).

Proper Stocking

Stockmen probably misunderstand this range management principle more than any other. Yet, it is so simple. A range in any given condition can only produce so much forage. If the demand is greater than the supply, then all animals will go to bed at night with empty or partially empty stomachs.

The short-term benefits of proper stocking are illustrated by Behment (1969). It can be seen in figure 4 that optimum livestock production is obtained with proper grazing. Utilization beyond this point results in a rapid decrease in production. Jones and Sanderland (1974) have calculated that zero gains per animal can be expected when stocking rate is double that required for optimum gains.

A long-term example of the adverse effects of overgrazing can be taken from Bentley (1898) who reported that stocking rates near Abilene, Texas, were 300 AU/section in 1898. Carrying capacities today are around 32 AU/section.

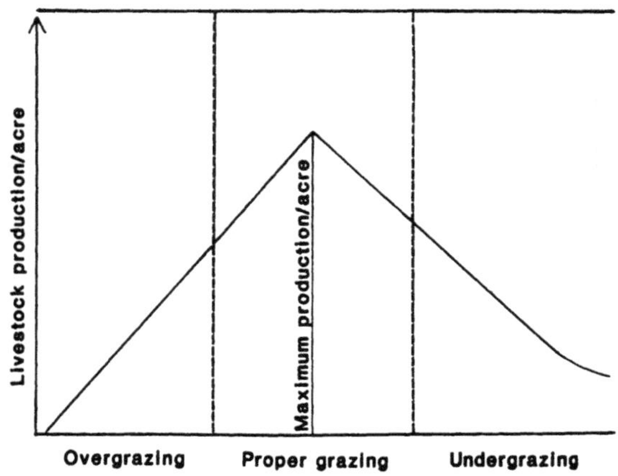

Figure 4. Relationship between livestock production and stocking rate. This relationship has been described by several authors (adapted from Behment, 1975).

Planned Grazing

The Society's 1964 definition of range management stated: "... the art and science of planning and directing range use to ..." These words cover the management in range management. They imply that maximum livestock production cannot be achieved haphazardly. A livestock operation must be planned and directed according to its needs. The realization and implementation of this factor alone could be one of range management's greatest contributions to the livestock sector in Latin America.

REFERENCES

ASRM. 1964. A glossary of terms used in range management. Soc. for Range Manage. 2760 West Fifth Av., Denver, CO.

Behment, R. E. 1969. A stocking rate guide for beef production on blue grama range. J. of Range Manage. 22:83.

Bentley, H. L. 1898. Cattle ranges of the southwest. USDA, Farmers Bull. No. 72.

Dyksterhuis, E. J. 1949. Condition and management of rangelands based on quantitative ecology. J. of Range Manage. 2:104.

Fernandez-Baca, S. 1975. Alpaca raising in the high Andes. World Animal Rev. No. 14:1.

Huss, D. L., M. A. Zavala, and S. Chazaro. 1972. The use of protein supplementation to increase livestock production in the Mexican tropics. Summary IX International Congress of Nutrition: 161.

Jones, R. S. and R. L. Sanderland. 1974. The relation between animal gains and stocking rates. J. Agr. Sci. 83:342.

FAO. 1977. Expert consultation for improving the reproductive efficiency of cattle in Latin America. FAO of the UN, RLAC, Santiago, Chile.

FAO. 1981. Production yearbook. FAO of the UN, Rome, Italy.

SRM. 1974. A glossary of terms used in range management. (2nd Ed.). Soc. for Range Manage. 2760 West Fifth Av., Denver, CO.

Stoddart, L. A. and A. D. Smith. 1955. Range Management. (2nd Ed.). McGraw-Hill Book Co. Inc., New York: p 1.

Stoddart, L. A., A. D. Smith, and T. W. Box. 1975. Range Management. (3rd Ed.). McGraw-Hill Book Co., New York: p 146.

34

MANAGING CATTLE IN A LOW-COST, EXTENSIVE RANGELAND SETTING

H. H. Stonaker

RECENT CHANGES IN THE NORTH AMERICAN HIGH PLAINS

Cattle management and production on the extensive rangelands of tropical Colombia contrast sharply with those on the High Plains of Colorado. Still, the extensive nature of the pasture resources for grazing beef cattle provides a common bond of interest to cattle ranchers -- South and North.

In the past 20 yr to 30 yr there has been a great deal of research and development in the natural grasslands. In the arid Pawnee Grasslands of eastern Colorado, integrated studies have been made of the total ecology, including flora, fauna, climate, and cattle productivity. The systems approach of the late Dr. George Van Dyne, who was one of the initiators of this work, has been widely appreciated. The land-grant universities in the region also have conducted cattle and range studies for many years. Within the High Plains states, these studies are well summarized at biennial meetings, such as range cow symposia.

The development of the Ogallala aquifer for pump irrigation changed much of the High Plains landscape from that of prairie to that of pivot irrigation. Vast acreages are now in corn, alfalfa, sugar beets, and beans grown on former rangeland that had required 15 to 40 acres per animal unit.

Winter feeding of cows on low-cost cornstalk residue pastures has required heavier stocking rates on native pastures during summer and less during winter, with less dependence on hay and other supplements in winter. Thus, much research has been devoted to better use of these crop residues to increasing their feed value through processes such as ammoniazation (researched largely by Dr. Klopfenstein of Nebraska) and to their proper supplemenation. Many protein supplements depend partially on low-cost urea. Rapid depletion of the aquifer and higher energy costs for pumping, however, portend further readjustments in crop and livestock use. Undoubtedly, cattle production will be affected.

A number of schemes have been set forth for grassland management. Decades of grazing-intensity studies on the short-grass Pawnee Grasslands have indicated that heavy grazing has produced more beef than have lighter grazing systems. Periodically, there is interest in reseeding dryland wheatland with wheat grasses, but whether there has been a successful effort to introduce or breed legumes for the High Plains is unknown. In contrast, it is a topic that has attracted much interest in Latin America and Australia.

In addition to the impact of crop residues, broad changes have been made through crossbreeding cattle and the introduction or reintroduction of many breeds. The amount of this being done in commercial herds through AI has reached a plateau, despite breakthroughs in estrus synchronization. Performance-tested bulls are widely available.

Dystocia, or calving difficulty, is a relatively recent cause of calf loss, which can be controlled. Calf scours remains another major cause of death loss. New electrolytes for restoring body fluids -- another product of university research -- have greatly improved the treatment of scours. The necessity for adequate colostrum intake at birth also is better understood today. The importance of energy feeds to condition cows for rebreeding is widely recognized. Marketing, too, has shifted to auctions and direct buying and, currently, there is great interest in the retained ownership of cattle through the feedlot. This appears to be the surest way for the rancher to gain the benefit of good management and genetically improved cattle. Having reputable, efficient feedlots nearby makes this easy.

Ranchlands have been attractive to investors and, although prices have dropped from all-time highs, these lands have been used as havens against inflation or for tax management. Thus, cattle operators -- especially younger ones -- are likely to lease pasture rather than own it. Such tenure conditions do not appear to be favorable to long-term capital improvements.

Despite the high dependence on the natural resources of native grasses and climate, techniques in cattle management have greatly changed in the past 3 decades. Productivity/unit of grassland on the experimental ranch at Akron, Colorado, has increased about twice in 20 yr. More efficient beef production means supplying an important food in a healthy, competitive industry that can offer a great amount of stable, high-quality employment. It offers the financial base for conservation and improvement of vast land areas. What more can we wish for? These types of economically viable increases in productivity are indeed awesome.

THE COLOMBIAN LLANOS

Our Latin American neighbors also have a great stake in the cattle industry, for many of the same reasons as our own. They produce a different type of beef more cheaply than we. Their resources of grass and climate make us "gringos" green with envy during our bitter and costly winter storms.

In brief, the following conditions or practices are common to cattle raising in the extensive grasslands of the Llanos -- the eastern plains of Colombia. Brahma or Zebu bulls are used. Bulls are kept with cows year-round. Cows calve first at about 4 yr. Expected calf crop born is about 45% to 55%, with a 40% weaning rate. All heifers are kept for breeding. The cow may have about four calves during her lifetime. Males generally will not be castrated and often are kept 4 yr to 5 yr before being trucked out of the Llanos to fattening pastures. There is little fencing. Although mineral supplementation is necessary, it is not used sufficiently. About the only pasture management is that of burning away the fibrous top growth of native pastures during the 4 mo to 5 mo dry season; however, more use is now being made of seeded, improved pastures. In contrast, in the richer grazing areas of the Magdalena River Valley and the north coast, different conditions prevail due to highly productive pasture. There, too, pastures are the exclusive basis of production because there is no storage of hay or silage and no use of grain.

In the 1960s and 1970s planners for improving national income and export income thought that improving cattle production should be a major cost-effective effort. Thus, a large international effort was initiated. The impact is not yet apparent because of the speed of other changes. For example, the export of flowers, a new industry, provides about twice the export value of beef. Grazing land prices, however, have increased many-fold within the last decade, and improvements are being made in Llanos cattle production.

Their potential for more dramatic increases in productivity seems even to exceed our opportunities in the Great Plains. Since I had the good fortune to be closely associated with two important research organizations in Colombia (CIAT and ICA), I would like to relate some of their results.

The basic objective of the experiments in Colombia was to develop more efficient and economical cattle production systems through study of the effects of mineral and urea supplementation, molasses grass pastures, early weaning, and crossbreeding.

The rationale for the experiments -- and for the establishment by ICA of the station, Carimagua, in the Llanos of Colombia -- stemmed from several conditions associated with the basic resources in Colombia and other countries in Latin America.

The primary product of the land in many of these countries is cattle, and beef is a basic food in the diet of the people. Approximately two-thirds of the beef in Latin America is produced in tropical regions (which contain 71% of the cattle population), and beef is an important commodity for export and foreign exchange in several of these countries.

Carimagua Ranch

Carimagua, an Indian word meaning water of the gods, is the name of a 22,000 ha ranch that is south of Orocue and bordered by the Muco River on the south. The Meta River, to the north, is the largest of several rivers draining the Llanos Orientales and during the rainy season is navigable by barges from Puerto Lopez, near the foothills of the Andes, to Puerto Carreno on the Orinoco River.

During the political violence of the 1940s, a reign of terror in the Llanos forced a great exodus of people from the region, including the owners of the ranch at the time. These owners were known as progressive ranchers and had made many improvements on the ranch, such as building barges to ship cattle and produce and planting new orchards and improved grasses (*Hyparrhenia rufa*, "puntero"). From the late 1940s until the ICA purchase of the ranch in October 1969, the ranch had been understocked and underdeveloped, which is not an unusual history for ranches in the region.

Brunnschweiler implied that the region of the Carimagua ranch was more heavily populated in Spanish colonial times. The capitol, Pore, in the Casanare region north of the Meta River, is now a small town of 1,000, but was once a city of 20,000 and the seat of government for approximately 130 villages. The missions established by the Jesuits had many cattle -- said to be of the San Martinero breed, probably a cross between Andalusian "Retinto" and the Asturian "Valle."

PROCEDURES

Experimental Design

The two experiments and corresponding treatments were arranged factorially as follows:

Mineral experiment	M_1	M_2
U_1	Herd 3	Herd 5
U_2	Herd 2	Herd 4

Pasture experiment	P_1	P_2	P_3
U_1	Herd 5	Herd 8	Herd 6
U_2	Herd 4	Herd 9	Herd 7

The treatments were:

M_1 = salt only; and M_2 = complete minerals

U_1 = no urea; and U_2 = urea supplementation seasonally

P_1 = native pasture continuously; P_2 = molasses grass, (*Melinis minutiflora*), continuously; and P_3 = native pasture in the dry season, molasses grass in rainy season.

Within herds:

Weaning: (W_1) five cows with calves weaned at 3 mo vs (W_2) 30 cows with calves weaned at 9 mo.

The five cows with calves to be weaned at 3 mo were randomly chosen from among the first cows to calf in each herd. That is, other cows could have been paired with them insofar as calving date was concerned. However, in the analysis, these five cows were compared with the 30 cows with calves to be weaned at 9 mo of age. Larger numbers and completely random selection of cows for early weaning were not considered feasible because of concern over costs and possible management problems.

Crossbreeding: (X_S) Alternate 28 days: 8 San Martinero bulls, one per herd, rotated among the herds in alternate 28-day periods.

(X_2) Alternate 28 days: 8 Zebu bulls of the Brahman breed, one per herd, rotated among the herds in alternate 28-day periods. Thus, bulls of a breed were in use 28 days and rested 28 days.

The first calf crop, making up almost one-third of the total data on calves, was not affected by early weaning or whatever cumulative effects long-term mineral and urea feeding may have had on the cow herd and subsequent calvings. They could have been affected by minerals, urea feeding, and crossbreeding.

Selection of Experimental Components at Carimagua

The herd-systems experiment included experimental, demonstrational, and commercial types of activities.

Native pastures for the experiment were located in areas where a combination of high land and low land could be incorporated within each of the native pastures. Empirically, 5 ha of native pastures were allocated for each animal unit. For seeded molasses grass pasture, only upland was used and this was allocated at the ratio of 1 ha per animal unit. Native pastures were further divided into three areas separated by firebreaks (disked bare soil). Each area within a pasture was purposely burned annually in a chronological pattern so as to provide a freshly burned area in each native pasture about every 4 mo. On the other hand, molasses grass pastures required protection from fire, a measure that was only partially successful.

To minimize any differences in productivity of the pastures within major nonpasture treatments, the herds were exchanged between the two pastures of their experimental treatments following each weighing period. Thus, Herds 2 and 3 alternated using the same two pastures, as did Herds 4 and 5, 6 and 7, 8 and 9. No pastures were exchanged between herds that received minerals vs those not receiving them; nor were there exchanges of herds requiring different types of pastures.

In the dry season, in addition to their own pasture, cows grazing molasses grass year-round had access to the molasses grass pastures vacated by Herds 6 and 7 when they were moved onto native pastures for the dry season. Thus, the stocking rate per hectare of molasses grass was approximately halved during the dry season.

MINERAL EXPERIMENT RESULTS

Characteristics of the Breeding Herd

Results obtained over the entire experimental period (1973-1977) are reported here in chronological order, with brief summaries of the treatment effects of minerals, weaning, urea supplementation, and crossbreeding. Other factors such as year, season, age, and reproductive status are included also. Particular attention is given to how nutrition is associated with conceptions, abortions, and calving rates. Also examined were effects on cow weights, death losses, and timing sequence of calving, calf weights, and mortality (see experimental design).

Observations on breeding cows reflect results of growth of their calves from 3 mo to 18 mo. At 18 mo, the calves were removed from the experiment. Death losses of calves also have been documented as to probable causes.

To simplify discussion, only the adjusted means for main effects and means for significant interaction effects have been presented.

BREEDING HEIFERS

Weights, Conceptions, Abortions, and Calving Rates

Figure 1 shows a rapid divergence of the growth curves representing purchased heifers in the mineral-fed herds vs those in the nonmineral-fed herds. By May 1973 (immediately previous to the first breeding season), the mineral-fed heifers weighed 31 kg (13%) more ($P<.01$) than did the herds fed salt only.

The original plan was to breed the heifers at 2.5 yr of age, in May 1972. However, examination at that time indicated immature ovarian development, and it was evident that the heifers were not cycling. Breeding was delayed until the following May when the heifers were 3.5 yr old and those on minerals weighed 257 kg (table 1). The conception rate was high in the next months, and it was apparent that the high level of breeding in May and June of the first breeding year would affect the remainder of the experiment by partially confounding results with seasonal calving. It also caused an atypical chronological pattern of breeding and calving, particularly in the native savanna herds.

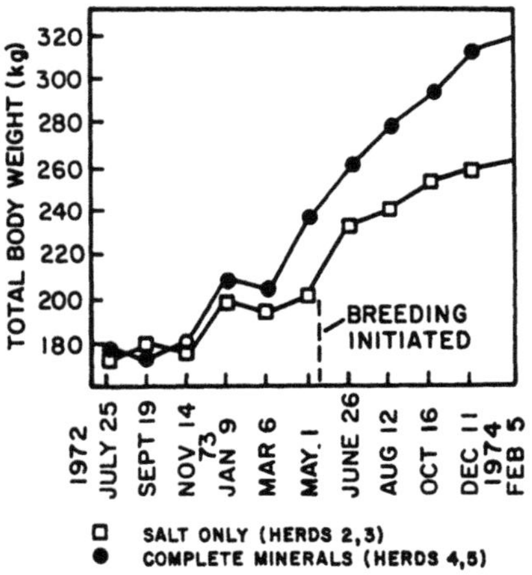

Figure 1. Weight changes of heifers on native pasture as related to mineral treatment.

TABLE 1. LEAST-SQUARES MEANS OF WEIGHTS OF EXPERIMENTAL HEIFERS AT BEGINNING OF BREEDING SEASON, MAY 1973 (MINERAL EXPERIMENT)

	Herds 4, 5, 6, 7, 8, 9			
	N	Constants	Least-squares means	Standard errors
Mean (MU)	140	242.1	242	2.09
Salt only	71	-15.3	227	2.93
Minerals	69	15.8	257	2.97

Surveys later indicated that, with uncontrolled breeding, there is a natural breeding season in the Llanos, usually during the dry season rather than the early wet season. As the experiment progressed through the years, the patterns of calving incidence associated with experimental procedures were diminished, and the calving dates began to cover a wider range over the other months. About 50% of all births occurred in the 4 mo between February and May.

First pregnancies. The mineral-fed heifers continued to outperform the nonmineral-fed herds in body weight (figure 1). The rate of conception in the first months of the breeding season was confirmed by palpation in October. Conception rates did not differ greatly between the mineral and nonmineral herds in the first year of the experiment. The large difference in calving percentage (table 2) could be attributed to a high abortion rate in the nonmineral-fed herds, a finding that was confirmed in subsequent years of the experiment.

TABLE 2. REPRODUCTIVE PERFORMANCE IN THE FIRST CALF CROP (CALVES BORN BETWEEN FEBRUARY AND DECEMBER 1974), BY HERD (MINERAL EXPERIMENT)

Treatment	Herd	No. of cows	No. of births	No. of abortions	Calving (C), %	Abortions (A), %	C+A	Time from beginning of breeding to conception, months
M_1 Native pasture	2	33	17	13	51.5	39.3	91	7.74
and salt	3	29	15	7	51.7	24.1	75	9.78
M_2 Native pasture and complete	4	31	27	2	87.1	6.5	94	4.27
minerals	5	33	30	1	90.9	3.0	94	4.74

Breeding Cows

Weights. Observations were made of variations resulting from effects of dry and wet seasons, minerals, urea supplements, early weaning, and reproductive status. Table 3 shows the resultant wide range of body weights. For example, over the years the cows increased in body weight by about 14%, plateauing at 339 kg in 1977.

TABLE 3. COW WEIGHTS (ADJUSTED) OVER ALL YEARS AS RELATED TO YEAR, REPRODUCTIVE STATUS, SEASON, UREA SUPPLEMENT, MINERALS, AND WEANING DATES (MINERAL EXPERIMENT)

Year, kg		Reproductive status, kg		Season, kg		Urea, kg		Minerals, kg		Weaning, kg	
1973	290	Open, dry	299	Dry	304	None	312	Salt	304	3 mo	317
1974	312	Pregnant 1-6 mo	320	Early wet	320	Urea	323	Minerals	331	9 mo	317
1975	314	Pregnant 6-9 mo	348	Late wet	329						
1976	332	Pregnant, lactating	325								
1977	339	Open, lactating	295								

PASTURE EXPERIMENT RESULTS

Only a few details of the pasture experiment are presented because the pattern of effects on cows and calves largely followed the results of the mineral experiment. The molasses grass pasture production more than tripled that of the native pastures. When grazed only in the rainy season (cattle grazing with savanna or native grass in the dry season) molasses grass gave excellent results. Molasses grass used alone and year-round, however, was a disastrous practice with high death losses of cows. Urea supplementation on year-round molasses grass was essential, but this apparently increased abortions. Molasses grass is not the best grass available for Llanos conditions today; *Andropogan guyana* is a new species that looks promising and *Brachyaria* is widely used at this time even though it is associated with calf nutrition problems. Both these grasses are resistant to fire, which molasses grass is not.

FIVE-YEAR PRODUCTION OF LIVE WEIGHT IN THE NINE HERDS

A Hypothetical Commercial Ranch Analysis

Table 4 shows the gross production of live weight beef per herd at Carimagua. For comparative purposes, Carimagua data may be compared with information from the Fondo Ganadero (Meta) related to the commercial production of leased herds in the Llanos. As a working rule of thumb, the Fondo Ganadero expects a doubling of the net weight produced per herd within a 5-yr period. Although results at Carimagua are not exactly comparable, weights were approximately doubled in Herd 1, the low-input (control) herd.

TABLE 4. LIVE WEIGHT BEEF PRODUCED PER HERD, 1972-1977[a]

Herd	M[b]	U[b]	Calves		Stockers		Cows, 1977		Heifers, 1972		Net wt produced per herd, kg	Wt ratio relative to Herd 3
			No.	Kg	No.	Kg	No.	Kg	No.	Kg		
1	1	1	10	1185	21	4066	24	7363	34	6468	6146	.75
2	1	2	27	2883	20	3197	31	9900	34	6606	9374	1.14
3	1	1	21	2124	27	4280	27	8378	35	6574	8208	1.00
4	2	2	32	4235	43	7852	32	11402	34	6467	17022	2.07
5	2	1	38	4317	38	7228	33	10862	33	6408	15999	1.95
6	2	1	30	3456	51	10248	33	10389	35	6709	17384	2.12
7	2	2	36	4929	45	9044	35	11858	34	6486	19345	2.36
8	2	1	28	3064	49	10032	25	7458	36	7013	13541	1.65
9	2	2	28	3699	48	10651	32	10277	37	7093	17534	2.14

[a] Includes live weight of all cows and calves in herds May 1977 and 18-mo wt of stockers leaving the experiment, less initial 1972 weights of breeding heifers entering the experiment.

[b] M = minerals, U = urea. Not fed = 1, fed = 2.
Mineral effect (M) = 7720 kg/herd.
Urea effect (U) = 2036 kg/herd.
Urea effect on molasses grass only = 3993 kg/herd.

Table 4 lists the numbers and weights of all calves below 18 mo of age at the end of the experiment in May 1977. Weights of all cows, as of May 1977, were added to the calf weights. The weights of the stockers (postweaned cattle) were added when they were removed from the experiment at 18 mo of age. The weights of the original heifers when they entered the experiment in May 1972 were then subtracted from these combined sums. Slight discrepancies in numbers are due to animals that were missing on the days when the herds were weighed.

The control herd performed poorly overall; cow survival rates were low, largely because of bone fractures. Bone fractures generally occurred when the cattle were being worked in the corrals and squeeze chute. Also, the total number of calves produced in Herd 1 was much lower than that of other herds, including the other nonmineral-supplemented herds.

Interestingly, the average weight of stockers at 18 mo was 194 kg in Herd 1 as compared to 159 kg in the other non-mineral-supplemented herds. Natural weaning may not have occurred until the calves were 12 mo to 14 mo of age, which explains the extra weight.

Herd 3, which received no mineral or urea supplement, also is useful for comparative purposes because it is representative of a number of herds under low-investment management in the Llanos. Except for Herd 8, a low-producing herd, all mineral-supplemented herds produced twice as much beef as did Herd 3. These ratios also are shown in table 4. Herd 7 was the most productive, yielding 2.36 times as much beef as did Herd 3. Herd 7 received mineral and urea supplementation and was grazed on native pasture in the dry season and molasses grass pasture in the rainy season.

The total mineral effect per herd was 7,720 kg body weight (Herd 3 vs Herd 4), which could be attributed to an estimated mineral consumption of about 3,470 kg during the 5 yr. Approximately 2.2 kg of beef were produced per kg of salt-mineral mix consumed (or, omitting the 47% salt, 4.2 kg of beef were produced per kg of mineral). On a market basis, the cost:benefit ratio could be as much as 1:20.

The total urea-supplementation effect per herd was associated with an additional 2,036 kg body weight per herd. Over the 4 yr of urea supplementation, the total seasonal consumption per herd was estimated at approximately 5,170 kg of molasses, 820 kg of urea, and 82 kg of sulphur. The cost:benefit ratio could be as much as 1:4. The greatest response to urea supplementation was on continuously grazed molasses grass pasture, where an additional 3,993 kg live weight were produced as a result of urea supplementation in Herd 9.

Herd 8 (on molasses grass continuously, with urea) did not perform as well as did herds on all other mineral-supplemented pastures, primarily as a result of heavy cow losses.

REFERENCES

Stonaker, H. H., N. S. Raun, and Juvenal Gomez. 1984. Beef cow-calf production experiments on the savannas of eastern Colombia. Winrock International. Morrilton, Arkansas. (In press.)

35

RATIONALIZING RANGELAND USE IN THE NEAR EAST BY MODERNIZING TRANSHUMANCE GRAZING

Donald L. Huss

The Near East is characterized by a harsh dry climate. Only 6% of the region is arable, 5% is forest, and the remainder is rangeland of which nearly 20% is irrigated. Unfortunately, policymakers too often consider these rangelands as either nonproductive or wastelands, and so they have a low priority rating in their development programs.

To the contrary, the Near East's rangelands are quite productive in their own rights. They provide a livelihood for millions of pastoralists as well as the bulk of the feed consumed by the region's some 300 million head of livestock. It was estimated in 1977 that these livestock annually produce around 1.5 to 2 million tons of meat valued at $1.5 billion to $2.0 billion (Huss, 1977).

Failure to recognize the importance of the rangeland resources and to make provisions for their proper care has caused a deplorable situation. Over vast areas literally nothing is left to graze, erosion scars are everywhere, and desertification is rampant.

Juneidi and Huss (1978) attributed this to overgrazing, untimely grazing, fuel gathering, and illogical cultivation. While many of the region's rangelands have been grazed for hundreds of years, destructive utilization has only recently become acute and widespread. Livestock numbers have increased several fold within the past few decades (FAO, 1972). The amount of shrubs that are grubbed or cut for fuel is astounding. In Jordan, for example, it is estimated that 182 million fodder shrubs annually are uprooted for cooking (FAO, 1976). This puts man and his animals in direct competition (figure 1). Overgrazing is also due to shrinkage of the rangeland area because of the expansion of cultivation to areas incapable of sustaining agriculture.

The problem, which goes beyond land misuse, involves a multitude of entangled sociological, economical, political, and marketing factors that cannot be unravelled easily. Many say that the problem is too entangled to solve. However, if man intends to live and depend upon these rangelands on a sustained basis, the problem <u>must</u> be solved.

Figure 1. Millions of valuable fodder shrubs, such as the sagebrush *(Artemesia herba-alba)* shown here, are annually destroyed for firewood thus putting man and his animals in direct competition (FAO photo).

GRAZING SYSTEMS

Although some livestock are grown under sedentary conditions on specialized farms, the vast majority are raised under a migratory system of production due to the region's ecology. Two migratory systems are recognized: nomadic and transhumance (FAO, 1974). The pastoral nomads move continuously with their families in search of water and pasturage for their animals. Nomadic grazing normally occurs in the very arid areas and is the only way in which the forage produced by sporadic rainfall can be utilized.

The transhumance system differs from nomadic in that the migrations are regulated by normal seasonal forage availability and climatic conditions. This is the major system used for millions of animals as they move between dry and wet zones and from winter to summer grazing areas in the whole Near East. A stockman involved in transhumance grazing is called a "transhumant" (figure 2).

Figure 2. A transhumant searching for forage for his sheep. His family may be the sole owners of the flock or he may be a shepherd for several different owners (FAO photo by author).

One solution to the range utilization problem is settlement, but attempts to do this usually result in failure because pastoralism is a way of life (Cole, 1975). Furthermore, with sedentarization, much of the forage produced on the seasonal rangelands would not be utilized and so some kind of migratory grazing system is needed. For this reason, migratory grazing itself is not the problem, but it must be modernized. Thus this article focuses upon the transhumance system.

THE MODERNIZATION APPROACH

Recent advances in an overall land-use policy can be used as a basis for modernizing transhumance grazing. These developments can be likened to pieces in a jigsaw puzzle as illustrated in figure 3 which, when put together, would form a beautiful picture of a modernized animal production system and rationalized range use.

THE PIECES AND PUTTING THEM TOGETHER

Homing Instinct

The social structure and organization of transhumant societies are based on a hierarchical system of family groups. A group of families constitutes a *hamula;* a number of *hamulas* form a clan, and a tribe is composed of a number of clans (Abdallah, 1978). All social relations are

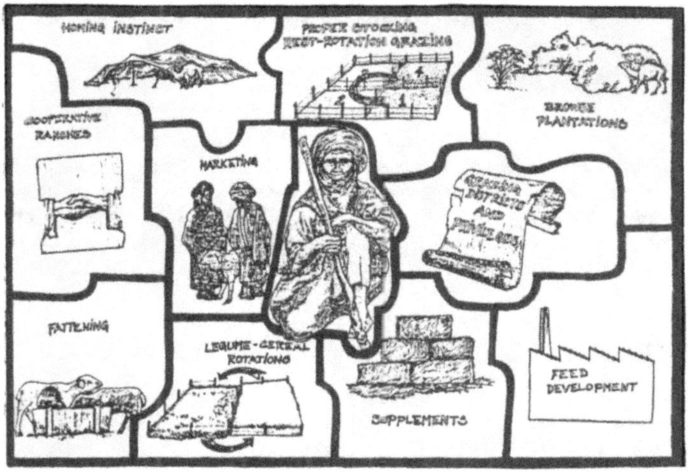

Figure 3. Recent isolated developments are like jigsaw puzzle pieces that can be put together to form a picture of a modernized transhumance livestock-production system and rationalized rangeland utilization.

governed by family ties and kinship, but each family has its own animals.

The families might separate during the migratory season or they may go as a *hamula*. Regardless, they have a homing instinct so that they return to the same grounds during the off season. This does not mean that they have control of the homing grounds because others also have the right to use them. As a result, the grounds offer poor fare to the animals. Nevertheless, the homing instinct and the social relationships connected to it are an important piece in the puzzle. The homing instinct now is being institutionalized by cooperative ranches formed for the exclusive use by the members, thus paving the way to modernization.

<u>Cooperative Ranches</u>

In ancient times, groups of pastoralists had grazing reserves for their exclusive use. These went by names such as *kose* in Kurdish or *hema* in Arabic (Draz, 1983). The purpose of a reserve was to defer grazing during the wet season in order to establish forage for use during the dry season. In effect, a *hema* was a cooperative ranch located on the people's homing ground. The means of harvesting or grazing the forage on the ranch was strictly controlled by the group so that vegetation would not be destroyed. Most of these ranches have been abandoned during the past 50 yr or so due to governmental programs that have made rangelands public property and open to grazing by all (Draz, 1976).

The *hema* concept has been successfully revived in Syria (Draz, 1983) and Jordan in the form of cooperatives. This requires that the government award a specific area to transhumants who have formed themselves into cooperatives, usually based on *hamula* or clan relationships, for their exclusive use. The responsibilities of ownership have been restored and at present the people are more amenable to cooperatives than to technical guidance. The next step to solving the puzzle is to fit together the homing-instinct pieces and the cooperative-ranches pieces as illustrated in figure 3.

Proper Stocking and Rest-rotation Grazing

Establishing cooperative ranches is not sufficient. They must be properly managed and improved to provide a greater supply of quality forage. This can be partially achieved through practices that permit natural regeneration of desirable species. Studies in the region have shown that rest-rotation grazing systems can result in several-fold increases in carrying capacities and animal production (FAO, 1976). These practices cannot be implemented on a widescale basis under present conditions, but they can be implemented on cooperative ranches. The puzzle can be further solved by fitting the proper-stocking-rest-rotation-grazing piece to the cooperative-ranches piece as shown in figure 3.

Browse Plantations

Various saltbush species (*Atriplex* spp) are being successfully used to establish plantations that provide supplemental feed during the dry season and some firewood. A spineless cactus (*Opuntia* spp) is also used in northern Africa to establish browse plantations. All of the browse plantation is rested during the wet season along with all of the rest of the ranch to permit recovery of the plants and to assure a dry season forage supply. By attaching the browse-plantation piece to the proper-stocking-rest-rotation-grazing piece (figure 3), the transhumants will have a fairly well-larded home to return to.

Forage Legume - Cereal Crop Rotation

Croplands can play an important role in the modernization process because of their feed and forage-producing potential. Currently, feed and forage crops are seldom sown. Traditionally, about 50% of the rainfed croplands are annually fallowed to maintain soil fertility. While fallow does maintain a fertility level higher than continuous cropping, fertility in general has declined and the decline continues.

The traditional farming systems can be largely eliminated by using forage legumes in lieu of fallow and in rotation with cereal crops. The various species of medic (*Medicago* spp) and vetch (*Vicia* spp) are suitable for this purpose in the Mediterranean climatic zone where rainfall exceeds 15 in. About 7.5 million to 10 million acres in the region could be annually sown to forage legumes (FAO, 1980). This cropping system is not only complementary to crop production (several-fold increases in wheat production have been observed), it also provides a badly needed and new source of fodder. Consequently, the forage-legume-cereal-crop-rotation piece of the puzzle is a most important one (figure 3).

Supplements

An adequate year-long, high-quality forage supply cannot be provided by rangelands alone. Supplementation to remedy the deficiencies of the range diet is necessary to obtain maximum livestock production. These needs can largely be met by processing the legumes in the previously mentioned crop rotations into hay. Thus, the supplements piece needs to be fitted to the forage-legume-cereal-crop-rotation piece as shown in figure 3.

Feed Development

An FAO study indicates that the region might be richer in feed supplies than it realizes (Abou-Raya, 1978). There are many agricultural and industrial by-products that could be utilized as feed. Since feed supplies are important to animal production in general, the feed-development piece must be fitted to the others related to croplands (figure 3).

Fattening Systems

The picture is far from being complete because of a gap between the cropland and rangeland sectors that will remain unless steps are taken to integrate them. This gap can be partially bridged with fattening systems (figure 3). This gap will permit the development of a land-use policy so that the rangelands can become the breeding grounds to produce offspring for fattening in the cropland area, which will siphon a greater number of animals from the rangelands at an earlier age than at present. Thus, another step is made towards rationalized grazing.

Syria has demonstrated that lamb fattening in the form of cooperatives is feasible and economical. These cooperatives purchase and provide feedstuffs at low cost and make loans from a revolving fund, but each member operates his own feedlot. The first 12 cooperatives were established in 1970 and had reached 55 by 1981 with a capacity to fatten about 1.5 million head annually (Draz, 1983).

Marketing Structures

The gap between range and croplands can be closed further with adequate marketing structures (figure 3). Historically, the transhumants have not had access to markets in which fair prices were given. Consequently, they are reluctant to sell their animals. The result is a prolonged finishing period on the range which in turn adds to overgrazing. Marketing structures that assure fair prices will serve as an incentive for the transhumants to sell. Marketing must not be a one-way street; provision must also be made for the sale of supplements and feeds at reasonable prices to the transhumants.

Grazing Districts and Privileges

Once the above parts of the puzzle are put together and become functional, rangelands outside the cooperative ranches can be brought into the picture so that pastoralists will have more confidence in modern technology and the technologists. The use of the rangelands outside the cooperative ranches can be rationalized by developing grazing districts, privileges, and permits. The picture will never be entirely completed until this piece is connected with the others (figure 3).

Man

All but one of the jigsaw puzzle pieces have now been put together. Rangeland misuse and low levels of livestock production are human problems, and it will take humans to solve them. Therefore, man is the most important piece in the puzzle. Man as the policymaker, planner, scientist, technician, and the public must become aware of the ultimate consequences of the continued destructive exploitation of the rangeland resources. And he must instigate remedial actions. The transhumants must also be involved in the formulation and implementation of these actions. The puzzle is solved when the piece related to man is inserted (figure 3).

REFERENCES

Abdallah, M. I. 1978. The nomadic pastoral system and its implications in the Near East. FAO of the UN, RNEA, Cairo, Egypt.

Abou-Raya, A. K. 1978. Preliminary survey of the feed resources of the Gulf and Arabian Peninsula countries along with possible means of developing them. FAO of the UN, RNEA, Cairo, Egypt.

Cole, D. P. 1975. Nomads of the Nomads. Aldine Pub. Co., Chicago.

Draz, O. 1976. Rangeland development in the Arabian Peninsula. EMASAR Rpt. FAO of the UN, Rome, Italy.

Draz, O. 1983. The Syrian Arab Republic -- rangeland conservation and development. World Animal Rev. No. 47:2-14.

FAO, 1972. Animal husbandry, production and health, fodder production and range management in the Near East and FAO's policies and plans for promoting the animal industry. FAO of the UN, Rome, Italy and RNEA, Cairo, Egypt.

FAO, 1974. The improvement of nomadic and transhumance animal production systems. AGA/Misc/74/3. FAO of the UN, Rome, Italy.

FAO, 1976. Desert creep and range management in the Near East. FAO of the UN, RNEA, Cairo, Egypt.

FAO, 1980. Prospects for increasing animal production in the region. NERC/81/4. FAO of the UN, Rome, Italy.

Huss, D. L. 1977. Importance of range development in dryland systems of farming for integration of crops and livestock husbandries in the Near East. 2nd FAO/SIDA Seminar on Field Food Crops in Africa and the Near East. Lahore, Pakistan. FAO of the UN, Rome, Italy.

Juneidi, M. and D. L. Huss. 1978. Rangeland resources of the Gulf and Arabian Peninsula countries and their managerial problems and needs. FAO of the UN, RNEA, Cairo, Egypt.

36

MODERNIZATION OF LIVESTOCK PRODUCTION IN INNER MONGOLIA: PROGRESS OR REGRESSION

Ervin W. Schleicher

INTRODUCTION

The autonomous region of Inner Mongolia is one of the major centers of the range livestock industry of the People's Republic of China (PRC), with rangeland comprising 74% of the area and some 40 million head of livestock. Located along the northern edge of the People's Republic of China, Inner Mongolia is shaped like a long (2,700 km), narrow crescent.

THE PROJECT

In 1979, the government of the PRC requested assistance from the United Nations Development Program/FAO in establishing a pilot demonstration center/ranch for intensive pasture, forage, and livestock production in the Wongnute Banner, Inner Mongolia. This cooperative effort is now commonly referred to as "The Grassland Project."

A representative site occupied by the Red Flag Brigade was selected for the project; this area is located approximately 350 km north of Chifeng and 20 km from Halisun on the Shilamulum River. The Red Flag Brigade consists of about 1,000 people operating 13,300 ha of land. The demonstration center occupies 4,493 ha of this land. Livestock included 3,000 cattle, 5,000 sheep, 4,800 goats, 900 horses, and 300 camels and donkeys. Annual rainfall in the region averages 340 mm (62 mm to 650 mm range). Climate, soils, and vegetation are similar to the Nebraska Sandhills. About 2,133 ha are classified as black, marshy, and sandy meadow soils while the hill land is white sandy soil. It is a grassland region characterized by frequent drought, continuous wind, low winter temperatures, and a short growing season.

The livestock industry of the area is characterized by heavy overgrazing, shortage of winter feed for the 7-mo feeding period, severe wind erosion with shifting sand dunes, and low production. Winter losses can exceed 10%. Because milk and cheese are important in the people's diet, most cows are milked following early weaning of the calves.

Breeding is done by artificial insemination. Native hay is cut by hand and carted to the farmsteads in September and October.

Objectives and Organization

General project objectives were to:

- Increase livestock production through modern methods of feeding, breeding, and management.
- Develop intensive fodder production using irrigation-silting and modern machinery.
- Improve productivity of range lands.
- Initiate soil conservation measures to control and stabilize shifting sand dunes.
- Construct essential improvements and infrastructure.
- Train personnel through study tours, work-study programs, fellowships, on-the-job training, and courses and seminars given by consultants.
- Increase the income and living standards of the people.

From the outset, it was clear that 2 yr could only be the first phase of a much longer project and that, if substantial progress was to be made to accomplish the objectives, much more time would be necessary.

Six expatriate consultants were selected for the project, including a livestock development specialist and mission leader, a forage production and pasture seed production consultant, a farm machinery consultant, a pasture improvement specialist, a ranch planning and management consultant, and an irrigation specialist. Later, additional consultants were contracted with expertise in soil conservation, sheep and goat production, forage and feed analysis, and land drainage.

This core group of consultants assisted the General Bureau of Animal Husbandry, officials of the Traudu League, the Wongnute Banner, and members of the Red Flag Brigade. The vice-chairman of the Revolutionary Committee of the Wongnute Banner was appointed project director.

Funding for the project was projected at $1.5 million for the UNDP contribution and $5.8 million from the Chinese government.

Development Plan

A development plan was designed to provide for the following specific activities:

1. Soil conservation, pasture, and fodder production
   - Fix sand dunes
   - Make land surveys and use capability map
   - Improve pastures by fencing and grazing systems, interseeding with introduced grasses and legumes
   - Improve fodder production with flood and sprinkler irrigation, silting controls, a

fertilizer program, introduction and screening of a wide range of plants for pasture and fodder production, and harvesting and storage methods to increase quantity and quality of storage

2. Irrigation and drainage
   - Emphasize the use of the shallow aquifer
   - Improve gravity irrigation techniques
   - Evaluate water requirements
   - Adapt structures to accommodate large machinery
   - Determine basic parameters for needed systems (A drainage expert was contracted because of insufficient knowledge of some parameters of major importance -- in particular the hydrodynamics of the soil and indications of severe salinity and alkalinity.)
3. Machinery
   - Acquire all necessary modern machinery and have on site by March 1981, using one manufacturer of imported machinery (John Deere)
   - Give highest priority to minimum tillage because of wind and soil conditions
4. Livestock production
   - Reduce numbers
   - Set up "selected pilot herds and flocks" to demonstrate optimum feed and management practices
   - Increase productivity on a per head basis rather than on numbers of animals
   - Maximize animal production per unit of feed available
   - Improve genetic potential by establishing a simple identification and recording system concentrating on a few important production traits: systematic culling; more efficient selection for economically important traits; selecting most appropriate lines/breeds to be introduced; examining possibility of using FAO "Artificial Insemination and Breeding Program" (AIBDP); studying advantages of keeping a few highly specialized dairy herds with the bulk of the cattle in the region on an extensive beef production system rather than all cattle being used for beef, milk, and draft; and improving veterinary control within the Brigade
5. Planning, facilities, and management
   - Prepare detailed topographic map for development planning
   - Plan and construct facilities which would permit efficient conduct of activities and, at the same time, be simple and low cost
   - Establish a recordkeeping system that would be an effective tool for evaluation, future planning, and application of results to other areas

6. Training
   - Visits by two study tour groups to Canada, Colorado, and Nebraska, U.S.A, with special emphasis on pasture/fodder production, farm management, and animal production
   - Visits by groups from mechanization section to John Deere, U.S.A., for mechanical and operational training
   - Consideration of group tours of other parts of the world, i.e., Europe and Australia

A tripartite review was set for the end of the initial project in 1981 to assess the progress of the project and make recommendations for the future.

OBSERVATIONS AND EXPERIENCES

Work had started in 1979 prior to the formal beginning of the project and was well underway when the core group first visited the project in April 1980. Each of the consultants made four, 30-day visits during 1980 and 1981: two visits in the spring and two in the fall of each year. Activities included reviewing progress to date, planning, actual field work, demonstrations, and teaching. A national staff of 21 members were assigned to the project, and labor was furnished by the Brigade.

The organization and management of the core group and other extra consultants were satisfactory for the first year; however, the project director felt that much more could be accomplished at less cost if two consultants could work full-time during the growing and harvesting season. Thus, an additional machinery specialist and a pasture and fodder crops specialist were contracted from June to October in 1981.

Following a project review in late 1981, it was agreed that limited UNDP/FAO participation would be continued for a second phase, 1982 to 1983. Two consultants (machinery and pasture and fodder crops) spent April to October of 1982 and 1983 on the project. Several other short-term consultants also were engaged (including a drainage expert) to finalize this phase of the project, which had been given limited attention.

RESULTS

Excellent progress was made in the development of the infrastructure, particularly in irrigation works, roads, buildings, machinery, the establishment of demonstrations, trials, and the screening of plant material. Much of the latter was done in cooperation with the Gerezhun State Seed Farm, which is near by.

Dune stabilization efforts using trees and shrubs on small areas slowed the advancement of the dunes from a rate of 3 m/yr to 2 m/yr. By 1982, 55 ha had been treated and, although findings were inconclusive, indications were that the best results were obtained by mixing locally adapted, low-growing grasses, legumes, and other plants with shrubs and trees. Twenty-eight kilometers of shelter belts were established. If appreciable progress is to be made, however, massive efforts must be mounted, requiring many years.

In range management, 2,000 ha were fenced for intensive development with more under construction. Trial reseeding was done in several locations with limited success. Four wells for livestock were scheduled for completion. Rotational grazing trials for cattle and sheep were completed, and electric fencing proved practical for both species.

Sixty-six kilometers of canals were constructed by hand: 41 canals, large and small, including the 21.4 km main-supply canal from the Halisun Diversion on the Shilamulum River. Bridges, sluces, drops, and other structures were completed; four deep wells were drilled and spray irrigation equipment was installed at one location. A total of 2,133 ha of black marsh and sandy meadow soils was designated for intensive irrigated pasture and fodder production. About 700 ha were readied for irrigation, drainage, and silting, and 133 ha of marshland were drained and 25 ha silted.

Trial plots and screening of plants to date have indicated that vital basic information obtained in other parts of the world could not be transferred easily to Wongnute. A very extensive testing program was initiated at the beginning of the project to test and screen various forage, grain, and pasture plants. A number of tillage methods were tried, especially those using minimum tillage, fertilization, and time and rate of planting.

Of 23 trials conducted, nine trials were concerned with screening and evaluation of the 150 introduced and local pasture and forage crop varieties; six trials demonstrated pasture seeding methods, three trials dealt with fertilizer, and two trials examined grazing management systems. The results of this work to date have provided invaluable information, with actual production comparisons to be made available after Stage II field sowings.

Preliminary indications are that silage production from imported *Sorghum bicolor* and *Sudax* can double dry matter production over that of locally grown crops. Results on silted soil at the nearby State Seed Farm indicate the production potential on comparable land at the Center to be 15,000 kg to 20,000 kg DM/ha. Fertilizer trials demonstrate significant increases. Native hay production, for example, can be doubled. Direct sowing of better pasture species in drained marshland indicated that *Medicago sativa* and *Lotus corniculatus* could be established using strategic mowing and herbicide to control resident vegetation. Brassicas and beets also showed some promise on the black soils.

A complete line of modern John Deere farm machinery is in place for the mechanization of the Center. Spare parts, a workshop, and storage facilities including a fuel depot make this an efficient operational unit.

Infrastructure of the Center was basically completed. A headquarters unit was constructed with offices and living accommodations for local staff and VIPs with a shopping center for the Brigade. A total of six bunker silos was constructed at several locations, along with several sheep-sheds, a dairy barn, hay storage, and other supporting buildings. A spray race and several feeding centers for cattle also were completed. A topographic map was prepared showing field designations, roads, electric lines, canals, and general land use. Roads and power lines were established to serve the three work-teams of the Demonstration Center.

In the livestock section, total numbers were reduced from 11,000 in 1979 to 9,700 in 1981, or a 12% decrease. As a result of improved nutrition, the mortality rate dropped from 8.5% in 1980 to 2.4% in 1982. Cattle and sheep numbers increased, but goat and horse numbers decreased. About 98% of the cattle were crossbred (Mongolian Red) and all sheep were crossed with fine wool rams.

Per capita income in 1979 was 91 yuan, 108 yuan in 1980, and 142 yuan in 1981 (1 yuan = $.68).

Breeding females in the cattle herd averaged 34.8%; the regional average is 40%.

Two study tour groups visited Canada and the U.S.A. in 1980 and 1981. These groups, including the project director, senior members of the Demonstration Center staff, and several officials of the General Bureau of Animal Husbandry, toured cattle and sheep operations ranging from extensive range cow-calf units in the Nebraska Sandhills to intensive feeding in northern Colorado and the Platte Valley. The tour groups all visited agribusinesses, educational institutions, and experiment stations.

Three members of the livestock production section spent 3 mo on a work-study program on ranches in Nebraska. Several groups of mechanics were hosted for mechanical training by the John Deere Company, and a group from the agricultural section visited Australia and New Zealand.

The core group of consultants conducted seminars and lectures while on their periodic visits to the project. They contributed informational materials from other countries including reference books, journals, and other publications.

NEW SYSTEM INTRODUCED

In late 1981, the government introduced the "Individual Responsibility System" for agriculture, which placed more emphasis on individual production efforts. Consequently, the Demonstration Center reorganized on the principles of state-collective joint management, unified accounting, two-level structure, setting prices, issuing shares, and distributing income according to labor expended. All property was turned over to the Center and shares were issued at 100 yuan each to the original owners based on the value of their assets. Total number of shares issued was 17,723, of which the state owns 9,216 (52%), collective 8,490 (47.9%), and individuals 15 (0.1%).

A management committee of seven was selected and eight special groups were set up: livestock, 3; feed production, 1; cattle finishing, 1; and dune fixation, 1. A production plan was drawn up that "established contracts and named rewards."

The following listings (table 1) show some of the production results obtained in 1981 and 1982 and progress compared to the previous year.

TABLE 1. PRODUCTION RESULTS FOR WONGNUTE DEMONSTRATION CENTER, 1982

Year	Gross increase	Breeding rate	Mortality	Sold for slaughter
	%	%	%	%
Livestock				
1981	18.4	55.7	2.7	25.8
1982	22.1	60.9	2.4	23.7

Total livestock numbers

Year	Cattle	Sheep	Horses	Goats
1981	3,024	5,101	741	3,900 (1980)
1982	3,385	6,057	384	1,955

Results of two groups of cattle fed in confinement on silage and grain were:

Group	Number	Days on feed	Daily ration: Silage	Daily ration: Grain	Gain/day
1	29	29	20 kg	2.5 kg	1.0 kg
2	53	55	20 kg	2.5 kg	1.1 kg

Two pilot herds of cattle totaling 157 head were set up for optimum feeding and management. Synchronized estrus was used on 890 of 970 breeding cows with good results; 53 ewes had winter lambs in 1981; 53 lambs survived and averaged 35.4 kg at weaning, 9.8 kg heavier than those born in the spring.

DISCUSSION AND SUMMARY

To date, crop yields have been low. The natural soils of the area are of low fertility and there is an alkali-salinity problem when water is applied. Drought was a most important factor in the first 2 yr, but this is probably a common occurrence. Irrigation has been slow for a number of reasons. Reliable power has not been available for the sprinkler area. Sand canals have been breached and the silt load has resulted in extensive canal silting. High winds can block the canals with drifting sand. Deposition of silt on the land to obtain the desired depth of 15 cm to 20 cm is also a slow process. Without the addition of silt, crop production seems likely to remain low. The problems of alkalinity, salinity, and drainage are most important. With special attention, however, the swamp areas can be immediately productive, with little extra effort.

Dune stabilization is a slow and extremely costly process, with no clear results to date. Indications are that several kinds of treatments should be applied. If the natural vegetation is destroyed, rehabilitation of the area will require massive efforts, accompanied by a long period of total exclusion of grazing.

Increases in livestock production, although significant, generally have been small, especially when measured on a per head basis. Inadequate nutrition is an extremely important factor and sophisticated techniques such as artificial insemination or estrus synchronization cannot overcome its effects in the breeding herd.

Animal numbers must be adjusted to the available feed supply. Fewer animals will produce more and reproduction and survival rates will remain low.

Changes in herd structure, stratification of the production cycle, specialization of function, and sale of younger animals can make significant contributions. Basic proven management and organizational practices are the foundation. Lack of timely financing is a major constraint.

The trial plots and demonstrations are of utmost importance to the total value of the Center as well as to the development of the country as a whole. Such work may be the major accomplishment of the entire project. Identification of valuable plants, their characteristics, and place in livestock production is a major step forward, as is development of systematic records.

The new state-collective joint management system should be evaluated regularly to make the Center an efficient organization.

Great strides have been made in the physical setup, organization of the ranch, acquisition of modern equipment, training and experience of staff, and establishment of trial plots and demonstrations.

Positive results have accrued with the feed production, irrigation and drainage efforts, and -- to a lesser extent -- with the improvement of livestock production.

Taking into account the experience gained and recorded, a meaningful assessment can be made of potential of the Center as weighed against the risks and high expenses involved. Dune stabilization, for example, has been very slow, expensive, but absolutely necessary, for the country because of its real threat to the future. The abuse and destruction of the natural vegetation in a fragile environment also may prove to be a major constraint to any development in the area.

The drainage problem of the swampy land has far-reaching implications, as do the irrigation by sprinkler and the accompanying salinity/alkalinity problems that could turn native grassland into a sterile desert.

Traditional social and political problems restrict the cost effectiveness of all animal production systems because they prevent stock reduction to a level commensurate with feed supply essential to an acceptable level of stock production. Malnutrition prevents the expression of the inherent ability of the present animals to produce, negates much of the superior potential of all breeding and management practices, and possibly produces animals less adapted to the rigorous natural environment.

The change in administration functions presents additional challenges. The operation of a tightly controlled Brigade varies greatly from that of a more liberal, individual-responsibility style. It requires much more understanding of the overall situation, the long-range implications of the Demonstration Center, and the actual responsibilities of the participants.

On one of the study tour visits to a Nebraska Sandhills ranch, a senior staff member of the Wongnute Center and a typical old-time Mongolian herdsman remarked, "These ranchers have the correct philosophy. Their land produces grass. They graze it with only enough animals to eat the proper amount, always having some to spare. Their animals are healthy and fat. All we look at is the number of animals we have -- the more the better. We are wrong. We need to pay attention to the grass we have first."

The progress that the Demonstration Center has made is very evident and is highly commendable. The hard work and dedication of the staff, the government, UNDP, and the Brigade has been outstanding. The future of the Center and its contribution to the improvement of livestock production hangs in a very delicate balance between the fragile environment of the area and the understanding and will of all the people involved to commit themselves totally to the goals of the project. The interrelationships of a multitude of factors, some of which cannot be controlled, are extremely complicated.

A build-up of livestock numbers before the problems of feed production are solved could result in lower total production and negate the progress to date. The effects on the grazing lands would be especially severe. The upgrading in the breeding program must be accompanied by better nutrition and careful management, or decreases in production will result. Production per unit of livestock, land, labor, and machine provides a meaningful and useful measure of progress or direction in which the program is going.

Government assistance in the way of financing and support must continue. One of the main contributions of the Center to livestock development is information; however, information that is not thoroughly tested and documented could have negative results in the long-term.

The information gained and the lessons learned at the Wongnute Banner Center will have widespread application, not only in China but throughout the world where there are similar conditions. The Demonstration Center can add materially to the store of knowledge urgently needed to advance the science of livestock and range production.

ACKNOWLEDGMENTS

The author acknowledges, with thanks, extensive conversations with William Hinton, machinery consultant, and Na Sen, Academy of Animal Husbandry, Hohhot, Inner Mongolia, PRC.

REFERENCES

FAO. 1980. FAO consultation. General report. Pilot Demonstration Center for Intensive Pasture, Fodder, and Livestock Production (Wongnute Banner). CPR 79/001.

FAO. 1982. Fourth FAO advisory panel mission. General Report. Pilot Demonstration Center for Intensive Pasture, Fodder, and Livestock Production (Wongnute Banner). CPR 79/001.

Harris, P. S. 1981. Consultant report of pasture and fodder crops. FAO. CPR 79/001.

Harris, P. S. 1983. Consultant report of pasture and fodder crops. FAO. CRP 79/001.

Kernick, M. D. 1980. First consultant report of pasture improvement and utilization. Pilot Demonstration Center for Intensive Pasture, Fodder, and Livestock Production (Wongnute Banner). CRP 79/001.

Wongnute Ranch Demonstration Center. 1981. Report on present construction of demonstration center.

Wongnute Ranch Demonstration Center. 1982. Making a start at transforming the backward situation in livestock production.

37

WHAT'S AHEAD IN HAY AND FORAGE PRODUCTION, STORAGE, AND HANDLING SYSTEMS

Allen R. Rider

One of the most important management decisions made by livestock producers is the selection of a forage system to match the current operation and planned goals. A wide variety of hay and forage equipment is available for today's producer. It is essential that the producer analyze each of the alternatives to determine if a specific system will provide the proper quality of feedstuff and match the labor and capital resources available on the farm or ranch.

This paper presents an overview of the major alternatives available for purchase by today's livestock producers. Data presented for each system will include field capacity and corresponding labor requirement, costs to own and operate machines, and management techniques to maximize performance of the various systems. Specific systems presented are the traditional baling system with hand hauling, automatic-bale-wagon system, round-bale system, large rectangular-bale system, loose-hay-stack-wagon system, and haylage system.

Field capacity and labor requirement data presented in tables 1 and 2 are based on university surveys made primarily on beef cow-calf and commercial hay operations in Oklahoma and Nebraska. It must be noted that throughput capacity of machines is significantly greater than the presented values; however, the data presented are based on actual field capacity, which includes lost time in the field (routine maintenance, repair of minor malfunctions, shutdown for breaks, etc.). Capacity on most operations will be within 25% of the tabulated values. The range is due to several factors, including field conditions, crop yield and variety, operator skill, and machine condition.

Cost data presented in tables 3 and 4 are based on techniques available from the American Society of Agricultural Engineers Machinery Management Standard for estimating fixed and variable costs for hay and forage equipment. Fixed costs include depreciation, interest on investment, taxes, insurance, and shelter. Variable costs include all operating expenses, which are labor, fuel, lubrication, repairs, and other supplies required to

TABLE 1. CAPACITY AND LABOR REQUIREMENT FOR HAY EQUIPMENT ON BEEF CATTLE RANCHES

	Capacity,[a] tons/hr	Labor requirement, man-hours/ton
Bale handlers (SP - 3 men)	5.0	.60
Bale mover (roll) - tractor mounted		
Haul - 500 lb	1.0	1.00
Feed - 500 lb	1.3	.77
Haul - 800 lb	1.7	.59
Feed - 800 lb	2.0	.50
Haul - 1,200 lb	2.5	.40
Feed - 1,200 lb	3.0	.33
Haul - 1,800 lb	4.2	.24
Feed - 1,800 lb	5.0	.20
Bale mover (roll - 1,200) - truck towed		
5 mi one-way haul	6.5	.15
10 mi one-way haul	3.7	.27
Bale wagon (PTO - 83 bale) - automatic	6.9	.14
Bale wagon (PTO - 104 bale) - automatic	8.0	.13
Bale wagon (SP - 160 bale) - automatic	13.7	.07
Baler (big rectangular - 1,750 lb)	12.2	.08
Baler (medium duty) - 14" x 18"	6.4	.16
Baler (heavy duty) - standard 14" x 18"	8.0	.13
Baler (round - 500 lb)	5.0	.20
Baler (round - 800 lb)	5.2	.19
Baler (round - 1,200 lb)	7.5	.13
Baler (round - 1,800 lb)	9.2	.11
Feed bales (with pickup)	1.0	1.00
Hand haul bales (3 men)	2.7	1.11
Mower (7 ft)	2.9	.34
Mower-conditioner (PTO - 9 1/4 ft)	4.1	.24
Mower-conditioner (PTO - 12 ft)	4.9	.20
Mower-conditioner (SP - 12 ft)	5.3	.19
Rake (single - 9 ft)	5.2	.19
Rake (tandem - 18 ft)	10.0	.10
Stack wagon (loose hay - 3 ton)	6.5	.15
Stack wagon (loose hay - 6 ton)	7.5	.13
Stack mover (loose hay - 3 ton)	5.0	.40
Stack mover (loose hay - 6 ton) - farm	10.0	.10
Stack mover (loose hay - 8 ton) - farm	15.0	.07
Stack mover (loose hay - 6 ton) - highway		
5 mi one-way haul	15.0	.07
10 mi one-way haul	7.0	.14
25 mi one-way haul	3.3	.30
Windrower (SP - 14 ft)	6.1	.16
Windrower (SP - 16 ft)	8.5	.12

Source: Oklahoma State University Cooperative Service Extension Publications.

[a] Capacities presented are based on typical field operations but may vary by 25% depending on actual field conditions, crop yield and variety, operator skill, machine condition, etc.

TABLE 2. CAPACITY AND LABOR REQUIREMENTS FOR HAYLAGE EQUIPMENT[a]

	50% moisture content		Dry hay equivalent	
	Capacity, tons/hr	Labor requirement, man-hours/ton	Capacity tons/hr	Labor requirement man-hours/ton
Feed haylage (belt feeder)	6.0	.17	3.8	.26
Feed haylage (chuck wagon or mixer-feeder wagon)	12.0	.08	19.2	.05
Forage blower	20.0	.05	12.5	.08
Forage harvester (PTO - small) w/pickup	7.7	.13	4.8	.21
Forage harvester (PTO - medium) w/pickup	10.0	.10	6.3	.16
Forage harvester (PTO - large) w/pickup	15.0	.07	9.4	.11
Forage harvester (SP) w/pickup	20.0	.05	12.5	.08
Haul haylage (w/forage wagon - 1 mi)	6.7	.15	4.2	.24
Haul haylage (w/truck - 5 mi)	16.0	.06	10.0	.10
Unload trench silo (w/tractor and front-end loader)	16.0	.06	10.0	.10
Unload trench silo (w/tractor-mounted unloader)	12.0	.08	19.2	.05

Source: Oklahoma State University Cooperative Extension Service Publications.
[a] Capacities presented are based on typical field operations but may vary by 25% depending on actual field conditions, crop yield and variety, operator skill, machine conditions, etc.

properly operate the machine, such as twine. Also, all pull-type implements received an appropriate charge for a properly sized tractor to power the unit. Costs are based on the most recent cost data available from hay and forage equipment manufacturers at the time of preparing the report. Labor was calculated at $6.00/hr. and diesel fuel was estimated at $1.00/gal.

TRADITIONAL BALING SYSTEM

The most common haying system for livestock producers continues to be the traditional baling system. The complement of operations for a traditional baling system varies for individual farms but generally consists of mowing-conditioning, raking, baling, and hand hauling of bales from the field to storage. Several mechanical attachments, such as bale throwers and multibale accumulators, have been used to reduce the labor requirement for removing bales from the field. But, most beef cattle ranches continue to rely on hauling bales by hand from the field to the storage site. This major labor consumer for the traditional baling system normally consists of a truck with attached bale loader and three-man hauling crew. Loading and hauling bales from the field to the storage area is often a custom operation. Custom rates vary considerably but typically range from $.35 to $.45/bale with an average rate of $.40/bale. The traditional baling system cycle is completed with the feeding operation, which is usually accomplished by hauling bales on a pickup truck from the storage site to the feeding area.

Some producers hire custom hay crews to windrow, rake, bale, and haul the bales to storage. The management decision to use custom operators, instead of purchasing and operating hay equipment, is usually based on the amount of hay harvested annually. The least-cost traditional baling system is to use custom crews for annual production of less than 200 tons and to own equipment for annual production exceeding that amount.

The custom rates for windrowing, raking, and baling range from $20/ton to $30/ton with an average rate of $25/ton. This custom cost combined with custom hauling to storage ($12.00/ton) and feeding bales ($10.00/ton) yields a cost for harvesting and handling bales of nearly $50/ton. Even though hay growers with low annual production levels cannot own and operate equipment for less, the convenience and capability to bale hay under proper conditions for high quality offsets the cost disadvantage. Custom operators are frequently busy when hay needs to be harvested and hauled to storage. This can result in lower-quality hay that is often unacceptable to the livestock producer.

Labor requirements for harvesting, handling, and feeding hay with the traditional baling system are nearly

TABLE 3. TOTAL COST PER TON OF OWNING AND OPERATING HAY EQUIPMENT[a]

Machine	Size	New cost, $	Tons per year 100	200	400	600	800	1,000
			---Cost per ton---					
Bale mover (round) Three-point hitch mounted - 2 moves								
	500 lb	500	$21.22	$21.13	$20.98	$20.93	$20.91	$20.89
	800 lb	500	14.23	14.09	14.06	14.04	14.02	14.00
	1,200 lb	500	11.53	11.30	11.24	11.22	11.21	11.20
	1,800 lb	500	8.57	8.30	8.21	8.19	8.18	8.17
Baler (big rectangular)	1,800 lb	65,300	76.77	40.78	22.92	17.04	14.15	12.44
Baler (round)	500 lb	9,600	14.96	9.73	7.17	6.36	6.25	6.13
Baler (round)	800 lb	12,200	17.73	11.07	7.81	6.77	6.43	6.33
Baler (round)	1,200 lb	14,700	19.97	11.90	7.92	6.63	6.00	5.63
Baler (round)	1,800 lb	22,500	28.33	15.96	9.84	7.84	6.86	6.28
Baler (traditional) medium-duty	14" x 18"	10,500	15.73	9.97	7.14	6.23	5.79	5.78
Baler (traditional) heavy-duty	14" x 18"	12,400	17.73	8.67	7.56	6.46	5.93	5.62
Bale wagon (PTO)	83 bales	12,300	16.13	9.39	6.07	5.00	4.48	4.30
Bale wagon (PTO)	104 bales	20,700	25.25	13.89	8.27	6.45	5.56	5.04
Bale wagon (SP)	160 bales	64,800	73.40	37.67	19.91	14.05	11.17	9.46
Mower-conditioner (PTO)	9 ft	9,700	14.93	9.67	7.11	6.48	6.40	6.33
Mower-conditioner (PTO)	12 ft	13,900	19.04	11.47	7.77	6.59	6.43	6.34
Rake (single)	9 ft	3,500	6.34	4.43	3.50	3.20	3.10	3.10
Rake (tandem)	18 ft	8,200	10.42	5.91	3.68	2.94	2.58	2.37
Stack mover (loose hay)	3 ton	4,500	8.81	6.36	5.16	4.78	4.73	4.69
Stack mover (loose hay)	6 ton	11,500	15.05	8.72	5.59	4.56	4.06	3.76
Stack wagon (loose hay)	3 ton	19,700	25.16	14.37	9.07	7.35	6.52	6.25
Stack wagon (loose hay)	6 ton	31,000	37.87	20.86	12.48	9.75	8.42	7.98
Windrower (SP)	16 ft	45,100	51.97	27.18	14.94	10.94	8.99	7.85

[a] All costs were calculated with the most recent cost data available at the time of writing this report. New costs were sourced from hay and forage equipment manufacturers. Ownership and operating costs were calculated using techniques from the American Society of Agricultural Engineers Machinery Management Standard.

TABLE 4. TOTAL COST PER TON OF OWNING AND OPERATING FORAGE EQUIPMENT FOR HAYLAGE[a]

			Tons per year of dry hay equivalent					
			100	200	400	600	800	1,000
			Tons of haylage					
Machine	Size	New cost, $	160	320	640	960	1,280	1,600
			Cost per ton					
Forage blower		3,900	$ 6.32	$ 4.17	$ 3.10	$ 2.75	$2.58	$2.48
Forage harvester	Small PTO	13,300	19.71	12.47	8.94	7.81	7.66	7.50
Forage harvester	Medium PTO	17,100	24.03	14.67	10.07	8.58	8.18	7.86
Forage harvester	Large PTO	30,000	37.80	21.22	13.01	10.33	9.01	8.24

[a] All costs were calculated with the most recent cost data available at the time of writing this report. New costs were sourced from hay and forage equipment manufacturers. Ownership and operating costs were calculated using techniques from the American Society of Agricultural Engineers Machinery Management Standard.

2.5 man-hours/ton. This labor requirement consists of approximately .4, 1.1, and 1.0 man-hours/ton for field operations, hauling, and feeding, respectively. With this high labor requirement, expanded hay acreage is normally limited and excellent management is needed to harvest the highest-quality hay. However, this system still provides a broadly accepted haying system that generally results in high-quality hay when fed. Also, the ability to control daily-feed-consumption rate with traditional bales is preferred by many livestock producers.

AUTOMATIC-BALE-WAGON SYSTEM

An automatic bale wagon with one operator can replace a two- or three-man bale hauling crew. Both self-propelled and tractor-powered bale wagons are available. Primary models available are 83-bale PTO, 104-bale PTO, and 160-bale self-propelled. A bale wagon is normally complemented with a windrower, tandem rakes, and a heavy-duty baler. A heavy-duty baler is needed to handle double windrows and produce a dense, uniform-length bale that can be handled and stacked with an automatic bale wagon.

A cost comparison between the traditional baling system and automatic-bale-wagon system reveals break-even annual tonnages of 350, 425, and 875 for the 83-bale PTO, 104-bale PTO, and 160-bale self-propelled bale wagons, respectively. If a producer stores 1-1/3 tons of hay per cow-calf unit, the break-even size of operation would range from a low of 250 head for the 83-bale PTO system to over 650 for the 160-bale self-propelled system. Feeding costs are assumed to be equal for the automatic-bale-wagon system and traditional baling system. Even though some automatic bale wagons have the capability to retrieve stacks and unload bales one at a time, feeding is normally accomplished with a pickup truck.

Due to the large break-even annual tonnages, bale wagons are not matched to small livestock operations. However, replacing a bale-hauling crew with an automatic bale wagon can reduce labor requirements substantially and provide an opportunity to generate additional income by custom hauling and stacking. Labor requirements for the 83-bale PTO, 104-bale PTO, and 160-bale self-propelled automatic-bale-wagon systems are 1.57, 1.55, and 1.49 man-hours/ton, respectively for harveting through feeding. High capacity hauling and stacking also protect the hay quality by minimizing weather damage in the field. This high quality can be preserved when properly stored; however, special barn dimensions are needed for inside unloading. To produce the highest quality hay and operate an automatic-bale-wagon system at maximum efficiency, the field manager must be capable of coordinating all harvesting and handling operations. In addition, a bale-wagon operator must be highly skilled (and generally performs most efficiently when paid an incentive wage).

ROUND-BALE SYSTEM

Round-bale systems are generally classified by the maximum-sized bale produced. Typical bale weights produced within the major categories are 500, 800, 1,200, and 1,800 lb.

Actual bale weights can be either lighter or heavier, depending primarily on baler dimensions, hay species, moisture content, and operator skill. The most common size for a round-bale system is within the 1,200-lb category; however, smaller round bales continue to gain in popularity because of their compatibility with smaller tractors and improved on-road transportation with two-bale-wide loading on trucks or wagons. Also, a baler that produces round bales nearly 8 ft long provides a hay package with excellent on-road transportation characteristics.

A typical complement of equipment for round-bale systems is a mower-conditioner or windrower, rake, baler, and bale mover. A rake is used primarily to combine windrows, which increases baler production and aids in forming uniformly shaped bales. A bale mover is needed to move bales from field to storage and from storage to feeding. Bale movers attached to a tractor's three-point hitch suit the smaller operations that do not move hay more than 2 mi or 3 mi from field to feeding. However, either tractor-mounted bale handlers for loading wagons and trucks or self-loading multibale movers are needed to haul bales long distances on roads and highways. Multibale movers towed by small trucks are also used to move bales from the field or storage site over public roads.

Economic evaluation of round-bale systems compared to the traditional baling system reveals that approximate break-even annual tonnage is 150 tons for the 500-lb category round bales, 125 ton for 800 lb, 125 ton for 1,200 lb, and 175 tons for the 1,800 lb round bales. These evaluations are based on a typical operation of mowing, conditioning, raking, baling, and two moves of round bales with a tractor-mounted, three-point hitch with attached mover. Two movers are required to relocate bales from the field into a storage area and then move bales from the storage site to a feeding location. Even though the 500-lb baler has a lower investment cost, operating costs for moving the smaller bales offset the lower ownership costs. However, the small round baler can be ideally suited to operations that have limited capital, small tractors, or need to load and transport bales on public roadways. Break-even tonnage of the 800 lb and 1,200 lb round-bale system compared to the traditional baling system are nearly equivalent; however, it must be recognized that as usage increases over 125 tons/yr, ownership and operating costs of the larger round baler becomes more economically viable because of the significant labor savings with a 50% heavier bale. The round baler capable of consistently producing 1,800 lb bales is primarily matched to large hay and beef cattle operations. In fact, this large baler becomes the most economical round-bale system with annual production levels in excess of 500 tons, assuming that bales are handled twice with a tractor-mounted single-bale mover.

One-man haying is feasible with a round-bale system. The bales are not only produced mechanically but also handled mechanically from the field to feeding. With this system, hand labor is eliminated; however, a mechanical device is always needed to move the bales. The total labor requirement for harvesting, raking, baling, moving, and feeding are approximately 2.3, 1.7, 1.2, and .9 man-hours/ton for the 500-, 800-, 1,200-, and 1,800-lb round-bale systems, respectively. The small 500-lb bale does not provide a labor savings compared to the traditional baling system unless a multibale mover is utilized. However, the convenience of feeding small round bales is preferred by many cow-calf operators.

Round-bale systems are well-suited to beef cow-calf operations not only because of cost and labor savings but also because of the reduced need for permanent structures and temporary covers to protect the hay during storage. The rounded shape, normally confined with twine, provides a weather-resistant surface that eliminates the need for covered storage. However, the rounded shape of round bales prevents efficient stacking; therefore, covered storage and long-distance hauling are expensive.

LARGE RECTANGULAR-BALE SYSTEM

A couple of manufacturers in North America are marketing balers that produce large rectangular bales. One unit produces a bale 38 in. x 48 in. x 8 ft, which typically weighs 1,300 lb to 1,400 lb in high-quality alfalfa. The most commonly used baler produces a bale 51 in. x 48 in. x 8 ft, which generally weighs approximately 1,750 lb. Bale length can be varied within the 6 1/2 ft to 8 1/2 ft range. Compared to traditional bales, these large bales tied with twine are well-suited for long-distance transportation due to the limited labor needed to load and unload trucks. Several commercial hay operations and large beef cattle ranches that must transport hay long distances have opted for this system.

A typical complement of equipment for this system is a self-propelled windrower, tandem rakes, baler, accumulator for the baler, and grapple fork attached to a tractor-mounted front-end loader. The baler has high capacity -- in excess of 20 tons/hr with good field conditions and heavy windrows. The accumulator that can carry up to three bales simultaneously is effective in grouping large bales in the field for more efficient removal. The mounted grapple fork is used to lift and load bales onto vehicles for transportation to the storage site or market.

Break-even annual hay production needed to economically justify ownership of a large rectangular-bale system is 900 tons compared with the traditional baling system. Based on the 1-1/3 ton/beef cow-calf unit, a beef cattle operation with over 650 head would be required to economically justify ownership.

The large rectangular-bale system has many features desired by both hay and livestock producers. The total labor requirement from field to feeding is approximately .6 man-hours/ton. Due to the large bale size, mechanical feeding, which also minimizes labor demands, is well-matched for large livestock operations. The feeding system will vary with the individual operation, but a self-loading feeder with hay processor is efficient for delivering hay directly into a bunk or on the ground. Also, the large rectangular bales can be fed whole (free choice) or broken into portions for hand feeding.

However, for the large rectangular-bale system, the break-even annual tonnage is too high for many hay and livestock producers. Despite the rectangular shape and good density that are conducive to stacking, protective storage either inside of a barn or with an upper surface cover is desirable to minimize spoilage due to weathering.

LOOSE-HAY-STACK-WAGON SYSTEM

Loose-hay-stack-wagon systems are normally categorized by weight of stack produced under typical field conditions. A wide range of wagon sizes and models ranging from 1 ton to 8 tons is available. Currently, the most common sizes used are 3-ton and 6-ton models. Also, the most popular units provide mechanical compaction of the stack by lowering a hydraulically controlled canopy several times during stack formation. This process increases stack density and forms a sloping thatched canopy to aid in minimizing weathering during storage.

A mower-conditioner or windrower and stack mover with stack wagons are normally used. Rakes are seldom used but may be needed to combine light windrows or turn hay during damp weather.

Compared to the traditional baling system, the break-even annual production levels for the 3-ton and 6-ton loose-hay-stack-wagon systems are 175 tons and 250 tons, respectively. These break-even tonnages correspond with cow-calf herd sizes of 125 head and 250 head, respectively. Loose hay stack feeding costs are assumed to be equivalent to the costs for moving a stack 1 mile. Feeding a stack free-choice or with feeder panels is acceptable to most livestock producers.

Loose hay stacks are not well-suited for commercial hay growers unless the market is nearby. Most loose hay stacks moved on roads are transported with a truck-mounted stack mover or truck and stack mover equipped with gooseneck hitch. Hauling capacity decreases rapidly with hauling distance. Due to low capacity, the costs become excessive when hauling loose hay stacks long distances. In addition, significant hay losses have been observed during on-road transportation. However, transportation losses can be minimized by securing the stack with ropes or a net.

Utilization of a loose-hay-stack-wagon system is encouraged by harvesting cost reductions, labor savings, and opportunity for a one-man haying system. Typical labor requirements for harvesting through on-farm feeding for the 3-ton and 6-ton loose-hay-stack-wagon systems are approximately .6 and .4 man-hours/ton, respectively.

Proper operation and management of a loose-hay-stack-wagon system will result in good-quality hay at feeding. One management technique for loose-hay-stack-wagon systems is stacking hay at a moisture content that is 3% to 4% higher than acceptable for baling. This technique reduces field losses while allowing hay to cure in the stack, generally without overheating or spoiling. Also, operator experience will improve the stack shape and density, which affects the storage characteristics. Loose hay stacks stored 3 mo or longer maintain more of the original quality if the stacks have well-shaped, thatched, sloping canopies to aid in shedding water. Wind and water damages are minimized with long hay interlocked within stacks. Also higher-density stacks generally feed more efficiently because the cattle tend to clean up loose hay on the ground instead of pulling more hay from a tight stack. If these operational and management techniques are followed, a stack wagon system can provide useful roughage from a variety of materials including high-quality alfalfa, grasses, forage sorghums, and crop residues, including small-grain straw and stover from corn and milo.

STORING AND FEEDING LARGE HAY PACKAGES

One key to success with large hay packages is proper management during storing and feeding. The cost advantage of large hay packages can be lost if the hay receives excessive spoilage during storage or is wasted during feeding.

The concept of storing hay outside without any protection is unthinkable to many hay and livestock producers. However, with good management practices, outside storage of hay has proven practical and cost effective, particularly in areas with annual rainfall under 40 in./yr. Based on public-service research and on-farm observations, application of the following management techniques will assist in minimizing losses during storage.

- Relocate package from the field to a storage site as soon as practical.
- Either select a storage site with excellent drainage characteristics or prepare the site by building a crushed stone base to hold packages off of damp terrain.
- Align packages to minimize area of exposure from prevalent winds and storms.
- Allow sufficient space between curved surfaces of large hay packages to facilitate shedding of water.
- Use the proper twine on bales to maintain shape throughout the storage season.
- Develop more than one storage site to prevent total loss of hay in case of fire.
- Select storage sites near the feeding area if practical.

With the acceptance of large hay-package systems, methods to feed hay efficiently are needed. The alternatives available to livestock producers are: 1) uncontrolled free-choice feeding, 2) controlled free-choice feeding, or 3) daily feeding. Research at several universities clearly indicated that uncontrolled free-choice feeding by placing packages on the ground without any protection from trampling resulted in excessive wastage, particularly if the feeding area had poor drainage. Controlled free-choice feeding by surrounding the hay packages with panels reduced wastage dramatically and still minimized labor during feeding. By feeding on a well-drained site and allowing access only to the amount of hay that can be consumed in one week, feeding losses are considered acceptable by most livestock producers. The daily feeding method by either manually or mechanically distributing hay from large packages results in the least waste. However, the reduction in feeding waste is often offset by the additional costs for labor and equipment to distribute hay on a daily basis. Each producer must decide which feeding method is best suited to the operation; but it is clearly evident that controlled feeding either with panels or daily distribution will reduce wastage significantly below uncontrolled, free-assess feeding of large hay packages.

HAYLAGE

Harvesting hay crops by chopping wilted windrows when the moisture content is near 50% and ensiling it is another alternative for providing a high-quality feedstuff for ruminant animals. In addition to high quality, a well-managed haylage system will minimize field losses, allow hay harvesting to occur during weather unsuitable for haymaking, and provide a completely mechanical system from field to feeding. Haylage is a labor-efficient system requiring approximately .5 man-hours/ton of dry hay equivalent, which is only 20% of the labor required for the traditional baling system with hand hauling and feeding.

Suitability of the haylage system, however, is often restricted due to the investment required for storage. An economic comparison of the haylage system with hay systems is not possible without defining the type of storage. Generally, a haylage system is not economically viable without harvesting at least 500 tons of chopped forage annually. A haylage system is usually best matched to an operation that also chops corn or forage sorghum for silage. With multiple use of the forage harvester and storage facilities, the cost per ton for owning and operating the system becomes economically feasible. To provide additional information for analysis of haylage equipment, tables 2 and 4 present field capacities and costs for owning and operating forage harvesters and related equipment, respectively.

FUTURE TRENDS

Over the past decades, hay and forage-harvesting systems have concentrated on improving quality of the product at feeding, increasing capacity, and reducing labor requirements. Future designs for hay and forage equipment will continue to concentrate on improving quality of feedstuffs and improving labor efficiency. Instead of a continuing increase in the size of harvesting equipment, most of the improved labor efficiency will be the result of better reliability. Electronic monitoring on farm equipment to assist the operator in maximizing throughput and minimizing downtime will become more available as electronic components continue to decrease in price and improve in reliability while operating in the harsh agricultural environment.

Several public service institutions are conducting research on equipment and techniques to improve hay and forage quality and production. Some of the most promising areas today are 1) chemical applications to enhance drying and preserve quality, 2) extraction of juice protein from alfalfa to facilitate more efficient feeding, 3) and biogenetics to alter plant characteristics for improved quality and production. The farm equipment industry monitors these types of public-service research results and also conducts research to ensure that the equipment of tomorrow matches the needs of forage and livestock producers.

38

UTILIZATION OF CROP RESIDUES BY BEEF CATTLE

Gary Conley

MANAGEMENT OF CROP RESIDUES BY GRAIN PRODUCERS

Crop residues have become a disposal problem for U.S. cash-grain farmers as crop specialization and farm acreages have increased. While most areas of the world use their crop residues, the recent trend in the U.S. has been away from farming systems that use crop rotation with livestock and toward intensive cultivation of a single crop. This has allowed a more efficient use of large-scale mechanization. At the same time, crop residues present a problem to seeding operations and create an environment conducive to increased insect and weed problems. Farmers are solving these problems several ways. Some farmers, using specialized equipment, are increasing their use of herbicides and insecticides while planting in high residue conditions; others are burning the residue to clear the land and kill insects and weed seeds. Maintaining residues increases both equipment and pesticide costs but does improve soil tilth and fertility. Burning reduces costs of cultivation and pesticides but can increase fertilizer costs; it damages soil tilth and eventually causes wind erosion and reduces water intake.

The grain producer may use ruminants to dispose of the crop residues. The residue may be harvested and stored for later feeding or may be grazed directly by livestock. Disposing of residues through grazing or by feeding them to ruminants will require additional management and an investment in both livestock and equipment. But, this system will increase cash flow from the operation while eliminating the machine and labor costs of the direct residue disposal. Soil fertility will be enhanced if the manure is returned to the soil. Also, available labor can be better utilized if livestock operations are timed to the best advantage of crop residues.

THE POTENTIAL FOR BEEF PRODUCTION FROM CROP RESIDUES

World population continues to increase while the area of cultivated land remains relatively the same. Population is estimated to increase 1.6% annually while cultivated land will only increase .2% annually through the year 2000. Currently there are 1.25 people per acre of cropland in the world, but it is estimated that there will be 1.65 people per acre in the year 2000, which means that each 100 acres of cropland will then need to provide food and fiber for an additional 40 people. A larger proportion of this cultivated land will be needed to produce food directly for human consumption rather than grain and forage for livestock consumption. This competition for resources will reduce the grazing area available to cattle, especially that which is now planted to high-quality forages such as alfalfa. Beef producers will then be forced to either reduce their herds or develop new sources of feedstuffs.

Crop residues available from corn, sorghum, and wheat in the U.S. have been estimated to exceed 200 million MT per year. Digestibility of these residues vary from 40% to 60%, depending on species, variety, season, fertility, region, etc. Assuming an average of 50% digestibility, beef cows will need to consume 60 lb per day of crop residues to obtain the energy necessary to raise a calf and rebreed. Thus, a cow will consume approximately 10 MT per year of crop residue. The potential for beef production, using residues, could increase beef cow numbers by 20 million head in the U.S. and forestall a shortage of animal protein as population pressure increases.

LIVESTOCK SUITABILITY FOR CONVERTING CROP RESIDUES

Ruminants can convert crop residues to protein and energy but some species are more efficient than others. Both sheep and cattle can convert residues to meat, milk, and offspring. Dairy cattle require a high energy intake for maximum and efficient milk production and so are not suited for a crop-residue program. Beef calves and yearling steers (a stocker program) need adequate concentrates or high-quality forages in order to grow and add muscle at an efficient rate. This leaves the adult beef cow and ewe as the animals best suited to low-cost conversion of residue to animal protein. The cow can consume adequate quantities of stover, stalks, straw, and other residues to maintain herself and to produce a calf once each year. If the management and labor skills are available, the ewe also can convert low-cost residues to animal protein.

TYPE OF BEEF COWS THAT ARE EFFICIENT CONVERTERS OF RESIDUES

Although cows must be able to consume more than 60 lb/day of crop residue, not all breeds or individuals within a breed are able to convert this large volume of material. Because of inadequate rumen volume, rate of rumen passage, or other factors, cows of certain genetic types may not be able to convert crop residues into adequate energy to provide milk for a calf and sufficient body condition to rebreed. Our experience indicates that cows with large frames and large rumen capacities are more likely to succeed as efficient processors of crop residue. The Holstein Freisian has developed a large rumen and frame, probably as a correlated response to selection for volume milk production, which causes the purebred to lose condition on a low-energy or crop-residue diet.

I would suggest a three- to five-breed composite animal be formed to optimize the conversion of crop residue. Each breed should be selected for one special trait along with its general adaptation to the environment. An example would be 1) Brown Swiss, selected for large rumen capacity, gentle disposition, excellent muscling; 2) Angus -- early maturity, easy breeder; 3) Hereford -- hardiness, good scavenger; 4) Holstein -- excellent udders, rapid growth, large rumen. When combined these breeds could increase variation in all traits and increase the effectiveness of selection in the specialized new environment of crop-residue conversion.

FARM SUITABILITY AND MANAGEMENT

Cash-grain farming is a high-risk enterprise. Weather and prices both create volatility of income. Two acres of crop residue from corn or sorghum will, on the average, support a cow and wean a calf. Gross income will be increased by the value of the calves weaned, which will reduce the volatility of the grain farm's cash flow. The majority of U.S. farm units are a combination of soil types and topography. Areas of wasteland and native grass unsuited to cropping are interspersed among the cropland areas. This noncultivated land provides ideal areas to maintain the cow herd during the calving season and during field operations. The capital requirements are not great when adding a cow herd to an existing grain farm. The cows are the major cost with other capital outlays, including hay harvesting equipment, corrals, working chutes, and fencing. These costs are very low when compared to the purchase price of grazing land in the U.S.

Each grain crop provides a residue that needs different management to maximize production. Applied research has shown that in most cases processing, packaging, and storing crop residue are not economically feasible. The value per ton is low and the cost of baling, stacking, chopping, or cubing is not balanced by an increase in value. Ammonia will add crude protein but the energy level in crop residue is inadequate for efficient conversion to digestible protein. NaOH alone, or in combination with CaOH, has increased digestibility by 10% to 15%. The chemical treatment has not been generally cost effective because the equipment and labor needed to treat the residues are too expensive to be included in the budget for the average cow herd operation.

Our management system has proved practical and cost effective. We use electric fences and divide the stubble-stalk fields into units that will place 10 cows per acre. The first 5 days the cows receive no supplement because they are recovering large quantities of grain (dropped ears or heads). Beginning on the 6th day they receive 1.56 lb of grain and .86 lb of 41% natural protein per day until that portion of the field is completely grazed out (usually 16 to 20 days). This process is repeated all winter. If the fields become muddy or the stalks are covered by snow, the cows are moved to an area of permanent grass and are fed baled or stacked stalks and supplement until the fields are dry and open again. Immediately following harvest, we bale 1 1/2 t of stalks for each cow, bull, and yearling calf that we plan to keep through the winter.

The crop residue from wheat, oats, or barley is managed much the same as that from corn or sorghum. We fence small areas and move the cattle frequently. Rather than use grain-protein supplement, we graze sudan, volunteer sorghum, and grass simultaneously with the small-grain stubble.

MANAGEMENT PROCEDURES FOR THE COW HERD

Timing of the cropping operations to benefit the cow-calf program is the most important crop-residue management consideration. Normal cropping operations of planting, spraying, and harvesting should be maintained. The calving season should be restricted to no more than 60 days and should be timed to occur in a slack season. Another consideration should be the availability of feed just prior and during the breeding season.

At Conley Farms the cows are put on fresh stalk fields beginning the first of November and are gaining weight and ready to conceive during the December 15 to January 15 breeding season. The other breeding season is June 15 to July 15. These cows have had access to the spring growth of volunteer wheat and grasses and are in a weight-gaining condition. These breeding seasons create calving seasons during the period just prior to the two harvests of wheat and corn or grain sorghum. This timing allows us to use existing labor to monitor the cows during calving. Our cows graze wheat stubble and volunteer sorghum or volunteer wheat in fallow fields during the summer months. In other programs it may be necessary to plant some area to hybrid sudan for summer grazing. This, however, depends on the acres of grass and straw available. If one acre of grass, sudan, or green volunteer is available for each two acres of straw (stubble), the cows can be maintained without energy or protein supplement. If grass or a green sudan-type crop is not available, it will be necessary to supplement the straw with 1.5 lb of grain and .90 lb of 41% natural protein.

The health program should be organized to fit each individual operation. In our seedstock operation the calves are identified (tatoo in left ear and ear tag with corresponding number) at birth. When the calves are 4 wk to 10 wk of age, they will need to be vaccinated, castrated and, if necessary, dehorned. While the cattle are in the corral, they are sprayed to remove parasites (lice, ticks, and other parasites). We vaccinate the calves to protect against *Clostridium chauvei*, *septicum*, *novyi*, and *sordelli*. Possibly a bactrin-toxoid should be given to stimulate the antibody system at this time. Two weeks prior to weaning, the calves should be revaccinated to protect against *Leptospira pomona*, *Haemophilus somnus*, IBR, PI-3, and the previous vaccines repeated. When the calves are weaned, they should be revaccinated against IBR, PI-3, and *Haemophilus somnus*.

The bull, steer, and heifer calves should be separated at weaning. All calves should receive a full feed of a moderate energy diet for at least 45 days and no longer than 60 days. This will prevent sickness, increase return (conversion is excellent at this stage), and allow accurate selection of replacements for rate of gain.

The spring calves, which are weaned in late October, may be grazed on fall-seeded small grains (wheat, oats, barley) following weaning. This grazing provides flexibility in the marketing of calves and low-cost grains. Fall calves will be weaned in late April and may be grazed on early spring grass or graze-out small grains.

PROFITABILITY FROM PRODUCING CALVES USING CROP RESIDUE

The quantity of supplement used in this system will average 270 lb of grain and 175 lb of 41% natural protein per cow. If we include bull feed and feed fed to postweaning calves, the quantity fed per calf will be 900 lb of grain with 360 lb of 41% natural protein. If we use average prices from 1983 and 1984, the supplement cost per calf would be $102.00. These calves averaged 635 lb after their 60-day feeding period, which gives a supplement cost of $.16/lb of calf. Approximately $.08/lb of calf is associated with the supplement fed to the breeding herd to produce a 500-lb weaning calf.

Labor costs are less than $.08/lb of calf produced because only those hours spent with the cattle, fencing fields, hauling feed, etc., are charged against the cow herd. If the crop residue had been shredded and incorporated in the soil, it would have been a labor expense to the farmer. If credit were given for this, the labor cost approaches $.05/lb of calf.

Health and death loss costs (which include vaccination, medication, calving assistance, and death) should be less than an average of $.025/lb.

The capital costs include such items as cows, hay equipment, corrals, vehicles, and spray equipment. Interest costs, repairs, and depreciation total $.18/lb to produce a 630 lb feeder calf. Assuming $.72/lb sale price for the 500 lb calf or $.68/lb for the 630 lb feeder calf, the profit per acre would be $88.75 and $91.35, respectively. These profits represent an excellent return to labor and capital from the utilization of crop residue by beef cattle.

REFERENCES

Bartle, S. J., J. R. Males, and R. L. Preston. 1984. Effect of energy intake on the postpartum interval in beef cows and the adequacy of the cow's milk production for calf growth. J. Anim. Sci. 58:1068.

Davis, M. E., J. J. Rutledge, L. V. Cundiff, and E. R. Hauser. Life cycle efficiency of beef production: IV. Cow efficiency ratios for progeny slaughtered. J. Anim. Sci. 58:1119.

Dickerson, G. E., M. Kunzi, L. V. Cundiff, R. M. Koch, V. H. Arthaud, and K. E. Gregory. 1974. Selection criteria for efficient beef production. J. Anim. Sci. 39:659.

Koch, R. M. and J. W. Algeo. 1983. The beef cattle industry: changes and challenges. J. Anim. Sci. 57(Suppl.2):28.

U.S. Meat Animal Research Center. 1974-1978. Germ plasm evaluation program. Reports no. 1-6. Report to the Agricultural Research Service, U.S. Department of Agriculture.

U.S. Meat Animal Research Center. 1982. Beef research program progress. Report no. 1. Agricultural Reviews and Manuals -- Agricultural Research Service, U.S. Department of Agriculture.

39

FORAGE HEIGHT AND MASS IN RELATION TO GRAZING MANAGEMENT

Iain A. Wright

INTRODUCTION

The aim of grazing management is to convert pasture resources to animal product in the most efficient way possible. This requires an understanding of the way in which animals respond to different sward conditions and the way in which the sward responds to grazing by animals. Although pastures vary enormously, this paper is concerned with the continuous grazing of sown pastures dominated by perennial ryegrass *(Lolium perenne)* with small proportions (10% to 30%) of other grasses, mainly annual meadowgrass *(Poa annua)* and white clover (0% to 10%) *(Trifolium repens)*. Different grass species and different climatic conditions from those in the United Kingdom may respond differently but the general principles should still apply.

MEASUREMENT OF GRASS HEIGHT

Grass height in this paper refers to the average height of the sward surface above ground level, measured by a specially designed instrument (Bircham, 1981). It consists of a graduated metal rod along which slides an outer sleeve with a rectangular perspex tongue 1 cm x 2 cm. The rod is placed upright on the ground and the outer sleeve moved down until the perspex tongue strikes the surface of the sward. The height can then be read directly off the graduated scale. Normally 40 to 50 such readings would be taken at random in a field and the mean sward surface height calculated. Other devices have been developed to measure grass height, but a simple ruler is probably adequate for many purposes.

SWARD HEIGHT — HERBAGE MASS RELATIONSHIPS

In general, the height of the surface of the sward is a good indicator of the amount of herbage. However, the density of the sward does affect the relationship. Figure 1 shows the relationship between sward height and herbage mass for two swards grazed by cattle. One was a close, dense sward while the other was a more open, less-dense sward. In general for the range of sward densities encountered in the U.K., open swards have to be 1 cm to 2 cm higher to achieve the same herbage mass as a close, dense sward.

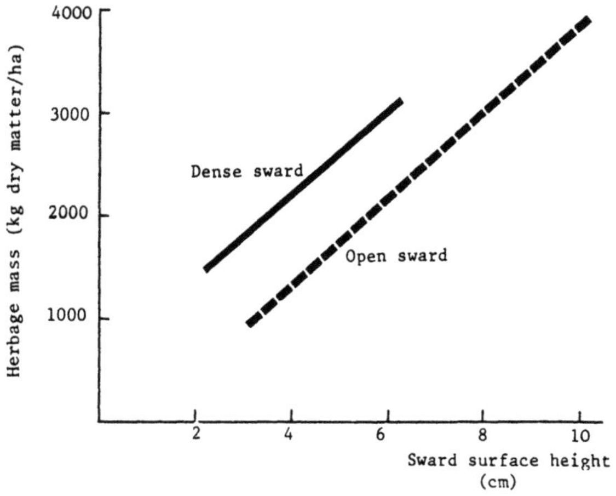

Figure 1. Examples of the relationship between sward surface height and herbage mass.

HERBAGE PRODUCTION IN RELATION TO HEIGHT AND MASS

Not all the herbage that is grown is available to stock for consumption. A grass plant is made up of a series of individual units called "tillers." During the growing season each tiller continually produces new leaves, and if these are not eaten, they eventually die and decay. Normally each tiller will have 3 to 5 leaves at any one time, each leaf a different age. The production of new tillers and new leaves constitutes "growth." The dying of old leaves is called "senescence." So the amount of new green material produced is the difference between growth and senescence and is termed "net production." Thus: -

NET PRODUCTION = GROWTH MINUS (-) SENESCENCE

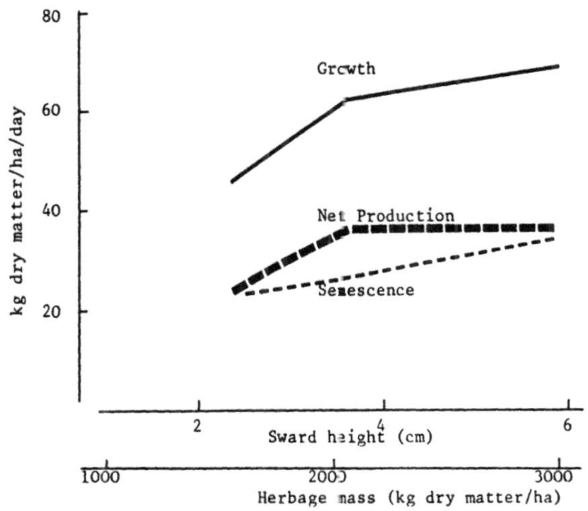

Figure 2. The relationships between growth, senescence, and net production and sward height and herbage mass.

The height of the grass has an effect on growth, senescence, and net production. Figure 2 shows the effect of maintaining swards at different heights by continuously grazing them throughout the grazing season. In the experiment (Arosteguy, 1982), three different swards heights were maintained on different plots by varying the number of yearling cattle. On three occasions throughout the grazing season, measurements were made of growth, senescence, and net production.

At low herbage height of 2 cm to 3 cm (corresponding in this case to herbage masses of 1400 kg to 1800 kg dry matter/ha), growth is depressed because insufficient leaf does not make the best use of available sunlight. At higher sward heights, the rate of growth increases, but the rate of increase is less. However, taller swards lead to higher rates of senescence. The combined effects of growth and senescence result in very similar rates of net production in swards of 4 cm or higher. Thus, over a fairly wide range of heights, the production from pasture is relatively constant. This conclusion was also reached by Bircham and Hodgson (1983) in pastures continuously stocked by sheep.

THE EFFECT OF SWARD HEIGHT ON ANIMAL PERFORMANCE

Beef Cows and Their Calves

The condition of the sward has a large effect on the performance of grazing cattle. In one experiment, cows and their calves were continuously grazed on perennial ryegrass swards at two different heights. The experiment was divided into two periods of 8 wk each. During Period 1 from mid-May until mid-July, 16 Hereford x Friesian and Blue-Grey (White-bred Shorthorn x Galloway) cows, which had calved in November-December, and their Charolais-cross calves and 28 similar cows and their calves that had been born in March-April grazed the short and tall swards. At the end of Period 1 the older calves and their dams were removed and the 28 spring-calving cows and their calves remained on the experiment for Period 2, which lasted for 8 wk more. The sward heights and masses are given in figure 3. During the first part of Period 1 the swards were taller than intended, but for the remainder of the experiment the heights were as planned.

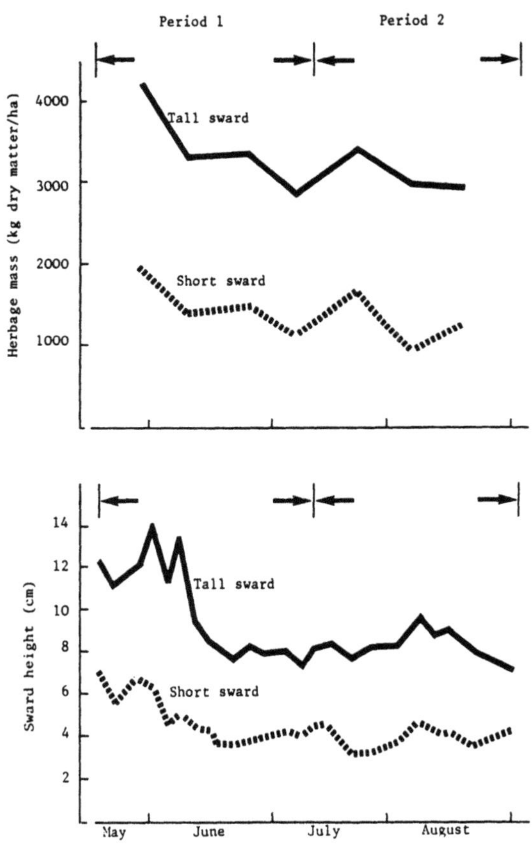

Figure 3. Sward height and herbage mass in the experiment with beef cows and calves.

The effects of sward height on animal performance are shown in table 1. The height of the sward had a large effect on all aspects of animal performance. In both periods, the taller sward resulted in greater live weight gains in the cows, and in the first period the response to sward height was the same for cows at both stages of lactation. Those spring-calving cows that were on the tall sward throughout the experiment gained between 1/1 and 1/4 of a condition score unit while those on the short sward gained less than 1/4. (Editor's note: Body condition score is explained in the paper entitled "The Relationships Between Body Codition, Nutrition, and Performance of Beef Cows.)" If full advantage is taken of the ability of beef cows to utilize their body reserves through the winter, it is vitally important that cows achieve a good level of body condition by the end of the summer.

TABLE 1. THE EFFECT OF SWARD HEIGHT ON PERFORMANCE OF BEEF COWS AND CALVES

	Short sward			Tall sward		
	Period 1		Period 2	Period 1		Period 2
	November/ December	March/ April	March/ April	November/ December	March/ April	March/ April
Cow live weight gain, kg/day	.53	.52	-.36	1.43	1.60	.52
Milk yield, kg/day	9.2	11.1	8.7	9.6	11.0	10.9
Calf live weight gain, kg/day	1.07	0.93	0.79	1.36	1.06	1.13

In Period 1, the milk yield of the cows was not affected by herbage height despite large differences in live weight gain, which shows the ability of cows to buffer milk production from the effects of nutrition. Only when herbage height got down to 4 cm and below was there a reduction in milk yield.

When the calves (born in March-April in Period 1) were 1 1/2 mo to 3 1/2 mo old, the effect of sward height was relatively small because at that age they eat a fairly small amount of herbage and are largely dependent on their mother's milk. However, the growth rates of older calves born in November-December in Period 1 and in March-April in Period 2) were affected to a much greater extent by sward height. In older calves herbage makes up a larger proportion of their total diet and so they are much more sensitive to variations in herbage height.

For cows and calves, herbage height should be kept about 8 cm so that cows make good live weight gains and put on reasonable amounts of body condition and the calves perform well.

Growing Beef Cattle

Sward height also has a large effect on the live weight gains of growing beef cattle. Charolais cross cattle of 14 mo to 18 mo of age were grazed on swards maintained at 4 cm to 5 cm or 6 cm to 8 cm in height. Table 2 shows that the shorter sward depressed live weight gain by nearly 40%.

TABLE 2. THE EFFECT OF SWARD HEIGHT ON PERFORMANCE OF GROWING BEEF CATTLE

	Sward height	
	4 to 5 cm	6 to 8 cm
Live weight gain, kg/day	.67	1.11

CONTROLLING HERBAGE HEIGHT

The previous discussion shows that herbage height and mass have a large effect on animal performance. For good animal performance on dense, continuously stocked swards, the sward surface height should be about 7 cm to 8 cm, and 1 cm to 2 cm higher on open swards. If it is lower, individual animal performance will suffer. If it is higher, it usually means that stocking rate is too low. When stocking rate is low, the cattle do not graze uniformly. Instead they graze some areas constantly, leaving the remainder ungrazed. These ungrazed areas go to seed and the digestibility and feeding value falls off dramatically. When grass growth declines later in the season and the cattle are forced to eat poor quality material, their performance drops. By maintaining the sward at a height of 7 cm to 8 cm, this problem does not arise. The sward stays fairly uniform with little seed head formation, the cattle graze highly digestible leaf throughout the grazing season, and their performance remains high throughout the summer.

But how can grass height be controlled when growth rates vary from year to year and within any one year? As herbage growth rates change, then the stocking rate must also change if grass height is to be maintained. Figure 4 shows the change in stocking rate that occurred in one experiment where the sward height was maintained at 6 cm to 8 cm. The stocking rate was changed as the grass-growth rate changed over the summer in the U.K. Theoretically the

stocking rate can be changed by varying the number of stock on a fixed area or by varying the area grazed by a fixed number of animals. In practice, varying the stock number on a given area is difficult, although there are situations when this is done. For example, if cattle reach slaughter condition at different times during the grazing season, the number can progressively be reduced, but it is often easier to vary the grazed area. In many systems of production, conserving fodder for winter feeding is a necessity. This need for
conservation can be incorporated into the grazing program. In practice only a part of the total area of grassland is grazed during the spring and early summmer, and this area can be adjusted to maintain grass height at the desired level. The remaining area is cut for silage or hay. After conservation, the grazed area can be increased to maintain sward height. This process can be repeated 2 or 3 times during the grazing season, depending on how many conservation cuts it is feasible to take. The latter part of the summer, after conservation is complete, the whole area of grassland is available for grazing.

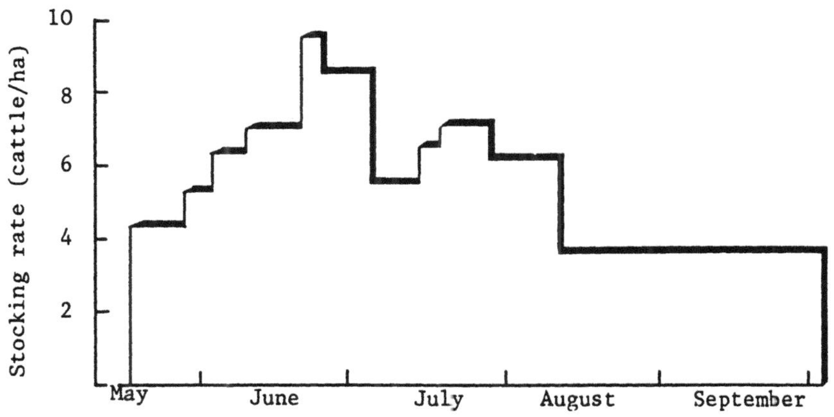

Figure 4. Stocking rate to maintain sward height at 6 cm to 8 cm.

The previous grazing system has led to very consistent animal performance from year to year. This leads to more predictable systems of management so that planning of marketing strategies becomes easier. Differences in grass-growth rates from year to year will result in slightly different proportions of grassland being cut and grazed; and so the quantity of winter fodder conserved will vary some. It is, however, relatively easy in the winter feeding program to adjust the quantity of fodder available with different levels of supplementary feeding.

REFERENCES

Arosteguy, J. C. 1982. The dynamics of herbage production and utilisation in swards grazed by cattle and sheep. Ph.D. Thesis. University of Edinburgh.

Bircham, J. S. 1981. Herbage growth and utilisation under continuous stocking management. Ph.D. Thesis. University of Edinburgh.

Bircham, J. S. and J. Hodgson. 1983. The influence of sward conditions on rates of herbage growth and senescence in mixed swards under continuous stocking management. Grass and Forage Science 38:323.

40

EFFICIENT FORAGE/LIVESTOCK PRODUCTION, NEW ZEALAND STYLE

Howard H. Meyer

INTRODUCTION

New Zealand Pastoral agriculture has made great strides in livestock productivity. This has not been so much by technological advances as by recognizing the need for concentrating on utilizing the available resources to maximize economic potential; in other words, doing what you do best.

Production emphasis is on output per unit input rather than on individual animal performance. A New Zealand dairy farmer is more interested in milk production per acre of land resource than in the herd average per cow. Beef and sheep farmers likewise concentrate on production per acre rather than average weaning or fleece weights.

The emphasis on productivity per acre reflects that New Zealand land costs are the biggest single cost item in livestock production, just as they usually are in the U.S. Return relative to other resource inputs is certainly not ignored; indeed, most production is gauged by marginal output response per unit of input whether those costs are labor, fertilizer, land development, or animal health remedies. This is not to say that every New Zealand farmer can predict the response to each unit of input, but that conscious assessment of investment alternatives play a large role in decision-making.

We have already noted that land is the largest single cost item in most sheep and cattle operations, but soil grows grass, not meat. Thus livestock production really consists of growing forage and using animals as the tools for converting it to usable products. In New Zealand, increasing livestock productivity has involved developing a better understanding of soil-plant-animal interactions, particularly in terms of optimizing forage production and utilization. This has included 1) choosing plants and animals that are suited to the environment and one another, 2) better forage and animal management practices, and 3) producing the end-product most suitable for the resources available.

FORAGE MANAGEMENT FOR OPTIMAL LIVESTOCK PRODUCTION

Since most livestock producers are already well aware of animal management principles, it may be worthwhile to first discuss several plant management principles and plant-animal interactions.

Get Grass Growing, Keep it Growing

All plants have annual growth patterns affected by season, temperature, and moisture availability. Apart form irrigation, there would seem to be little we can do to influence forage production, but in many cases this is not true. Left undisturbed, plants go through their regular cycle from germination through reproduction. Once a plant enters the reproductive phase, its growing or vegetative stage generally ceases, regardless of temperature and soil moisture conditions. Frequently forage production ceases not because soil moisture or temperture become limiting but because plants are allowed to go to seed. This is repeatedly demonstrated on lawns and roadsides where mown areas remain green while adjacent areas contain brown, rank growth. One area has been kept vegetative while the other has been allowed to quit producing.

Plants allowed to mature not only become poorer in quality for livestock use but may undergo a substantial depletion of root reserves as well. This causes slower recovery and regrowth when conditions again become favorable for growth.

The solution is to keep the forage growing, either by grazing management or mechanical removal of surplus growth, preferably as conserved feed. Allowing forage to mature before cutting it as hay often defeats the purpose of the exercise because plants have already ceased growing and the quality of the harvested product is poor. In areas where unpredictable weather makes early harvest of hay difficult, producers might consider silage as an alternate method of forage conservation. The fast rate and amount of regrowth after early forage conservation is often quite surprising to producers who have traditionally harvested hay later in the season.

Grazed pasture plants undergo regular and predictable patterns of recovery and growth. During the early period after grazing, the plant must utilize root reserves for initial regrowth. As leaf area increases, growth accelerates and root reserves are replenished from the enlarging upper plant. If growth is allowed to continue long enough, shading occurs, deterioration of lower leaves begins, and the net growth may become zero. Desirable plants, such as clover growing close to the ground surface, may suffer considerably and even be lost from the pasture.

Maximizing forage production by grazing means keeping the plants big enough to grow rapidly, but not allowing them to reach the stage where deterioration becomes too high.

Since livestock are our harvesting tools, the timing and severity of grazing will determine the utilization level of forage present and also greatly influence future pasture forage production.

Plant Resistance and Susceptibility to Damage

Pasture typically consists of a variety of plant species ranging from highly desirable to undesirable. Desirability is generally judged by the grazing preference of animals. Thus our most desirable pasture species are the ones most vulnerable to overgrazing and eventual disappearance unless forage grazing is properly managed.

All plants have some point in their annual cycle at which they are most susceptible or resistant to damage. In fact, this is often a major factor in determining when best to control brush or undersirable weeds by herbicide application. Likewise undesirable plant species can often be controlled or eliminated by strategic high-intensity grazing. Livestock can be very effective tools for changing pasture composition, either for better or worse, depending on the manner in which they are grazed. Given the chance to continually graze one area over a long period of time, livestock will repeatedly defoliate preferred plant species and ignore unpalatable ones. Is it any wonder that such pastures eventually consist of little but brush and unpalatable weeds?

Desirable plants should be grazed heavily when they will suffer least from defoliation. Undesirable plants should be grazed heavily when they will suffer the most. In this way, we can swing the balance in favor of the preferred plant species. Not every undesirable plant is controllable by grazing alone, but by using short-term high-density grazing and utilizing the different grazing and browsing patterns of different livestock species, many noxious weeds can be controlled or eliminated.

Weeds -- A Symptom, Not the Problem

The replacement of desirable forage plants by undesirable ones results in lower productivity. The typical approach to "solving" the problem is either herbicide treatment of "weeds" or pasture renovation by cultivation and re-establishment of desirable plant species. Success is usually short-lived and the exercise must be repeated.

Pasture deterioration is usually the result of pasture mismanagement. If we view the problem as one of absence of desirable species rather than presence of weeds, it is clear that the goal is to increase the proportion of desirable plants. Accordingly, we must correct the cause of the problem by promoting the desirable species. This solution may consist of many factors including improving soil fertility or alternating time and(or) intensity of grazing.

Forage Varieties to Suit the Need

It goes without saying that forage species adapted to the local environment will be more productive than unadapted ones. However, consider using combinations of species that work together. Choose a forage that is the most useful to the livestock being produced, and other forages to fit particular needs.

Producing legumes and grasses together. Inadequate soil nitrogen is a major factor limiting grass growth in most pastures. Nitrogen fertilizers are typically expensive to apply in adequate quantities. Legumes, on the other hand, are net producers of soil nitrogen, which becomes available to other plant species. While this fact is well-known to most farmers and ranchers, New Zealand farmers put this knowledge directly to use. They consider clovers as an essential part of the pasture and manage accordingly. Their management is based upon a fertilizer program to feed the clover, and they then count on the clover to produce nitrogen to feed the grass. In such pastures the bulk of livestock forage is produced by the grass varieties, but this is possible only because of clovers present in the pasture to produce nitrogen.

Choosing the forage most suited to stock being produced. Various grass and legume species are suited better to different types of livestock. A typical example is tall fescue, which produces a large amount of dry matter and grows well into the summer in dry regions where most perennial grasses become dormant. While this forage is excellent summer cattle feed, particularly for beef cows, it is quite unpalatable to sheep and is eaten only when alternative forges are very limited.

Another factor to consider is the ratio of legume to grass. Clovers tend to produce less dry matter than grass but the feed quality is higher. Thus a clover-dominant pasture may be ideal for fattening lambs but a grass-dominant pasture is more useful for growing cattle or maintaining ewes. In New Zealand the clover:grass ratio in the typical white clover-perennial ryegrass pastures is manipulated through the use of beef cattle and sheep. Cattle grazing over a period of time favors development of clover-dominant pastures while heavy sheep grazing reduces clover-incidence and results in grass-dominant pastures.

The number of forage species in a pasture will influence the manageability of forage production. While including a wide array of forage types in a seeding mixture may intuitively seem to be insurance against variation in weather, soil type, etc., it makes forage management very difficult. Just as it means that the best species for a given set of conditions is present, it also ensures that numerous less-desirable species are also present. The most manageable pastures consist predominantly of one grass and one legume. It is better to sow the different grass varieties in the areas of the farm best suited to their growing needs rather than spreading all varieties everywhere in a shotgun fashion.

Considering special forages. While permanent pastures with low annual maintenance costs are the foundation of livestock production, there is certainly a place for special forages in many production systems. These may be either annuals or perennials grown for a variety of reasons: 1) to produce conserved feed 2) to overcome green-feed shortages, or 3) to have a high-quality forage available for growing lambs or weaned calves. Costs per pound of dry matter may be higher for some of these forages, but returns may be sufficient to warrant their use.

LIVESTOCK AS TOOLS FOR HARVESTING AND MARKETING FORAGE

The primary role of sheep and cattle in New Zealand is to convert forage into meat, just as is done in the United States. Since forage is the resource to be converted, many choices are possible as to the most efficient type or types of livestock to be used, the optimum level of production, and the target product to be marketed.

Livestock as Forage Harvesters

New Zealand farmers manage livestock as forage harvesters to a much greater extent than do U.S. producers. Sheep and cattle are managed under so-called controlled grazing systems designed to optimize forage production and utilization.

Pasture subdivision is an essential component of controlled grazing systems. There is little advantage to increasing forage production unless the additional forage is utilized. The intent is to use animals to harvest what the farmer wants harvested at the time he wants it harvested. This may be particularly critical when livestock are to be used for controlling less-palatable forage species.

Subdivision is also essential in feed budgeting, the other primary ingredient of efficient forage management. At high stocking rates it is usually necessary to budget or ration the forage available to make it last throughout feed shortfall periods. Obviously this cannot be done without regard for the nutrient requirements of stock. Thus it is important that the stock used to clean up unpalatable forage are not at a production level requiring high nutritional inputs. Also the time of lowest feed availability should coincide with a period of relatively low nutritional requirements. Breeding ewes, for instance, are used best for forage grazing control following weaning. The typical winter feed shortage and rationing of limited feed supplies should coincide with early gestation before nutrient requirements for pregnancy rise.

Species, Breeds, and Combinations of Livestock

New Zealand farmers are typically referred to as sheep and beef farmers because few farms contain only one or the other species of livestock. Over the past 2 decades only once have beef cattle been as profitable as sheep to the producer. One beef cow in New Zealand is considered to have annual forage requirements equivalent to 3 1/3 ewes. Hence 10 cows are the equivalent to 35 ewes. From the standpoint of net return, farmers would appear to be better off if they ran fewer cows and more ewes. However, cows are a very important grazing-management tool used for removing surplus forage (in essence, lawnmowers) to keep pasture plants in the productive vegetative state. Beef cows typically constitute about 30% of the total stocking rate in ewe equivalents. This is considered to be about the minimum necessary for optimizing total farm productivity; therefore, cattle are essentially a "necessary evil" and a liability to the sheep operation.

Despite beef cattle being less profitable than sheep, New Zealand farmers still pay considerable attention to raise the type of cattle most profitable to their operation. Most beef cattle are Angus, Hereford, or crosses of the two breeds. European exotic breeds have been tried extensively and largely found to be unsuitable for the New Zealand production systems because of greater winter maintenance feed requirements, poor reproduction, and increased physical damage to wet pastures during winter.

Choice of sheep breeds is also determined by economic factors, both in terms of best utilization of forage resources and production of most salable products. Nearly all commercial ewes in New Zealand are dual-purpose whiteface ewes. These are comprised largely of Merino in dry areas and coarser, wooled breeds like the Romney in wetter areas. About 55% of the gross income from sheep is derived from wool sales, including wool and pelt allowances on slaughter animals.

The Romney breed, which once made up about 90% of the New Zealand national ewe flock, has made way for large numbers of Coopworths and Perendales, breeds developed in New Zealand from crossing Romneys with Border Leicesters and Cheviots, respectively. Such crosses, combined with strenuous selection for productivity, have resulted in more productive sheep under many conditions and have given Romney breeders considerable impetus to improve their breed.

Producing at the Optimum Level

Under most production systems, the maximum level of production is seldom the optimum in economic terms. Increasing the output by one extra pound of meat or wool is not of benefit if production cost is more than the price received.

New Zealand farmers have found intermediate levels of individual animal performance to be most profitable and so have concentrated on increasing stock rates to an optimum level, which is much higher now than was common a few years ago. Since both costs and returns go into determining profit, expensive technological approaches have not been widely put into practice; hence the minimization of forage conservation in favor of managing sheep for year-round grazing and the use of livestock rather than machinery and chemicals for forage management.

New Zealand farmers have selected cattle for increased growth rate and selected sheep for increased twinning. In both species, selection emphasis has been redirected as farmers saw the mature size of cows increase and as ewe twinning rates reached levels considered optimal for labor, forage, and management resources.

Replacement beef heifers are mated to calve for the first time at 3 yr of age, and most ewes first lamb at 2 yr old. While the unachieved potential reproduction may at first seem inefficient, producers have found that the extra feed necessary to achieve earlier reproduction can be more profitably invested in maintaining additional mature breeding animals. The same is unlikely to be true for U.S. producers with different resources and operating under considerably different input and output price structures.

Producing the Most Profitable Product

Just as maximum production is not always the same as optimal production, the product receiving the highest price does not always give the greatest net return. Slaughter lambs in New Zealand are killed at 65 lb to 70 lb live weight, not because they can not be grown heavier, but because that is the weight of market lamb most suited to the seasonal forage production systems. Choice of breeds and selection programs have been planned around producing adequately finished young lambs for slaughter directly from spring pasture. Unfortunately for New Zealand producers, the changing consumer demand in favor of leaner meat has made existing lamb carcasses fatter than desired, and overfat lamb is now the major single problem of the New Zealand sheep industry.

It is important for both producers and the meat industry to decide what their marketed product should be. Current U.S. trends toward slaughter of lambs at increasingly heavier weights is a serious disadvantage to producers whose resources are more suited to marketing adequately finished 100 lb lambs. Requiring additional weight on such lambs often means producing it at an economic loss and ending up with a product less desirable to consumers.

Farmers have always taken pride in heavy weaning weights or selling a high proportion of animals for slaughter rather than to feedlots. The prestige involved has often encouraged lower stocking rates to increase individual animal performance.

Individual U.S. producers must decide what they can produce most profitably, whether it be beef vs lamb, slaughter animals vs feeders, or breeding stock vs meat animals. Many farmers struggle each year to produce fat lambs for sale early in the season before market prices drop. Repeatedly they lamb before spring forage growth is available, experience increased lamb mortality due to inclement weather, feed purchased nutrition to achieve maximal lamb growth, and still have only a portion of lambs ready for slaughter before summer drouths arrive. Too often the attraction of "topping the market" leads producers away from doing what they do best with available resources.

SUMMARY

The basis for New Zealand's very efficient system of meat production is recognizing the resources available and using them to maximum advantage. Forage is grown and managed as a crop with livestock used as the harvesting and marketing tools. Adjusting the forage and livestock species to suit one another and managing the whole system rather than the individual components has allowed for dramatically increasing the output per resource unit invested.

41

UTILIZING IMPROVED PASTURES

Gerald G. Bryan

A majority of the U.S. beef production and much of the world's livestock production is from improved grasslands. Indeed, improved grasses and legumes contribute much to the total meat and milk produced throughout the world. Each day we are more aware of the need for practical and profitable information about improved pastures and their value and usage. The emphasis on pasture and forage in this series of seminars reinforces this point.

In our area of Oklahoma and Arkansas, we find that most of us are good agronomists. We can grow vast amounts of many different kinds of grasses and legumes, but we are not as efficient in our utilization of what we grow. Being able to grow large amounts of forage (4 t to 5 t/acre) on good pastures is wonderful, but it is not always profitable. To be profitable, forage must be efficiently utilized to produce meat. Utilization -- with production -- is the key. In the last few years, the extension and research people have realized that the stockmen need to know not only how to use the newest and best forage varieties and how much and what kind of fertilizer to use, but also how to use forage production within a total program that produces a profit. In other words, the stockman needs to know how to utilize what he produces.

Improved pastures in our area of Oklahoma and Arkansas are primarily Bermudagrass and tall fescue. These grasses account for approximately 90% of the improved pasture acreage in this area. Native grasses are confined to undeveloped areas, hay meadows, and forest lands. Legumes are used in combination with fescue and Bermudagrass to varying degrees. Annual pastures -- wheat, rye, annual ryegrass, sudan, and millet -- are used mainly for stocker grazing or hay production on tillable land. Even with many acres of improved pastures, there is still great potential for improved production per acre and per animal. Forage production can be dramatically improved by use of newer varieties and proper fertilization and weed control. However, the greatest improvement needed is in the area of utilization (table 1).

TABLE 1. ESTIMATED PRODUCTION OF PASTURES IN SOUTHEAST OKLAHOMA

Type of pasture	Pounds of forage/acre	Acres per cow
Timberland	35- 300	20-100+
Native and unimproved pastures	200- 100	15-25
Improved pastures	1,500- 4,000	2-4
Potential forage production of area	12,000-20,000	1 or less

Bermudagrass and tall fescue are most often used for cow-calf operations where they provide large amounts of forage over a relatively long grazing season. Most grazing is done continuously with a low level of management and low-input costs. The normal stocking rate is a cow per 3 to 4 acres or more. Recently, higher land costs and higher interest rates have forced producers who wanted to expand their operations to consider intensifying to increase incomes. To become profitable, intensifying requires not only increased cattle numbers per acre and more forage production per acre so that more pounds of beef are produced at the same or a lower cost, but also increased management inputs. A current demonstration at the Kerr Foundation addresses the question: What level of intensification is the most profitable? More important, is intensified cow-calf production profitable? Admittedly, there are times when cattle prices and input costs make less management more profitable; but generally, if beef can be produced within an acceptable price range, profits can be realized. To intensify an operation requires a total forage-livestock management program: pastures and fertility program must produce adequate forage; cattle must be genetically capable of profitably converting forage into beef; and management must be upgraded to a level sufficient to manage the operation.

The Kerr Foundation intensification project has utilized four levels of intensification in an attempt to determine the level of intensification that is best for our area. In four projects with variable levels of management, land areas, and stocking rates, 25 head each of Brahman x Angus cows were mated to percentage Simmental bulls for February, March, and April calves. Combination pastures of tall fescue and Bermudagrass were crossfenced for controlled rotation and were fertilized and managed to provide year-long forage either standing or baled.

Treatments were:

- Project 1-A. A high-intensive system with a stocking rate of one cow per acre on 25 acres. This system has utilized a combination bermuda-fescue-clover pasture, with fertility rates designed to produce required forage. This project is designed to explore the potential of area soils and environment.

- Project 1-B. A medium-intensity system with a stocking rate of one cow per 2 acres, pastured on 50 acres of Bermuda-fescue-clover pasture using a medium fertility program. Use of deferred fescue and rotation grazing is stressed.
- Project 1-C. A low intensity system of 75 acres using Bermudagrass and fescue pastures stocked at a rate of one cow per 3 acres. A protein supplement is fed to supplement hay and a low fertility program is used.
- Project 1-D. The same treatment as C except that calves are carried through the stocker phase and are sold at 600 lb to 800 lb during the following spring. Fifteen acres of fescue-ryegrass-red clover is used in this system for the above purpose.

Weaning weights (table 2) are vital economic criteria that can be affected by level of intensification. With heavier stocking rates, the cow herd tends to utilize more of the available forage, with less forage available for the calves. The lightest calves are produced when the stocking rate is one cow per acre, and the heavier calves are produced at a stocking rate of one cow per 2 to 3 acres. Maximum utilization of available forage occurred with the one cow/acre stocking rate, but "maximum" is not always the most profitable, or even the most productive, with all measurements considered.

TABLE 2. CALF WEANING WEIGHTS FROM VARIOUS STOCKING DENSITIES

	1982 adjusted wt		1983 adjusted wt	
Project	(254 day)	Top group, %	(240 day)	Top group, %
1-A (cow/acre)	565 (-32)	94.6	543 (-72)	88.0
1-B (cow/2 acres)	585 (-12)	98.0	615	100.0
1-C (cow/3 acres)	597	100.0	600 (-15)	97.5
1-D (cow/3 acres w/stockers)	540 (-57)	90.5	584 (-31)	95.0

Table 3 lists the economics of production. Utilization of the total pasture resource was greatest in the one cow per acre treatment, because more pounds of beef were produced per acre. This level produced the highest return per acre. However, the cost per pound of beef was highest with the lowest return per cow. Optimum utilization appears to be a stocking rate of one cow/2 acres. In times of exceptionally low cattle prices, the lighter stocking rates may be the most profitable if land ownership costs are not considered. If land costs are considered, the higher levels of intensification are the most profitable use of the pasture resource.

TABLE 3. ECONOMIC SUMMARY OF INTENSIFIED COW-CALF STUDY

Total investment	Cow/acre 1-A $30,165		Cow/2 acres 1-B $44,525		Cow/3 acres 1-C $57,195		Cow/3 acres w/stockers 1-D $57,280	
	1982	1983	1982	1983	1982	1983	1982	1983
Pounds of beef/acre	441.7	521.4	245.6	258.2	160.6	183.4	164.4	211.7
Pounds of beef/cow	460.1	521.4	511.7	561.2	502.8	600.1	472.7	584.3
	---------- $ ----------							
Income								
Calf sales	6957.20	7535.29	7595.42	7565.92	6901.59	7667.80	7030.41	8279.28
Stockers (net)	--	--	--	--	--	--	--	1851.72
Hay	700.00	1075.00	1440.00	3675.00	900.00	1050.00	1400.00	1800.00
Total income	7637.20	8610.29	9035.42	11240.92	7801.59	8717.80	8430.41	11931.00
Total operating expenses	5531.42	6134.31	5299.66	6352.65	3226.32	4189.68	3807.36	6611.00
Return to land, labor, capital, management								
Total	2105.78	2475.98	3735.76	4888.27	4575.27	4528.12	4623.05	5320.01
Per head	87.74	99.04	155.66	212.53	198.92	205.82	192.63	212.80
Per acre	84.23	99.04	74.72	97.77	63.55	62.89	67.00	77.10
% return on investment	7.10	8.21	8.30	10.98	7.93	7.92	8.14	9.29
Breakeven operating costs								
Per head	230.48	245.37	220.82	276.20	140.27	190.44	138.64	264.44
Per acre	221.26	245.37	105.99	127.05	44.81	58.19	55.18	95.81
Per pound of beef	.50	.47	.43	.49	.28	.32	.34	.45

Fescue is a very versatile forage. Its cool-season growth lends compatibility to the warm-season growth of Bermudagrass for a well-rounded forage program. Because of fescue's cold tolerance, it can be used to provide green grazing in the winter in a conventional grazing program or deferred and grazed on a part-time basis as a protein supplement. Cow overwintering costs can be dramatically reduced by deferring grazing during the fall growing season until forage is needed in winter and then grazing 3 to 4 head of cows/acre on alternate days for 2 hr (table 4).

TABLE 4. TYPICAL SOUTHEAST OKLAHOMA SPRING CALVING OPERATION

Typical feeding program	Cost/cow	Winter forage program	
4 lb of 20% CSM[a] cubes/day for 80 days @ $200/t	$32.00	One acre of fescue should provide enough protein for 4 dry cows when grazing is deferred until December and then grazed 2 hr every other day.	
5 lb of 20% CSM cubes/day for 40 days @ $200/t	20.00		
Hay - 30 bales @ $1.50	45.00	Cost	Cost/cow
Total winter feed cost, 20% CSM cube-and-hay program	$97.00	Fertilizer 75-50-50 = $38/acre	$ 9.50
		Emergency feed	7.50
		Hay - 20 bales @ $1.50	30.00
		Total winter feed costs, forage program	$46.75

[a] CSM -- cottonseed meal.

Some of the most dramatic improvements in utilization of improved perennial forages has been with light weight, 400 lb to 500 lb stocker cattle. Bermudagrass and fescue both have traditionally been cow grasses. Stockers have generally failed to perform well on these grasses due to their lower feed quality. Heavy stocking rates to utilize Bermudagrass in its early and more nutritious growth stages has produced outstanding results in several studies in several states.

Fescue, however, has been another story. It has consistently produced poor animal gains. An endophytic fungus was recently discovered that apparently reduces animal gains and causes other problems associated with fescue grazing. New stands of fungus-free fescue produce dramatically improved animal gains (table 5). Economic utilization of existing infected (55% to 85% infection level) stands of fescue is a real problem, especially in areas difficult to re-establish. How can an existing infected (55% to 85% infection level) stand be utilized economically?

A trial at the Kerr Foundation using deferred stockpiled fescue, grazed by yearling steers with various levels of supplemental feeding (table 6A), indicated that protein supplementation (41% cottonseed meal) during the winter grazing period could be highly efficient and profitable. Feeding later in the spring after fescue greenup was not efficient or profitable. However, as reflected in tables 6B and 6C, supplemental feeding was not economical at any time when fescue forage was of sufficient quality to produce satisfactory gains. Energy supplementation was always less efficient and less profitable than was protein supplementation. The data in tables 6B and 6C also show the value of protein supplementation when fed with high-quality fescue forage. The value of maintaining a short, high-quality forage, e.g., as shown by Oliver (1978) with Bermudagrass grazed to obtain maximum use of the forage, is also evident with the fescue in these data. The calves on this trial were rotated on closely grazed fescue, whereas calves in tables 6A and 6B were grazed continuously on the same pasture. The high-quality forage in this trial seemed to offset the effects of the fungus in fescue plants.

TABLE 5. GAINS OF STEERS GRAZING *A. COENOPHIALUM* -INFECTED AND NONINFECTED TALL FESCUE PASTURES, MARION JUNCTION, ALABAMA 1978-81

Tall fescue pasture	Animal days/acre	Beef gain, lb/acre	Avg daily gain, lb	Gain/steer, lb
Noninfected	240	426	1.83	318
Fungus-infected	311	301	1.10	185

The use of implants and other growth promotants is a well-utilized practice in the livestock industry. Monensin has been used in feedlots for several years. Recently it was approved for use by stockers grazing pastures. The use of monensin to improve gains of steers grazing Bermudagrass was tested in the summer of 1981 and was shown to be a profitable aid (table 7). In a similar trial in 1982, monensin improved gains even more markedly because the cattle were gaining at a much faster rate than in the 1981 trial.

Proper use of improved forages can be much more profitable than is conventional, low-management use of improved grasses or native forages. Utilization of high-quality and productive forages can produce beef with a low break-even point that can improve profits for the producer. A grazing study on ladino clover (table 8) in the mid-1970s illustrates this point very well. Stocker steers and heifers were grazed on ladino clover, with poloxalene used to control bloat. The cattle gained at very acceptable rates to produce a high return per head and per acre based on both $28.00/cwt cattle in 1975 and $60.00/cwt cattle in 1982.

Even with today's current price squeeze, good management of improved forages can be very profitable. Improved grasses and legumes have a high-yield potential; therefore, it is up to the good manager to take advantage of this potential by using livestock in an effective, efficient manner to produce a profit.

TABLE 6A. SUPPLEMENTAL FEEDING OF STEERS GRAZING HEAVY STOCKPILED FESCUE (1982-1982)

	Control	(Daily) 1 lb of 41% CSM	(Daily) 3 lb of 14%	(Daily) 1 lb of 41% CSM
Daily gain (12-23 to 3-9)[a], lb	.14	1.09	.74	1.12
Ratio feed to gain	-0-	1.027	4.582	1.512
Feed cost/lb of gain, $	-0-	.11	.38	.17
ADG (3-9 to 5-4), lb	1.11	.59	.81	1.0
Total gain (12-23 to 5-4), lb	71	116	101	140
ADG, lb	.54	.88	.77	1.07
Additional ADG over control		.34	.23	.53
Ratio feed to gain		2.862	12.376	2.814
Feed cost/lb of gain, $		.33	1.03	.32

[a] Numbers in () indicate dates of feeding period.

TABLE 6B. SUPPLEMENTAL FEEDING OF STEERS GRAZING GROWING FESCUE (1982-1983)

Daily gain (11-30 to 3-15), lb	.45	.60	.64
Ratio feed to gain		6.56	18.53
Feed cost/lb of gain, $		.74	1.54

TABLE 6C. ROTATION AND SUPPLEMENTAL FEEDING OF STEERS GRAZING FESCUE (1983-1984)

Daily gain (12-15 to 3-13), lb	1.49	1.61
Ratio feed to gain		8.46
Feed cost/lb of gain, $		1.02

TABLE 7. EFFECT OF MONENSIN ON GAINS OF STEERS GRAZING BERMUDAGRASS IN LATE SUMMER (1981)

	Type of supplement	
	Monensin block	Salt and mineral
Gain in 56 days, lb	14.56	8.19
ADG, lb	.300	.146
Cost of supplement/hd, $	1.66	.68
Monensin/lb of gain	15.4	

TABLE 8. THE PERFORMANCE OF ANGUS AND CROSSBRED HEIFERS AND STEERS GRAZING LADINO CLOVER AT KERR FOUNDATION, POTEAU, OKLAHOMA

	1975	1976	Average 1975-76
Days on pasture	189	183	186
No. hd/acre, avg	3.3	2.8	3.1
Wt on pasture, avg[b]	472	568	520
Wt gain/hd, avg[b]	308	273	291
Avg daily gain/hd	1.63	1.51	1.57
Live weight gain/acre, lb	1,025	809	917
Grazing days/acre	637	537	587
Net income/acre, $	385.33	286.33	335.83

[a]Bloat controlled by feeding 1 lb of 12.5% protein pellet containing 5 g of poloxalene per pound daily in 1975 and 1.25 lb in 1976.
[b]Cattle weighed on and off pasture after a 24 hr shrink period.

REFERENCES

Bacon, C. W., J. K. Porter, J. D. Robbins, and E. S. Luttrell. 1977. Epichloe typhina from toxic tall fescue grasses. Appl. Environ. Microbiol. 35:576.

Bryan, G. G. 1984. Kerr Foundation data, reported in Kerr Foundation progress report. Kerr Foundation, Box 588, Poteau, Oklahoma 74953

Hoveland, C. S., R. R. Harris, E. E. Thomas, E. M. Clark, J. A. McGuire, J. T. Eason, and M. E. Ruf. 1981. Tall fescue with ladino clover or birdsfoot trefoil as pasture for steers in northern Alabama. Alabama Agri. Exp. Sta. Bull. No. 530.

McMurphy, W. E., G. W. Horn, and J. O. O'Conner. 1981. Gains of stocker cattle on midland and hardie Bermudagrass pastures. A 5-year summary. Okla. Agri. Exp. Sta. Misc. Pub. No. 112-127.

Oliver, W. M. 1978. Management of coastal Bermudagrass being grazed with stocker cattle. Proc. Forage-Livestock Conf. Intensive Forage Utilization. The Kerr Foundation, Poteau, OK.

42

PUTTING IT ALL TOGETHER: FORAGE AND CATTLE

P. D. Hatfield and Connie Hatfield

BACKGROUND

We make our living converting desert bunchgrass into beef. Some years the desert cooperates better than others, but always we are adjusting our cattle and ourselves to better fit the desert. Our ranch still contains the remnants of old abandoned homestead cabins -- testimony to those who failed trying to change a hostile environment to fit their needs. Our program is to adapt our cattle to what we have; not change what we have to fit a type of cattle that may be popular at the moment.

A more appropriate title for this talk would be "Manipulating Grass, Time of Calving and Breed of Cattle So Your Cows Will Perform on What Your Ranch Can Produce." The discussion will illustrate our philosophy on a cattle ranching operation: Produce the optimum forage from your land base, then harvest that crop with cattle. No other environment will be exactly like ours and most will be much different. However, the principles for matching cattle to forage are the same everywhere. Try to apply our philosophy to your ranch environment as the discussion progresses.

Each year on this ranch we have the use of 14,000 acres of soil, approximately 12 in. of moisture, below-freezing temperatures every month, and abundant sunshine. Our primary job is using these raw materials to produce year-round digestible forage in a manner that improves the environment for the future. This forage is then harvested with cattle. There is no farming, haying, or irrigating, and our cattle are on grass 12 mo of most years.

Production-related work on the ranch involves two major activities: 1) manipulating bunchgrass to produce forage with year-round value and 2) breeding and selecting for trouble-free, athletic cattle for which the nutrient requirements fit that grass.

Thanks mainly to the homesteaders, our ranch is divided into 23 units. These units support some plant growth most years from March through July; however, the majority of our grass crop for the year is grown in May and June. During May and June, grazing management is highly intensive. A major goal of that management is to prepare high-quality regrowth for later use during the dormant period.

During this critical period of fast growth, all cattle run together in one herd and move to fresh pasture every few days. A key consideration is to move them to fresh pasture soon enough to avoid eating regrowth from plants that were grazed on the first days of grazing. The decision as to which pasture to move to next is based on what type regrowth we can expect to produce, while at the same time meeting the physiological needs of the plant.

We have a general idea where our next moves will be, but routinely change the original grazing plan when weather dictates a move to a different unit that would benefit both the plants and the livestock.

During the fall, winter, and early spring, the cattle are divided into 3 or 4 herds. During the spring they may stay in the same unit for several weeks as they graze regrowth.

If we are to do the best possible job of converting our grass to beef with a cow-calf operation, figure 1 suggests that fertility is our top concern. For any cow to efficiently perform in our environment, she must calve every year in March or April. The very best of time for us to have a calf born is the last week of March. A cow that is calving at that time has had grass supplying most of her nutrients for one month.

After calving (when her nutrient requirements double), the grass also supplies her needs. By mid-May through July, when the feed is at its best, she has enough nutrients to support maximum milk production, to gain weight, and to breed back -- all with only a little mineral supplement. The calf also is big enough by then to eat a lot of that hard-desert grass and make maximum weight gains. By September, when the grass starts to get tough, the calves are big enough to wean. The cow can go on making her living on the grass with only minor supplementation because this is the time of year when her nutrient requirements are lowest.

Maximum nutrient value in grass is produced by grazing it at a time when full regrowth will not quite occur. A rough picture of the difference in grass quality between our managed and unmanaged grass is shown in figure 2.

After we have done our best to provide quality year-round forage, we select cattle (through artificial insemination and ruthless culling) that have nutrient requirements similar to the year-round nutrient values of our grass. It isn't the ranch's job to produce what the cow needs to perform; it's the cow's job to perform on what the ranch can best produce. She needs to perform with a bare minimum of outside supplemental feeding and labor.

With this philosophy, performance information and breed characteristics are looked at in the light of how they fit the year-round forage resource. Beef cows exist to convert forage to food. A cow in a hothouse situation where everything she needs is served with a pitchfork and bucket may

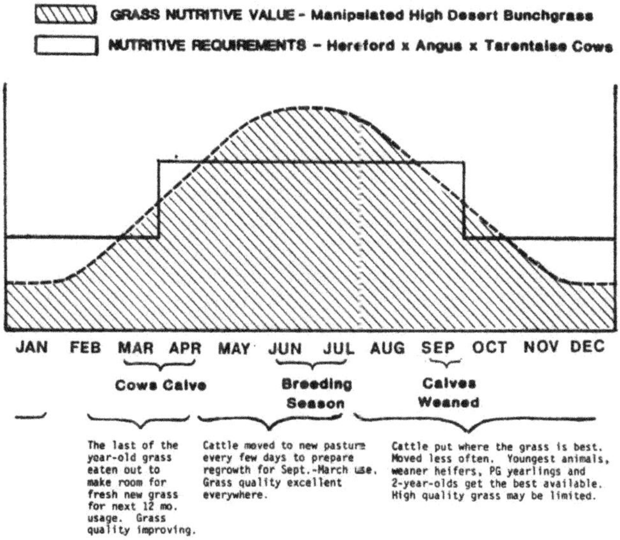

Figure 1. Converting grass to beef.

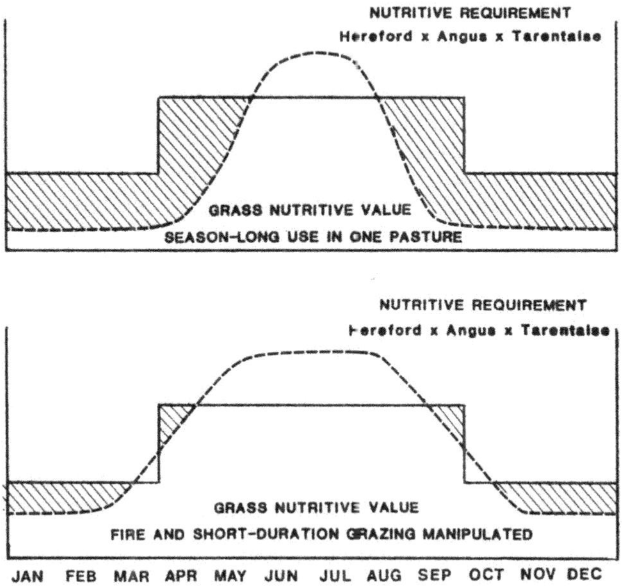

Figure 2. Differences in grass quality between managed and unmanaged grass.

have little bearing on profit in the real world. High milk production and extreme growth rates are of value to a cow-calf operator only if an inexpensive feed source is available to support that performance.

BEEF SELECTION PROGRAM

What type cow do we need to harvest this grass? The following theoretical or impossible program is not meant to be funny or cute. Its purpose is to point out the constant trade-offs involved in any beef selection program. Think along with the curve charted in figure 1.

From weaning time in September until calving time in March, the ideal cow would be a 900-lb 1950-model, Hereford-Angus cross with lots of hair and 3 in. of backfat. Wintering costs are the biggest out-of-pocket expense that ranchers in our area have to pay. This type of cow could winter with very little supplemental feed in our harsh area.

At calving time in March, the ideal would be to drop the Hereford and switch to a Tarentaise-Angus cross. We also would need to modernize the Angus half to about 1970. It would be optimum to stay with this Tarentaise-Angus cross until she was bred in early June.

After breeding, the ideal for one month from mid-June to mid-July would be a Simmental or Brown Swiss crossed with a 1980-model Angus. At this time of year, we have the strongest feed in the world and this breed combination could really convert it.

From mid-July to weaning in September, we would need to switch back to the Tarentaise crossed with the 1970 Angus -- to better match the declining forage quality, but still keep milk production at a high level.

At weaning, we would have gone full circle and would be back to the 1950-model Hereford-Angus cross for maximum wintering efficiency.

With this cow program, we could consistently wean 600-lb calves at 180 days of age, winter their mothers for about $20/head in supplement costs, and quite easily have over 95% conception in a 45-day AI period.

Unfortunately, we can't switch breeds and types within breeds at a moment's notice. After experimenting with many breeds and types through our 100% AI program over the past 17 yr, we have settled on the following combination to tailor a cow for our environment. We use Hereford, Red Angus, and Tarentaise, staying away from the more modern type Hereford and Angus.

We use these breeds and types for the following reasons:

The Hereford -- has the most generations of adaptability to range conditions, a good hair coat, and the ability to lay on external fat for wintering ability. Late puberty, average milking ability, and eye problems are its main drawbacks.

Red Angus -- calving ease, early puberty, and adequate milk production are strong points for our operation. Disposition and an occasional dumpy throwback are the main negatives for us.

Tarentaise -- introduced into the United States in 1972 from the French Alps. They are the smallest, most athletic European breed -- about the same size as our better Hereford and Angus cattle. They are cherry red in color with black pigmentation around the eyes and body openings.

They were originally imported for their maternal traits. They have the ability to breed at an early age and to be mature and in almost full production at 2 yr of age. They are able to breed back under practical range conditions, and have significantly improved teats, udders, and milk production.

The Tarentaise also show ability to increase the rate of gain of their offspring over that of the British breeds (but not to the extreme of the larger European breeds), and to improve carcass quality by reducing fat and increasing length.

Negatives for the breed would be their unconventional conformation and inability to put on condition in the summer for use in the winter. Tarentaise are by far the hardiest of the European breeds we have used; but they, like the other European breeds, have been barn-wintered for generations and lack the fat cover and hair coat that give the Hereford-Angus cross that competitive edge for winter hardiness.

Our goal from this breeding program is a 1,000-lb to 1,050-lb cow that will produce a 475-lb to 550-lb steer calf at 6 mo of age. We have weaned a few calves at 600 lb, but have often culled their mothers for failing to breed back in our 60-day AI season. We don't have the environment to support that level of growth. If all our cows produced that heavy a calf, we would soon be out of business. Anything below a 450 lb calf is hurting us as those mothers are loafing and changing our bunchgrass into fat on their backs instead of milk for their calves.

Other environments will obviously support heavier weaning weights and more growth than ours, but with the introduction of the large European breeds, we have all the growth and milk that most practical cow-calf environments will support today. Unfortunately, most purebred breeders and AI organizations have not yet recognized this fact as they charge forward to win the size and milk contest.

MATCHING COW-CALF PRODUCTION WITH FORAGE

The theme of this talk is matching cow-calf production with forage. Optimum performance for this segment of the industry in any environment will not necessarily produce the fastest-gaining, highest-yielding carcass steer for the feedlot segment. Our breeding program produces a 500-lb weaner steer calf at 6 mo of age that will make a low-choice, yield grade 2 or 3, 1,200-lb steer at 14 mo to 16 mo of age under feeding programs in common use in our area. That is an acceptable commercial product for today's market. Through our AI program, we could immediately increase the feedlot performance, but in the process would produce a cow unsuited for our environment.

One of our main programs is to produce replacement heifers. We winter and breed all heifers in a 45-day AI program, with no cleanup bulls.

Again, to fit our winter feed resource for both the weaner heifer and her mother, we need to wean a 450-lb to 525-lb heifer calf at 6 mo of age. A weaner heifer of this weight can be wintered on our grass from October through March with approximately 3/4 ton of purchased alfalfa hay as protein supplement. We must have a heifer that will reach puberty at 600 lb, which is our target weight for May. We have the month of May for a breeding flush on our top quality grass before we start AI in June.

By September, our heifer will be a bred heifer weighing 800 lb to 850 lb. Our area is good for yearling gains, but if the calving season is to match the feed supply, the yearling heifer has to make most of her gain after she is bred. This requirement effectively eliminates the Brahman composite breeds and European carcass breeds from our program because these breeds need to weigh over 600 lb to be in heat at 15 mo of age. We could haul in feed and make our heifers weigh 800 lb at 15 mo of age, but then <u>any</u> breed would show heat in our AI season. That would be against our philosophy of selecting cattle to work on what <u>our</u> ranch can produce. We are in the grass raising and harvesting business, not the feedlot business.

The second critical performance trait for our operation is the ability to raise a marketable calf as a 2 yr old and breed back for a calf as a 3 yr old, with a 12-mo calving interval. At this point we lose the large European dual-purpose breeds. They are milking heavily and still growing. We cannot economically provide adequate supplement to support their growth and milk. Too many of them never show heat in our 60-day AI season with no cleanup bulls.

Our operation is simpler than most cow-calf ranches. Due to our environment, we do not have the option of economically raising anything but grass. Most ranches also raise hay, silage, winter grain crops, etc. Basically though, all ranches are the same in that most of the products to support a beef animal must come from the soil. The cow's only logical function is as a harvesting machine for the optimum crop that ranch can produce.

Our environment is harsher than most for a good bit of the year -- but for a short time, none is as good. Everyone's criteria will not be the same as ours, but by setting standards, selecting breeds that have a chance to meet those standards, and ruthlessly culling those that don't, real progress can be made. We have developed a herd of cattle that look surprisingly alike, even though no effort has been made to select on visual appraisal.

Part 5

PRODUCTION AND MANAGEMENT OF SHEEP, GOATS, AND CERTAIN OTHER SMALL ANIMALS

43

MANAGEMENT TOOLS FOR GREATER SHEEP PROFITABILITY

Howard H. Meyer

INTRODUCTION

Sheep producers have a wide array of management tools to increase production efficiency. Some of these are used extensively, others are not. This is partly due to the level of producer awareness, cost of implementation, and expected return to investment.

Recent scientific and technological advances have put new tools in the hands of producers to use along with many well-tested age-old practices. No single management tool is likely to be universally beneficial to all producers. While it is important for producers to be aware of the options available, it is also essential that producers carefully consider each item as it applies to their particular production system.

MANAGEMENT PRACTICES

Flushing

Increasing ewe nutrition levels prior to mating has long been known to increase flock reproductive rate, primarily through increased twinning. This is due to higher ovulation rate of ewes in good body condition.

The length of flushing necessary to increase ovulation rate is dependent upon initial body condition. Flushing of thin ewes causes them to ovulate as though they were in better body condition. Ewes already in good condition are unlikely to show a flushing response because they are already near their peak ovulation rate. Many producers erroneously think it is beneficial to reduce body condition in order to get a response to flushing. All they are doing is getting the same ovulation rate but with thinner ewes at the end of mating.

Three weeks of flushing prior to mating will usually give as good a response as longer flushing. Since flushing serves only to increase ovulation rate, once ewes are known to have mated, they can be put on lower nutrition to save feed costs. While not all ewes will conceive at first service, approximately 80% should be pregnant, and so it is seldom beneficial to continue flushing the whole flock for the possible benefit to the 20% returning to service. Under conditions of autumn forage shortage, New Zealand farmers use marking crayons on all rams and draft (sort) off marked ewes as often as twice weekly. Marked ewes are retained on lower feed levels and run with rams carrying a different crayon color. Feed savings achieved in a large flock can be quite substantial.

Since response to flushing is inversely related to ewe condition, flushing feed can most profitably be allocated to the thinnest ewes. By condition scoring his ewes a farmer can better utilize the flushing feed available.

Likely responses to flushing of moderately thin ewes is 10 to 20 more lambs born per 100 ewes lambing. By estimating the net value of extra lambs produced and dividing the value by the number of ewes flushed, a producer can quickly determine the cost of feed he can afford to put into flushing each ewe.

Condition Scoring

Condition scoring is a simple procedure of feeling the animal's body to determine its condition, much as is done by a lamb buyer in selecting lambs suitably fat for slaughter. Condition scoring the ewe flock is useful for several purposes:

- To identify thin ewes requiring special attention
- To select thinner ewes for flushing if the whole flock is not being flushed
- To estimate the average condition of the ewe flock to decide on feeding levels to get ewes ready for mating, lambing, etc.

Estimating the average condition can be done by selecting a random group of 20 or 25 ewes from the flock, and can be very useful in following average flock condition over seasons and years relative to feeding management and(or) productivity.

Condition scoring is done by feeling over the top and side of the loin. Scoring is typically based on a scale of 1 to 5 with 1 representing emaciated ewes with backbones sharply prominent and 5 representing overfat ewes in which the backbone cannot be felt. Most scorers find it useful to include half points on the scale for greater precision. Good-condition ewes are about 3.5 on the scale. At scores of 4 or above, ewes are unlikely to give any response to flushing. Ewes of 1.5 or below are in bad shape and require attention.

Although subjective and fairly imprecise, condition scoring is more useful than weighing sheep to estimate body condition because ewes within a flock may differ considerably in frame size (particularly if the flock contains several breeds), and ewes of the same weight might be quite different in condition.

Grazing Management to Assist in Controlling Internal Parasites

Young animals, especially lambs, are more susceptible to intestinal parasites than are mature animals. Grazing weaned lambs ahead of yearlings or ewes in a rotational pattern is useful in reducing infection of the lambs. It is the logical grazing pattern if we want good lamb growth while only maintaining ewe weights. By using the ewes as grazing tools to clean up unpalatable forage,the ground is exposed to the sun and to air circulation to assist in killing parasite larvae hatching on the pasture. Adding cattle to follow sheep in the system makes it even better since the grazing patterns are complementary and the time period between ewe and lamb grazing is increased.

Grazing management alone is unlikely to remove parasite worries, but it can reduce the need for parasite control remedies while increasing stock productivity.

Overmating

Producers desiring concentrated lambing patterns to fit labor resources and allow better grazing management might consider overmating, i.e., mating more ewes than they intend to lamb. By using marking harnesses and changing color regularly, a producer can identify ewes marked late in the mating period and can expect them to either lamb late or possibly be dry. He then has the option to sell them at the end of mating when prices are usually quite good and smaller producers are often looking for a few additional ewes. In this way, lambing will be concentrated and feed is not wasted carrying late lambing or dry ewes for several extra months.

MANAGEMENT TOOLS

Ram Harnesses

Ram harnesses containing a marking crayon are widely used in Australia and New Zealand to identify mated ewes. By periodically changing color of crayons, producers can use approximate mating dates to estimate lambing dates. This allows for better allocation of nutrition to pregnant ewes according to their stage of gestation. It also identifies ewes that have cycled repeatedly through the mating season and are likely to be dry.

Marking harnesses are also useful management tools for ram management. Purebred breeders can quickly identify both low-libido rams that are failing to mark ewes and working rams whose ewes are showing high incidence of returns to estrus. In the group-mating situation, nonworking rams can be identified by lack of wear on their crayon. Simultaneous use of different crayon colors in multiple-ram groups is not a reliable means of determining sire parentage of lambs since most ewes will be mated by more than one ram and co-twins sired by different rams are common.

Farmers using marking harnesses must check the harnesses are properly adjusted to keep them from tangling or injuring the ram over time. Crayons must be checked periodically. A harnessed ram who has worn away his crayon will lead to considerable confusion and uncertainty of mating dates and returns to service.

Eartags

Eartags have long been essential to purebred breeders for easy identification of individuals and of performance recording. Commercial farmers can also benefit from use of eartags. The durable, large, flexible plastic eartags can be a valuable recording tool to identify the most productive ewes and select best replacements. The eartag can become the record sheet carried around by each animal. Individual numbering of eartags is not important if written records are not being kept.

By using different colored tags each year, a ewe's age is immediately obvious. With a bit of ingenuity, a simple hole punch can be used to record each years lambing performance or virtually any other useful information. For instance, by reserving four areas around the perimeter of the tag and two in the center we have room to record six years of lambing information. Multiple births could be recorded by a single hole near the edge, failure to lamb by a notch at the edge. Singles need not be recorded. In this way we would know that a five-year-old ewe expected to have lambed four times and carrying a unblemished eartag has produced a single lamb each time. With slightly more complexity we could note lambing assistance required or failure to wean one or more of the lambs born.

Ewes that commit a "sin" worthy of culling by the producer (e.g., deserting their lambs) can be marked immediately by cutting the eartag in half. Such ewes will be obvious at culling time and culling decisions will be made on past performance rather than present appearance.

Record-keeping eartags on ewes allow the producer to know the mother's past history when handling newborn lambs. In this way we will know if a twin ewe lamb was the ewe's first multiple birth or if she twinned regularly, and he can mark the lamb accordingly.

Marking Newborn Lambs

Many producers lambing indoors routinely mark lambs to identify them with their dams before mixing groups together. This has usually been done with number brands and suitable branding paint. Now aerosol scourable raddle (marking pigment) is available to conveniently mark lambs born outdoors as well.

A simple system of marking with aerosol raddle can greatly assist in selection of replacements without individual identification and record keeping. For instance, by changing colors weekly, the age of a lamb is readily seen, and selection within color code will quite effectively adjust for age at weaning.

Location of the mark (e.g., shoulder vs rump) can be used to indicate whether the dam was a young or mature ewe and allow that adjustment at weaning. If the dam is carrying a record tag as mentioned previously, lambs from high fecundity mothers (e.g., over 150%) can be marked on the head. This information is more important than birth rank of the lamb alone, since a twin birth says nothing about the lamb, just that the dam had twins at her last lambing. In choosing between two lambs that both were born to five-year-old ewes, a single born to a ewe having twinned at every previous lambing is expected to have higher fecundity than a twin born to a ewe producing only singles previously.

Marking the lamb at birth with such information will allow for better selection decisions at weaning without the need for individual identification and record keeping. While such a system will not allow the same accuracy and progress expected from accurate records, it does allow a commercial producer to apply considerable selection based on actual performance.

Teaser Rams

The first seasonal estrus of ewes can be moved forward by 2 wk to 3 wk by the introduction of rams. Vasectomized teaser rams introduced before mating can result in more ewes cycling at the start of mating and a shortened lambing season as a result. The ram effect occurs only in ewes about to cycle. Ram introduction in deep anestrus has no effect on initiating earlier breeding.

Harnesed teaser rams can also mark cycling ewes to hand mate to a specific ram or to select cycling ewe lambs to be mated later in the season. Some producers use teasers to identify nonpregnant ewes at the end of mating, although an intact ram could be used just as well if the ewes were to be sold.

Chemical Reproductive Aids

Numerous products are used for synchronizing ewes, increasing ovulation rate, or inducing parturition. These may all have a role in particular operations.

A new product has just become available in New Zealand and Australia for increasing ovulation rate by immunizing ewes against some of their own hormones. Two injections are required to get results in the first year, but only one booster thereafter in each year that a higher ovulation rate is desired. The treatment boosts ovulation rate by 20% to 30% and is equally effective on genetically high or low fecundity ewes and whether or not they have been flushed.

SUMMARY

Numerous management techniques and tools are available to sheep producers. Some of these allow more accurate assessment of animal performance, others can be used to reduce feed costs by better allocation of feed resources relative to nutrient requirements. While some are tailored to intensive production systems, others are labor-savers under extensive conditions. All warrant consideration by producers for making their operations more efficient and profitable.

44

IMPORTANCE OF SHEEP'S TEETH IN PRODUCTIVITY

Howard H. Meyer

INTRODUCTION

Positioning and soundness of a sheep's teeth (particularly incisors) receive much attention both from purebred breeders in the selection of future breeding animals and from commercial producers in culling decisions. The prevailing attitude is that without sound teeth, production will suffer. The degree to which this is true depends on both the severity of the "fault" and the conditions under which the animal is to perform. It is important for producers to consider both factors.

INCISOR DEVELOPMENT, ERUPTION, AND WEAR

As with all ruminants, sheep have no upper front teeth. They bite or tear off forage by grasping it between the incisors of the lower jaw and the dental pad of the upper jaw. There are eight incisors in the front of the lower jaw and a full set of molars (12 each in the upper and lower jaw) in the rear of the mouth for grinding food.

Sheep incisors are much like human incisors in that there are sets of both "baby" and "permanent" teeth. The baby teeth fall out and are replaced by permanent teeth according to a somewhat predictable schedule. The permanent teeth are well-formed at birth and "erupt" from the jaw rather than grow.

Baby teeth are replaced in pairs so that a sheep's age is often determined by the number of permanent teeth present. Hence the terminology of two-tooth, four-tooth, six-tooth, and full-mouth ewes. The first pair of permanent incisors is generally in place at about 17 mo of age with successive pairs erupting at roughly 9-mo intervals.

As the cutting edge wears away, teeth continue to erupt and maintain a firm contact with the pad. This leaves less root of the tooth in the jaw. As wear and eruption continue, the tooth eventually has inadequate root left to anchor it, and so it loosens and falls out.

The majority of incisor wear occurs not while the animal is grazing but while chewing its cud. The incisors are repeatedly rubbed against the pad thousands of times a day as the cud is chewed and rechewed. Dust or grit ingested while grazing tends to accumulate on the pad, forming a fine, abrasive paste that the incisors repeatedly grind against. The effect of soil ingestion on teeth wear has been clearly demonstrated in trials with confined fistulated sheep being maintained on hay. Small quantities of soil placed directly in the rumen resulted in a dramatic increase in teeth-wear rate. Wear rate was related to both the quantity and abrasiveness of soil added.

RESEARCH INTO TEETH ERUPTION AND WEAR

Despite the importance that breeders and producers attach to tooth and jaw soundness in sheep, little research has been directed toward the subject. Researchers concerned about excessive wear rates (Barnicoat, 1957, 1959; Cutress and Healy, 1965; Healy and Ludwig, 1965) have tried to relate the problem to either soil ingestion or mineral deficiencies in the diet. Physical characteristics of the diet have also been implicated with the suggestion that some newer varieties of pasture grasses may require more cud chewing and result in greater teeth wear (Barnicoat, 1957). Observations that teeth wear is greatest at the time of highest feed requirement for production (Barnicoat, 1957) suggests that the increased quantity of feed ingested may contribute to increased teeth wear. Little evidence is available to suggest that occlusion (teeth placement) is related to wear rates.

GENETIC VARIATION IN ERUPTION AND WEAR

Recently reported extensive studies (Aitken and Meyer, 1982; Meyer et al., 1983) are underway to examine genetic variation in teeth eruption and wear and to relate teeth measurements with one another and with productivity. Considerable variation was found in the age of the ewe at permanent incisor eruption, both between and within breeds. Among the numerous genotypes present in the 1,350 ewes repeatedly observed for tooth eruption, Suffolks were found to have early tooth eruption while Southdowns were very late. A line of Romneys, with a long-term history of selection for increased litter size, demonstrated considerably earlier eruption than did other Romney lines in the trial. The Romney is also known for early onset of puberty, so it appears that selection for prolificacy has accelerated maturity for both traits. Within breeds, however, ewes showing earlier puberty did not acquire their first permanent incisors earlier.

Growth rate and body weight had a slight effect on teeth eruption; heavier ewes got their permanent incisors at a bit younger age. Observed heritability of age at eruption was .18 ± .06, reflecting moderate within-breed genetic variation.

The variation in the age at which teeth erupted is sufficiently great that sheep cannot be accurately aged by dentition. Studies are continuing to determine if age at eruption has an effect on age at eventual tooth loss.

Among the ewes observed for teeth eruption, about 800 were marked to study tooth-wear rate. This was done by using a dental drill to place a mark at the gum line of ewes about 21 mo old. The distance from the mark to the tooth-cutting edge was recorded and remeasured 12 mo later to determine the amount of tooth disappearance through wear.

Both the breed of ewe and the sire within breed were found to significantly affect teeth-wear rate. The heritability estimate for wear rate was .46 ± .13, indicating that the trait is very heritable. The major nongenetic factor influencing wear rate was the number of lambs reared. Mean wear rates for ewes weaning 0, 1, and 2 lambs were 3.5 mm, 3.7 mm, and 3.9 mm, respectively. This agrees with the earlier suggestion that increased food intake may lead to greater tooth wear.

Teeth placement had no effect on amount of wear. This supports other studies (Purser et al., 1982) that found no relation between teeth placement and productivity of Scottish Blackface ewes. Ewes that got their teeth at a younger age showed slightly more tooth wear over the 12-mo study period than those whose teeth came in a little later. Since teeth initially have a very hard enamel surface above the underlying dentine, these ewes may have already been grinding on the softer material by the start of the trial.

BREEDER CONCERNS AND SELECTION EMPHASIS

As ewes age, their teeth typically become long, loosen, and fall out. Long teeth are generally seen protruding forward beyond the pad, and some observers believe this to be the expression of previous malocclusion. This is not necessarily true. As long as teeth squarely meet the pad, the pressure will keep them repressed in the lower jaw and wearing against the pad. Eventually, the amount of anchoring root will be insufficient to hold the tooth in place, and as it leans outward to miss the pad, eruption will continue until the tooth falls out.

There is evidence that the size of the pad may change due to seasonal and(or) hormonal influences. Hence an animal judged to have slightly protruding teeth at one observation may be normal at a later check. Breeders may be wise to check teeth over a period of several weeks before making final selection.

At this point there is a lack of information relating early tooth placement to longevity. Certainly sheep with severe malocclusion should be culled, but culling high-performing sheep because of minor malocclusion may be counterproductive.

Loose or missing teeth will make ewes less competitive than full-mouth ewes. However, even under conditions of very intensive grazing, broken-mouth ewes perform well if managed as a group with similar ewes. Producers may wish to consider adjusting management to cater to a group of such ewes where conditions allow.

SUMMARY

Sufficient variation exists in timing of incisor eruption to make dentition an unreliable means of accurately determining the age of sheep. Studies have shown that loosening and eventual loss of teeth is due to excess wear and insufficient remaining root to adequately anchor the tooth. Neither time of eruption nor tooth placement relative to the pad are closely related to wear rate. Heritability of tooth-wear rate is quite high but objective selection of animals would be difficult. It is suggested that minor malocclusion is not likely to warrant selection pressure relative to other performance traits.

REFERENCES

Aitken, W. M. and H. H. Meyer. 1982. Tooth eruption patterns in New Zealand sheep breeds. Proc. of the New Zealand Soc. Anim. Prod. 42:59.

Barnicoat, C. R. 1957. Wear in sheep teeth. New Zealand J. Sci. and Tech. 38A:583.

Barnicoat, C. R. 1959. Wear in sheep teeth. VI. Chemical composition of teeth of grazing sheep. New Zealand J. Agr. Res 2:1025.

Cutress, T. W. and W. B. Healy. 1965. Wear of sheep teeth. II. Effects of pasture juices on dentine. Ibid. 8:753.

Healy, W. B. and T. G. Ludwig. 1965. Wear of sheep teeth. I. The role of ingested soil. Ibid. 8:737.

Meyer, H. H., W. M. Aitken, and J. E. Smeaton. 1983. Inheritance of wear rate in the teeth of sheep. Proc. New Zealand Soc. Anim. Prod. 43:189.

Purser, A. F., G. Wiener, and D. M. West. 1982. Causes of variation in dental characters of Scottish Blackface sheep in a hill flock, and relations to ewe performance. J. Agri. Sci., Cambridge 99:287.

45

WOOL AND MUTTON PRODUCTION AND MARKETING IN SOUTH AFRICA

Hans E. Nel

The sheep and wool industry has always been one of the key industries in agriculture in South Africa. It is the livelihood of a large percentage of the rural population, and wool is the most important agricultural export product to earn foreign currency. Wooled sheep contribute an average of 10% to the value of agricultural production in South Africa. Wool's contribution to the export value of agricultural products was about 15% the last number of years.

Wool producers are encouraged to obtain the highest possible income from sheep and wool within the framework of optimum land utilization. Financial encouragement comes from both the government and the wool growers' own organization, the Wool Board. In 1969 the South African Department of Agriculture started the Stock Reduction Scheme whereby ranchers were paid to take stock off the range until the stocking rate met the carrying capacity of the natural grazing. The idea was to prevent overgrazing, destruction of the "veld," and soil erosion. Good management of rangeland through the application of prescribed rotational grazing policies was an inherent part of this improvement scheme. A system like this can work in South Africa because about 99% of all grazing land is privately owned.

Other financial encouragement came from the National Wool Growers' Association that taxed itself during good years in order to be largely self-sufficient during difficult years. They built a Wool Stabilization Fund that is used to buy producers' wool whenever the market price is lower than the so-called "floor price," which is decided on from year to year. Furthermore, wool growers accepted a one-channel marketing system and the pool system of payment, which will be discussed later.

SHEEP BREEDS

The sheep industry in South Africa started in 1652 using the indigenous fat-tailed and fat-rumped sheep of the native inhabitants. These sheep were reasonably suitable for the provision of fresh mutton to colonists and for supplying ships passing by the Cape of Good Hope on their way from Europe to the East.

Merino

Today, the Merino is the country's most important breed of sheep. It forms about 62% of the total sheep population of just over 30 million. On the average, the Merino is smaller than the American Rambouillet and has more wrinkles and a lower fertility rate. It produces a high quality fine wool. Most Merinos are to be found in the Karoo, an exclusive sheep-producing area in the Cape Province where an annual rainfall of 10 in. to 15 in. gives rise to a mixture of grass and low-growing bush. This area not only contributes a large percentage of the total wool clip, but the majority of the Merino purebred operations. The districts of Somerset-East, Cradock, Graaff-Reinet, Murraysburg, and Richmond can be regarded as the heartland of the Merino industry.

Wool-mutton Breeds

Apart from the Merino, South Africa's wooled sheep population consists of smaller numbers of different breeds and types. The South African Mutton Merino (formerly known as the German Merino) has increased in popularity in recent years and, while still in relatively small numbers, this breed is now found to be widely distributed, especially in the higher-rainfall regions.

In order to produce slaughter lambs of high quality in the higher rainfall areas, a few dual-purpose sheep breeds have already been developed in the country. One example is the Dormer, which was developed from a cross between the Dorset Horn and the German Merino. The Dormer was especially bred for the production of slaughter lambs in the winter rainfall area of the Western Cape Province. Good conformation and a relatively high fertility of 150% are typical characteristics of this breed. Dormers are normally kept on improved pasture for at least the lambing season. Another dual-purpose breed is the Dohne-Merino, which was developed from a cross between the Merino and the German Merino on the Dohne experiment station in the Eastern Cape Province. This breed is particularly well adapted to the sour grasslands of the summer rainfall areas along the southeastern seaboard. Comparable to the Dohne-Merino are other smaller breeds like the Letelle, the Walrich, and the Grassveld Merino.

Indigenous fat-tailed sheep are disappearing from the South African scene. The Namaqua, the Afrikaner, the Blackhead Persian, and the Van Rooy are examples of these breeds.

Dorper

Because of the low economic value of the indigenous sheep types, and also because wool cannot always be produced successfully in the dry regions, it has been necessary to develop a type of sheep capable of producing slaughter lambs under dry conditions. Thus, the Dorper mutton sheep was developed from a cross between the Dorset Horn and the Blackhead Persian. This work was done in the 1930s at the Grootfontein College of Agriculture in the Karoo. During the past few decades, this breed has made considerable progress and is now distributed throughout the country. Although most Dorpers are blackheaded, a White Dorper strain was also developed over the years.

A Dorper ram

Afrino

When Dorper rams are used on Merino ewes for the production of slaughter lambs, the black and kempy fibers contaminate the Merino wool and produce problems. This was one of the reasons why the South African Department of Agriculture decided in 1969 to develop a new white-wooled, dual-purpose sheep breed adapted to the arid Karoo conditions. This work was done at the Carnarvon Experiment Station in the Karoo. After evaluating several crosses, it was decided to develop this new breed with a genetic composition of 50% South African Mutton Merino, 25% Afrikaner, and 25% Merino. By choosing this cross, no colored or kempy fibers were brought in, but at the same time the adaptability of the Afrikaner was combined with the excellent wool characteristics of the Merino and the fertility and good mutton qualities of the South African Mutton Merino. While guarding against colored and kempy fibers, as well as horns, most of the selection pressure goes to two characteristics, namely a high production ability under arid, natural grazing conditions and a good weight for age. This new breed has been named the Afrino.

Karakul

The Karakul is the only sheep in the pelt or fur sheep category. Although Karakul sheep are regularly shorn, and a fair amount of Karakul wool is marketed annually, the main product of this breed is the much soughtafter and well-known pelt of the newborn lamb. Karakul sheep are restricted to the dry northwestern Cape Province where they are well adapted to the prevailing dry climatic conditions. Rainfall in these parts varies from 4 in. to 8 in. a year. In years of good rainfall, ranchers do not slaughter ewe lambs but allow them to run with the breeding sheep in order to raise replacement ewes. When rainfall is low, all lambs are usually slaughtered, and the dry ewes receive supplemental feed in order to survive on the poor grazing until the following season. It is a well-known fact that feeding has a great influence on the thickness of a lamb's skin at birth. When feed is scarce and rainfall low, the skin is thinner and the wool shorter. The industry prefers a thin-skinned, short-haired pelt since a lightweight form of clothing can be made from it. South African and Southwest African pelts are sold under the trade name of SWAKARA and are 97% black in color. In the past, black was popular for ladies' evening wear, but color variations, and especially light colors, now attract the greatest attention in fashion markets. As a result, more attention is being paid at research stations to the breeding and selection of animals with lighter-colored pelts. Today about 40% of SWAKARA pelts are color-treated. The bleaching process can, however, damage a pelt by lowering the quality of the fiber and affecting the texture, luster, and durability of the wool. For this reason the process is not carried out to the full. The black wool is now only bleached to a lighter shade, or to an indeterminate color, and then dyed a bright color. Research stations have already made substantial progress with the improvement of pelt quality and color.

Breeders' Associations

All the different breeds have their own associations for promotion, advertising, and organizing annual ram sales. Prices for rams are, on the average, much higher in South Africa than in the U.S. The record price for an individually registered Merino ram is over $40,000.

THE WOOL INDUSTRY

South Africa is the world's fifth largest wool producer, following Australia, the Soviet Union, New Zealand, and Argentina. It normally produces around 230 million lb of wool per annum, with an average clean yield of approximately 57%. More than 80% of the total production consists of Merino wool, and more than 80% of the clip has a spinning count of 60s and higher. Therefore, it is mainly suited to the worsted wool industry.

The Wool Producer

The South African wool producer himself oversees the shearing, grading, baling, labeling, and dispatching of his wool for marketing. A wool producer, or a prospective wool producer, can learn all these steps in the wool industry by undergoing training in the theory and practice of wool production at agricultural high schools, agricultural colleges (junior colleges), and universities. An extension service, which is an arm of the Department of Agriculture, also advises producers continually on the latest and most effective production techniques.

The National Wool Growers' Association (NWGA)

The NWGA, affiliated with the South African Agricultural Union, is the official mouthpiece of the wool producers. It functions in much the same way as its U.S. counterpart.

The Wool Board

The South African Wool Board was launched as a board of control under the Marketing Act on July 1, 1972. It consists of sixteen members, ten of whom are wool producers, nominated by the National Wool Growers' Association. The other members represent the wool trade, the processing industry, the brokers, and the Department of Agricultural Economics and Marketing; two members are appointed for their special knowledge of marketing in the textile industry.

Under the "Wool Scheme," the Wool Board is empowered to acquire and market all wool produced in the Republic of South Africa and Southwest Africa. The Wool Board is authorized to buy and sell wool and dispose of it as desired. It is a corporate body and may acquire and dispose of properties. It may undertake research and promotional work, develop the demand for wool, and cooperate with other bodies, locally and overseas, in the best interests of the wool industry. The Board also has the authority to place a levy on wool, to build and administer a reserve fund, and to appoint agents to perform certain functions for the Board.

Pool system. The Board administers a pool system on behalf of producers. The pool system ensures that every type of clip is assessed individually on merit, independently from other types and according to its own selling performance. This enables the Board to group certain types, where desirable, at the end of the season when marketing performances are known. This is considered to be the fairest basis for administering a pool system for the producers, and furthermore, it reflects clearly what types of wool are in greatest demand.

In practice, this means that the entire clip is turned over to the Board, where it is evaluated objectively and sorted. An advance payment is paid to producers for each of the more than 1,200 types. Wool is then pooled by type and sold at public auction with the proceeds reverting to each pool. An important aspect is that wool is not sold for the individual account of each producer. Remuneration consists of an advanced payment and then a final payment calculated on the basis of the average proceeds attained by each type at the end of the season.

Provision is made for a guaranteed price or floor price so that each producer is assured a minimum price. Should the price for wool fall below this floor price, the difference is paid to the producer from the stabilization fund.

All producers of similar quality wool, therefore, receive a similar price during that season, regardless of the actual time of delivery of their clip. Among other things, the scheme maintains stability for the individual producer by ensuring him the average price for the season and avoiding price fluctuations in the wool market.

Another important goal of the Scheme is to enable the Wool Board to market its commodity in a modern and competitive manner, including international cooperation with other wool-producing countries. This can only be done provided the Wool Board controls the entire South African clip.

In administering the Woolmark (the international quality symbol for articles of pure, natural wool) and also the Woolblendmark (for wool-rich blends), the Wool Board ensures that products displaying these symbols comply with the strict requirements attached to the symbols.

Other functions of the Wool Board are to:

- Undertake country-wide trade promotion in cooperation with manufacturers and retailers
- Provide educational information to schools, colleges, and other organizations
- Supply the press and radio media with a comprehensive news and fashion service, and also publish a magazine Golden Fleece, which is directed at the interests of the industry
- Make available publications on all aspects of wool
- Assimilate statistics for domestic and overseas use
- Support research by means of bursary awards and significant contributions to the South African Wool and Textile Research Institute of the Council for

Scientific and Industrial Research and to the Animal Disease Research Fund

At the Board's experimental laboratory in Port Elizabeth, the latest research data are processed into a usable form and made available to the industry.

Funding. The Board is financed by means of a levy paid by each producer according to his production and his interest from certain investments and income from the leasing of buildings. The government is not obliged to contribute any specific amount toward the financing of the Wool Board but does, in fact, provide some financial support for the execution of the Board's task.

LAMB AND MUTTON

The South African Meat Board controls and promotes the production and marketing of meat, including lamb and mutton, in approximately the same way as the Wool Board does for wool. Carcasses, slaughtered in modern abattoirs near the big cities, are auctioned according to grade in the nine controlled areas and are subject to guaranteed minimum prices for the producer -- except in the lower grades. There is no control at country auctions in rural areas. Lamb carcasses are graded in much the same way as in the U.S.

IMPLEMENTATION OF TECHNOLOGY

Research findings from the Department of Agriculture, and from universities, are applied in varying degrees by sheep and wool producers. Most of the big ranchers follow an annual routine of drenching and vaccinating to prevent major diseases like bluetongue, pulpy kidney, and others. Because the profitability of sheep production is directly correlated with the number of lambs weaned per ewe mated, most producers strive for a high weaning percentage. The rainfall pattern for specific years can, however, influence the weaning percentage drastically in most parts of the country.

Performance Testing

There seems to be an increasing awareness among South African sheep producers of the value of performance testing as a means of identifying superior breeding animals. Many breeders participate in the South African Performance Testing Scheme for wooled sheep. This is a government-aided scheme that assists breeders by making certain calculations for them and keeping them informed on the techniques of scientific breeding principles. At present, fleece weight, clean yield percentage, average fiber thickness staple length, and the number of crimps per inch are measured. Body weight and growth rate are measured on request, while fertility is measured according to a record of the lambing performance of individual ewes. The breeder is responsible for recording fleece and body weights and for forwarding a 150 g wool sample to the laboratory at Grootfontein for analysis. Any registered breeder of an acknowledged breed of wooled sheep may join the performance testing scheme, and any wool producer can send wool samples in for analysis, free of charge.

Group-breeding Schemes

Several group-breeding schemes for sheep of different breeds are operating in South Africa. The purpose of these schemes is to increase the size of the genetic pool, to allow a two-way flow of genes between different flocks, and to combine performance testing with subjective evaluations of animals.

REFERENCES

Department of Agricultural Technical Services. 1975. Wool-sheep are winners. Government printer, Pretoria.

Hugo, W. J. 1968. The small stock industry in South Africa. Government Printer, Pretoria.

Van Rensburg, C. 1978. Agriculture in South Africa. Chris van Rensburg Publication (Pty.) Ltd., Johannesburg.

46

SHEEP PRODUCTION AND MANAGEMENT IN A DRY CLIMATE

George Ahlschwede

INTRODUCTION

Profitable sheep production in a dry climate depends on a number of factors, but it is most directly influenced by the amount and quality of lamb and wool that can be produced per ewe from the resources that are available. Range management, supplemental feeding, sheep selection, wool preparation, and preventive health programs all regulate lamb crop percentages, lamb weights, wool quality, and wool weights. The management of the resources, which vary from ranch to ranch, provides sheep producers with daily challenges to enhance their opportunities for profit.

We did not need the extreme drought of the past year as an example for my use to prepare for this paper and this presentation. Sheep numbers in West Texas have been greatly reduced and some counties report reduction of more than 50% of their ewe numbers. Such is the case for sheep production in a dry climate where producers must depend on their range resource for the maintenance of the ewe flock. Supplemental feed is available to improve sheep condition and to help meet nutritional needs at critical stages of the production cycle, but producers cannot purchase roughages to maintain their ewe flocks.

RANGE MANAGEMENT AND SUPPLEMENTAL FEEDING

Range management is an important key to profitable sheep production. New range management techniques are continually being tested and evaluated. The ability of a rancher to blend his livestock with his range for optimum production remains at the heart of the system that allows our range producers to survive. Their ability to monitor range conditions and the condition of their sheep allows them to make supplemental feeding decisions that can enhance their productivity.

Adequate nutrition is essential during three critical periods of the reproduction cycle to help maximize lamb-crop percentages: 1) prior to breeding when ewes should be in a moderate condition and should be gaining in weight; 2) just prior to lambing when the ewe is developing her fetus, getting ready to milk, and maintaining her body and wool production; and 3) during lactation when the nutritional requirements are really the highest for the ewe.

In a dry climate, a combination of range management and supplemental feeding must be used to satisfy the nutritional needs during these three periods. Fresh and rested pastures can be utilized to help improve nutrient availability, but supplemental protein and energy also can be added. The choices of supplements vary in feed type and in physical form. Oil meals have long been utilized to provide protein. Corn and grain sorghum are excellent energy sources. More recently many producers have used whole cottonseed as an excellent protein and energy supplement. In periods of marginal range forage availability, alfalfa hay has been used as an excellent supplement. The physical form of the supplement and the method of feeding in large rough pastures creates many management opportunities. For many years, salt has been used to limit feed intake. Salt has been mixed with oil meal and grain sorghum in a ratio of 1 part salt, 1 part meal, and 3 parts grain sorghum and fed in a self-feeder. Hand feeding on a two- or three-times per week basis requires more labor. The new blocks are more expensive but require less labor. The choice of the supplemental feed and the method of feeding varies greatly among ranchers, depending on their labor, facilities for feed storage, feeding equipment, and the cost of the feed delivered to their ranch. Regardless of such choices, it appears that the most profitable sheep production systems in our dry climate utilize supplemental feeding for their ewe flocks.

One of the other feeding opportunities that many of our ranchers enjoy is the possibility of developing a creep-feeding program for their lambs. Young lambs are very efficient and utilize a high-concentrate ration quite well. Creep feeding helps to increase market weights off the range, allows for earlier weaning of lambs, allows the ewes to get back in shape for the next breeding faster, and helps get the lambs to market before the price breaks and the summer heat arrives. The quantity of feed needed for a creep-feeding program is not great and it allows the lambs to express their superior growth potential that has been developed in the breeding program.

SELECTION AND BREEDING PROGRAMS

The development of a sound sheep-breeding and selection program is another of the keys to profitable sheep production in a dry climate. The most desirable program utilizes the selection of replacement females that are born and raised on the ranch. The selection of the best females from this group incorporates the ability of these females to produce and survive within that particular ranch situation. Over a period of years, the most adaptable females are those that are retained, and they are the ones that can produce under the particular feed, climate, and management system of that given ranch. Ideally, records of productivity would be utilized to make selection determinations, but in most ranch situations in West Texas, no identification is available. The lambs from multisire breeding groups are weaned and a preliminary sort is made of the ewe lambs. Off-type ewe lambs are culled and the remainder are kept as possible replacements. This group is then evaluated for potential as yearling ewes (observing body structure, size, and wool traits, including length of staple, fineness, density, face cover, and uniformity of wool cover).

Adaptability is so important that ram selection should incorporate some of the same detail employed in selecting females for a given ranch situation. A new stud ram should be selected by using as many records as are available. Ram tests can provide a great service in this respect. Ranchers can test some of their best range rams and compare their performance to other rams in the same test. Using these data, a prospective stud can be selected that complements, yet exceeds, the performance of the ranch rams on test. This new stud then can be mated to a select group of the top ranch ewes to produce rams that can be mated to the group one and group two ranch ewes. This allows the producer to pick up as much adaptability as possible in developing his breeding program.

The choice of a breed for sheep production in a dry climate can probably be debated -- but, in my area of the country, fine-wooled sheep seem to be the most adaptable. They excel in fine wool production and seem to survive and stay productive longer than do other types. The selection program just described works well with fine-wooled sheep.

Some sheepmen in dry climates may wish to purchase replacement females and crossbreed for market-lamb production. This approach reduces the management program required for developing a selection and breeding plan, focusing more sharply on lamb production. Even in this situation, it is advisable to select replacements that are as productive as the ranch's management programs can handle. Potentials for lamb and wool production (the genetic ability for wool and lamb production) are purchased when replacements are purchased. Such sheep producers can increase overall productivity only through superior management of the environmental factors that the producer controls.

WOOL PREPARATION

Lamb production in a dry climate is dependent on many factors, some of which have been discussed and some, such as rainfall and predator control, are very unpredictable. It is certainly most important to optimize lamb production, but wool production cannot be ignored. Range lambs in a dry year may net producers only $25 to $30 per ewe. Wool production in that same year may reach the same level of net return, including the incentive program now in effect. This opportunity makes it essential to spend some time on wool preparation.

Uniformity is the key word in wool preparation. An ideal program starts with the breeding program and the elimination of ewes with off-type fleeces. Wool production will be enhanced by both the supplemental feeding program and the health and parasite control program that I will discuss next. Management at the shearing pen also is necessary to make sure that the fleeces are kept clean and are packaged in the best possible manner. Sorting of the fleeces for grade and staple length will make each wool bag more uniform and will increase the buyers' confidence in the clip. The producer's presence at the shearing pen is essential to help convey a positive, interested attitude to the shearing crew and to the helpers. This alone will improve the quality of the shearing operation and will result in a more uniform, higher-valued clip.

PLANNED HEALTH PROGRAM

A planned health program must accompany a successful sheep production enterprise in a dry climate. This program must be designed as a preventive program to guard against disease and infections that can cause decreases in lamb and wool production. Productivity can be reduced severely by ineffective health; parasite control programs and the annual prevention design are essential to maintain production at an optimum level. The on-going health program may receive slight changes and improvements from year to year, but the basic program should be routinely followed to prevent interruptions in production.

The health program will vary depending on whether the producer is raising or purchasing his replacements. The health program will be an additive program for the ewe lambs and rams kept for replacements, while purchased rams and ewes need an isolation program coupled with the vaccination program of the existing flock. Producers should make no assumptions health-wise when receiving newly purchased replacements. Treat these newcomers as if they have never received any preventive treatments. Replacement ewes should not be mixed with the existing flock until 1) they have received the routine vaccinations already carried by the existing flock and 2) have been observed for any disease or infection outbreak. A strict isolation program should be followed while new rams are brought into the flock. New rams should be run separately and should be given their vaccinations. New ram additions should not be mixed with existing rams, even during breeding season (if possible). This procedure will help guard against any possibility of spreading a reproductive-type disease to the entire ram battery.

The health program for a dry climate should include internal or external parasite control, enterotoxemia and soremouth vaccinations, and examinations of the ram battery for ram epididymitis. Dry areas such as West Texas generally do not foster the magnitude of health problems sometimes found in more humid areas. Foot problems, such as foot rot, do not exist unless brought into the area on newly purchased sheep. Once these initial infections are controlled and successfully treated on imported sheep, the dryness of the area generally provides protection from further infestations.

Enterotoxemia is responsible for a majority of the sheep health death losses in our area; the ewe flock should be vaccinated annually about 1 mo prior to lambing. Most of the producers in our area are using the new combination of Type C and Type D vaccine for the annual vaccine. This annual vaccination protects the ewes from sudden changes in feed supply, such as feeding mistakes in the supplemental feeding program, and also provides immunity to the newborn lambs. Winter and spring rains can cause drastic changes in forb production that greatly affects milk flow. The annual vaccination of the ewes before lambing will help prevent baby lamb losses. Purchased replacement ewes and ewes that previously have not been vaccinated for overeating will require two vaccinations about 3 wk apart to build up the needed immunity. The last vaccination should be given about 1 mo before lambing, at the same time the annual booster is given to the vaccinated flock. Newborn lambs should receive their first vaccination for overeating at about 1 mo of age. Prior to this time, they will be protected from the disease by the antibodies they receive from their mother's colostrum.

Soremouth can cause severe decreases in the performance of young lambs. There also may be some link between soremouth in lambs and infected and sore udders in ewes. Our Texas soremouth vaccine manufactured at the Sonora Experiment Station has many strains in the live vaccine. It is a good policy to revaccinate the ewe flock every second or third year to help build and check their immunity. All lambs should be vaccinated at 2 wk to 4 wk of age, and all newly purchased replacements should be vaccinated immediately.

The control of internal and external parasites is important to successful sheep production in a dry climate. Variations in rainfall make it necessary to continually monitor the condition of the ewe flock and to determine its need for stomach-worm control. Fecal egg counts also can be utilized to help determine when drenching is needed. A variety of products are available for this purpose, and their timely administration seems to be as important as the choice of the product. Dipping vats for the control of external parasites have nearly vanished from the scene in

Southwest Texas. They have been replaced by sprayers of all types. The best control can be obtained by spraying out of the shearing pen, using adequate pressure and making sure that the sheep are thoroughly covered. A light spray over the back will not get the job done. New control compounds and some pour-on type products appear to be most effective.

Other health problems occasionally crop up in a dry climate; the most prevalent are probably those caused by poisonous plants on a dry range. Most such problems occur when sheep are forced to eat these plants because of lack of desirable forage. Dry, hot, dusty conditions help promote polyarthritis and pneumonia. Bluetongue is also a problem some years. An experimental vaccine has been developed, but as yet no company has signed on to produce it commercially for our people in the Southwest.

Ram health programs are similar to those of the ewe flock, with the exception of ram epididymitis. The testicles of breeding rams should be examined before breeding season to check for abnormalities. With any type of abnormality, analyses should be made of the semen and blood analysis for disease and fertility. Ram testicles also should be palpated following the breeding season to again check for any infections that might have developed during the breeding season. Rams with abnormal testicles should be removed from the ram battery to prevent the spread of any disease during the nonbreeding season.

A planned, preventive program to maintain health is one of the keys to a successful sheep production program in a dry climate. Isolation of new breeding stock, plus a strict vaccination and parasite control program should allow sheep in a dry climate to function and produce without health-related restrictions. Remember that a small amount of prevention easily outperforms any kind of a treatment program.

MARKETING

The subject of profitable sheep production in a dry climate cannot be completed without briefly addressing the marketing opportunities. Wool marketing should be enhanced by a good wool-preparation program. The improvement of the uniformity of the clip will improve the confidence of the buyers and help the producer receive the full market value. Lamb marketing is much more complex. Lamb marketing actually starts when the rams are put with the ewes. This beginning dictates to a large degree when the lambs will be ready for market. Under dry conditions, the majority of the range lambs will not be fat enough to go to slaughter and must enter the feeder market. Creep feeding has the potential to adjust weight, condition, and marketing date. Marketing trends and patterns have traditional ups and downs and producers should understand when and where their lambs will need to enter the system. If range conditions allow, some sliding of marketing dates may be possible. It is important to note that range operators in a dry climate usually have limited flexibility in marketing their lambs.

47

SHEEP PRODUCTION AND MANAGEMENT IN A HOT-HUMID CLIMATE

Leroy H. Boyd

Sheep of the traditional wooled type generally have not flourished in hot-humid climates. This lack of adaptability is primarily because of 1) heat stress produced when the animal's wool coat slows the discharge of body heat, 2) parasitism, and 3) serious foot problems. In such environment, there is a decline in the number of lambs born per ewe and fleece weight is below average; sheep are raised primarily for meat and wool is secondary.

I will draw from my experience in managing sheep in the southeastern U.S. to outline some of their limitations as well as opportunities for their genetic improvement. Also, I will report the findings of a study done in northern Mississippi to select for adaptability.

Regardless of geographic location, the genetic makeup of the animals must be compatible with the environment to ensure maximum production. One positive advantage of raising sheep in a hot-humid environment is that the challenge readily identifies the genetically strong and weak. Fertility, prolificacy, and growth are excellent measures of adaptability.

LIMITATIONS AND OPPORTUNITIES

Climate

Annual rainfall is above 40 in. to 45 in. in the southeastern U.S., which is seasonally both hot and humid. Thus expanding sheep from temperate areas to this more hostile environment could prove to be costly in both death losses and reduced production for the reasons outlined below.

No sheep diseases are unique to the hot and humid regions. The climatic conditions encourage the growth of internal parasites and mosquitoes are a menace when sheep are raised along rivers and marshy areas on the gulf coast. Foot scald and foot rot can be kept under control with proper sanitation, careful hoof trimming, and regular use of the foot bath. When grazing under thorn-bearing trees, portions of the thorn can become lodged in the hoof pad and cause lameness.

Parasites

I learned early about genetic resistance to internal parasites. After purchasing a group of bred purebred ewes in February 1964, they were lambed out in April and by August half were dead. When the registration certificates of the living and dead were separated, it became clear that all the daughters of one ram were alive, all daughters of another were dead, while the third ram had some in both groups. A closer check revealed that for the third sire, his living daughters were produced by the living ewes and the dead were from the dead ewes.

Our producing ewes are drenched just before breeding, after lambing, at weaning (60 days), and twice more during the year.

Lambs are drenched at weaning, and if kept for breeding they will be drenched four additional times during the year. Those lambs retained for breeding that show any genetic weakness for parasitism (continually scouring and unthrifty) or any heat stress during the summer months are slaughtered.

Predators and Controls

Chief predators are the domestic dog and coyote with some loss to fox and bobcat. Fencing, either net wire or energized high tensile, is used to contain sheep and discourage predators. Most sheep are confined at night in nearly predator-proof enclosures. Sheep quickly adjust to the night confinement. More labor is required daily to move sheep in and out but certain advantages are offered, including the easier administration of health care.

A good-working Border Collie is invaluable in gathering and working the sheep.

Heat Stress

Sheep with advanced symptoms of heat stress will "hump up" in the back, break behind the shoulders, and are reluctant to stand and move. After prolonged standing, the backlegs will tremble uncontrollably and in a short time the sheep will fall to the ground. Further physical stress at this point should be avoided. Occasionally these symptoms are diagnosed as blue tongue. However, immediate shearing of the heat-stressed sheep will result in a remarkable recovery in 24 hr to 48 hr. Heat stress during pregnancy results in abortion or fetal dwarfing. (Survival of fetal dwarfs is rare.)

Shearing

Twice-a-year shearing is recommended. Shearing usually occurs in early spring and again in midyear. Some commercial producers shear only once per year and that occurs just before initiation of the breeding season in midsummer. A short fleece is desirable for the cold-humid winter and the short fleece holds a minimum of the moisture, which comes as rain at this time.

Diet

Depending on plant species available, grazing can be extended up to 10 1/2 mo in the deep South. The major growing season occurs from early March to late October. Depending on the year, limited grazing of cool season plants is available from late fall through winter. Cool season grasses most frequently available are annual ryegrass, wheat, oats, and fescue, used mostly for maintenance of mature sheep. Legumes suited for the area are subterranean (Mt. Barker variety which is low in estrogen) ladino, crimson and red clover. Warm season grasses grown in the region are common Bermuda and other improved Bermuda varieties, such as Dallisgrass and bahiagrass. Sheep will eat the ergot-containing seed produced on Dallisgrass during midsummer. Side effects are nervousness and an uncoordinated gait. Within two days after removal from the pasture, affected sheep will fully recover. Mowing the pasture will prevent a reoccurrence. Temporary grazing and a source of hay can be provided by the hybrid sorghums. Properly cured ryegrass hay is fed successfully to sheep. Short-term grazing is provided by tyfon, purple top, and other varieties of turnips. Corn is the grain most readily available.

Body weight can be maintained more easily by regulating grazing time, which allows the producer to increase the stocking rate. Approximately 7 hr of grazing time are required each day.

SELECTION FOR ADAPTABILITY

The main challenge to increasing sheep numbers in a humid environment is identification of suitable breeding animals. Detrimental effects of high ambient temperatures (such as fetal dwarfing) on reproduction in sheep have been reported by Yeates (1958). Fetal dwarfing can be characterized by abortions, premature lambs or, in some cases, lambs born alive but so small and weak they die within one week (Goode, 1964). Hafez (1968) reported that in hot climates the later developing parts of the animal's body, such as loin and rump, are stunted. Hammond (1932) reported that greater development of the later-maturing parts of a sheep body was associated with a more physiologically complete animal. Bonsma (1966) reported that tropically adapted animals have more body cooling because of more vascularity in their hides than nonadapted animals.

In 1978, a selection study was designed to objectively identify Polled Dorset sheep that were adapted (A) or not adapted (NA) to northern Mississippi. Past performance data of ewes and their lambs were used to assign them to a selection group. From the original 37 ewes, 15 were selected for the A group and four for the NA breeding group. Eight bred ewe lambs were purchased for the NA flock. The remaining ewes were identified as intermediates. No sheep were transferred between groups. Out-of-flock rams were purchased and used only in A or NA flocks, along with rams raised in the flock. Currently, the A flock contains 88 ewes with over 96% tracing to one ram and 85% to one ewe. Ewes were given an opportunity to lamb three times in two years. Ewe lambs were exposed to fertile rams after reaching 9 mo to 13 mo of age. Breeding periods were April 15 to June 1, August 5 to September 5, and October 15 to December 1. Fall lambing ewes were bred during lactation. Those breeding periods allowed the ewes to conceive or be in gestation or the young lambs to be growing during a period of high temperature and humidity.

The objective of this study was to determine the relationship of ear thickness (ET), chest depth (CD), rump length (RL), body length (BL), and body height (BH) with growth and reproduction in purebred Polled Dorset sheep. The adapted (A) flock was selected for and the nonadapted (NA) flock against greater body measurements.

Calipers were used to make the external body measurements. Ear thickness was measured at the tip of the ear. Chest depth was measured just behind the front legs. Rump length was measured from just in front of the point of hip (hook) to the pin bone, on each side and just below the dock. Body length was the distance from the base of the neck (head up) to the pin bone. Body height was the distance from the floor (concrete) to the top of the withers. Body weight and all the external body measurements were made at birth on ewes and their lambs. Lambs were weighed and weaned at 60 days of age. Lambs were remeasured for all traits at 120 days of age.

Although management practices varied between seasons, they remained relatively constant within seasons. During the hot and humid summer season the A ewes frequently could be found grazing during the middle of the day, while the NA ewes were lying in the shade. Concentrate feeding was begun (corn-soybean meal, 16% CP) two weeks before the fall lambing season started. Daily intake level was .5 lb per ewe and was gradually increased to 2 lb and to 3 lb, depending on forage quality and availability. In the fall, both pregnant and lactating ewes grazed (6 hr to 7 hr daily) common Bermuda-Dallisgrass pasture. Limited grazing of Mt. Barker subclover-annual ryegrass pasture occurred during the fall and winter, with peak grazing during the spring. Concentrate feeding was not begun until after lambing for those ewes on subclover-ryegrass pasture. During the winter, medium-quality alfalfa hay was fed, if suitable grazing was not available. Large round bales of grass hay and a protein supplement-salt mixture (2:1) was available for all dry ewes. One week prior to weaning, concentrate feeding was stopped. For the last 48 hr prior to weaning, all feed and water were withheld from the ewes, yet the lambs could continue to nurse. All sheep were allowed free access to a mixture (equal parts) of salt, feed-grade ground limestone, and soybean meal.

Lambs had access to a creep area where an 18% (winter) and 16% (summer) CP concentrate ration (corn-soybean meal containing .5% ammonium chloride) and alfalfa hay were available free choice. Fall-born lambs were maintained in drylot after weaning. Winter-born lambs were placed on pasture after weaning. Spring-born lambs were placed on pasture at 3 days to 4 days of age. Lambs were vaccinated for overeating, soremouth, and pasteurellosis.

Ewes were drenched prior to breeding, after lambing, at weaning, and two other times during the year. Lambs were drenched at weaning and at 4- to 6-wk intervals. Adult sheep were sheared in the spring and just before the summer breeding period. Lambs were sheared after weaning and two other times during the year.

All external body measurements (except ear thickness) of producing ewes were significantly different ($P<.001$) between A and NA groups. Respective average body weight and external measurements values for A and NA ewes were: 152.58 lb vs 120.47 lb; ET -- 2.27 in. vs 2.07 in.; CD -- 12.76 in. vs 11.66 in.; RL -- 10.00 in. vs 9.00 in.; BL -- 31.21 in. vs 28.80 in.; and BH -- 27.03 in. vs 24.37 in. External measurements were taken each time just after lambing. Usually the first set of data recorded was used for statistical analysis.

With two or more opportunities to lamb, the increased reproductive performance of the A ewes (1.96 vs 1.10 lambs per ewe per year) can be accounted for by lambing rate (+13%), livability (+11.7%), and a shorter lambing interval (-3.8 mo). The A ewes average age was 8 mo younger than NA ewes. At 120 days of age, 156 lb of lamb had been produced per A ewe per year compared to 65 lb for NA ewes.

Adapted lambs were heavier ($P<.01$) at birth, 60-day weaning and 120 days of age. With the exception of ET at birth, all external body measurements were different between A and NA lambs. The longest RL measured at 120 days for an A ram and ewe were 10.2 in. and 9.0 in., respectively. The shortest RL measured in an NA ram and ewe lamb were 5.9 in. and 5.6 in., respectively. At birth all lambs were taller than long but this relationship is reversed at 120 days. Within a selection group, ram lambs had the longer RL.

Correlation coefficients are lower for ET in relation with all other measurements. The highest correlation coefficients between CD, RL, BL, and BH were obtained at birth and 120 days. Body composition does not influence the skeletal reference points, RL, BL, and BH, to the degree found with CD. Body length and height are nearly equal in importance in regard to skeletal development. (By 120 days of age, body length should exceed height.) These data clearly point out that when body height exceeds length, development is more immature and lamb-like.

CONCLUSIONS

Based on the findings of this study, those sheep with a thicker ear, deeper chest, longer rump, longer body, and greater height at the withers are heavier, produce more lambs more frequently, and have greater lamb livability and gain. This greater production is achieved in an environment that challenges the sheep with higher ambient temperature and humidity and where internal parasites could have been a problem.

Selection of breeding stock should be made for RL and done when the lambs are 120 days of age. Selection for longer rumps in sheep is positively correlated with increased skeletal height, length, and depth. With increased skeletal extension came an increase in body weight, without apparent changes in body composition.

REFERENCES

Bonsma, J. C. 1966. The Santa Gertrudis, past, present, and future. Santa Gertrudis Recorded Herds, Vol. 3.

Goode, Lemuel. 1964. Effects of summer confinement upon ewe productivity. J. Anim. Sci. 23:906 (Abstr.).

Hafez, E. S. E. 1968. Adaptation of Domestic Animals. Lea & Febiger, Philadelphia, PA.

Hammond, J. 1932. Growth and the Development of Mutton Qualities in the Sheep. Oliver and Boyd, Edinburg:- Tweedale Court.

Yeates, N. T. M. 1968. Fetal dwarfism in sheep -- an effect of high atmospheric temperature during gestation. J. Agric. Sci. 51:84-89.

48

ANGORA AND MEAT GOATS IN SOUTH AFRICA

Hans E. Nel

The number of goats in South Africa has varied from 5 million to 7 million over the last 2 decades. Most of these were meat-type goats. Monetary return per animal unit is, however, much higher in Angora goats. Because goats are predominantly browsers, they are often used to graze in combination with sheep and cattle. If properly managed, goats can be a great asset in controlling the intrusion of undesirable woody plants, especially the *A* . When goats stand on their hind legs, they have a reach of up to 5 ft.

ANGORA GOATS

Turkey, South Africa, and the United States (Texas) are the biggest mohair producers in the world. As far as quality is concerned, the South African product compares with the best, partly because of the methods used for grading and preparation for market. Most Angora goats are found in the eastern part of the Cape Province where 95% of the mohair clip is produced. The major production districts are Jansenville, Steytlerville, Willowmore, Somerset East, Pearston, Cradock, Uitenhage, Bedford, and Adelaide. Small numbers of Angora goats are also kept with relative success in other parts of the country that have a mild, dry climate and an altitude not much above 4,000 ft. Angora goats are rather sensitive to climatic extremes; therefore, sudden cold, rainy, windy weather may result in heavy losses.

The interests of the Angora industry are promoted by several organizations. The South African Mohair Growers' Association, with local branches in most of the districts, serves as spokesman for the mohair industry, promotes marketing, and organizes annual flock competitions. The Angora Goat Ram-Breeders' Association strives to improve the quality of breeding material. It is responsible for the inspection of rams eligible for registration, holds a number of ram sales every year, and publishes *The Angora Goat Journal*. The Mohair Board promotes the industry through research and publicity and controls the marketing of mohair.

To improve the quality and quantity of mohair, Angora goats are being graded and selected according to breed characteristics and an Angora Score Card. This is mainly a subjective evaluation that takes into account both conformation and mohair characteristics such as length, fineness, density, and character and style.

Producers look for a good, uniform length throughout the fleece because this affects both the quantity of mohair produced and the value of the fiber for processing. A well-bred Angora should produce at least one inch of hair per month. Angoras are normally shorn twice a year. Fineness and uniformity of fineness are regarded as the most valuable characteristics of mohair since they not only determine the fineness of the yarn but also the weight and attractiveness of the woven material. Fineness is directly affected by the age of the goat. Kid hair, produced during the first 6 mo of life, is the finest mohair produced. Fineness tapers off with age. Mohair with good character and style is kind, but firm, to the touch. It has just enough protective yolk, is bright and lustrous, and forms a blunt-tipped, twisted, wavy, solid staple that is nearly uniform in size all over the body. An animal shows a good density when all parts of the body are evenly and well covered.

Mohair has several desirable characteristics that make it very valuable in the textile industry. It is a firm, strong, and smooth fiber with a high luster and low felting capacity, which makes it especially suitable for woven material in which luster, durability, and firmness are required. Because of its structural smoothness and high luster, mohair does not readily collect dust or absorb the smell of smoke, and it shows up color shades very well. It is excellent for curtains, pile upholstery fabrics, and lightweight, crease-resistant summer clothing for both men and women.

Grading and Marketing

In South Africa, the mohair fleece is sorted into more than 70 grades, or types, when animals are shorn. The purpose of this grading system is to present the raw material for sale in an attractive, standardized form so that buyers' and manufacturers' confidence in South African mohair can be gained and retained. When the grading system for mohair was put together, characteristics considered were length, fineness, quality, clean yield, vegetable matter, stains, and soundness of the fiber. This grading system makes provision for three main divisions; namely, kids, young goats, and mature goats. Each of these divisions is subdivided into classes according to combinations of fineness, length, and quality.

The entire South African clip is marketed through the Mohair Board in much the same way as the pool system used by the Wool Board. (See "Wool and Mutton Production and Marketing in South Africa.")

Research and Technology

Research on Angora goat production and mohair processing is undertaken mainly by three institutes. At the Grootfontein College of Agriculture, much attention has been given to the problem of low reproduction, including abortion, and the general physiology of the reproductive organs. Two types of abortion have been identified, and many producers who followed proposed guidelines have succeeded in getting rid of abortion problems in their flocks. Other research areas at Grootfontein include selection and breeding systems, histology and morphology of the skin and its covering, feeding, and grazing habits.

The mohair fiber (and its response to several manufacturing processes) is investigated at the South African Wool and Textile Research Institute in Port Elizabeth. Grass seed contamination in mohair caused a considerable number of problems a number of years ago. As a result of research carried out at this institute, a method has been developed for processing mohair with a seed content of up to 15% into tops.

The Leather Research Institute at Grahamstown is conducting studies on goat skins and the usefulness of skins of widely differing qualities.

MEAT GOATS

The meat goat is a traditional favorite with nomadic people, especially in developing countries. In South Africa, most meat goats are owned by people of the black tribes. A small number of meat goats are often kept on ranches as slaughter animals, mostly for farm labor consumption. If a meat goat is slaughtered at an early age, the meat is flavorsome and tender. The meat of older goats is often coarse-grained. A very good-quality leather is made from goat skins.

Meat goats in South Africa are sometimes classified into five groups: 1) the common boer goat, 2) the long-haired boer goat, 3) the polled boer goat, 4) the Bantu goat, and 5) the improved boer goat.

The Improved Boer Goat

The improved boer goat was developed from indigenous goats over the last 4 decades. It was not developed on an experiment station but by dedicated producers who followed a prescribed breeding and selection system to develop it into one of the world's outstanding goats.

The South African Boer Goat Breeders' Association was formed in 1959 and is doing an excellent job to promote this goat. They formulated a breeding policy to develop a uniform breed with significant meat potential, rapid growth rate, and high fertility, without losing its natural hardi-

Boer goats browsing a woody shrub

A Boer goat buck

ness and adaptability. The Association also regulates annual sales and runs an inspection system according to a set of phenotypic breed standards.

The improved boer goat was developed in the eastern part of the Cape Province, especially in areas where thornbush, *Acacia karoo,* and "spekboom," *Portulacana afra,* predominate. It is an excellent animal to use and control this type of brush.

This goat was selected mainly for a meaty conformation and the characteristic color pattern of a white body and a red-brown head with a blaze. A white body is desired to ensure easy visibility for rounding up goats in dense brush areas. A dark pigmented skin on the hairless parts is considered absolutely essential for resisting sunburn that could lead to skin cancer. A loose, supple skin and short hair are important characteristics for adaptability in the warm and sunny climate of South Africa. Short-haired animals can also readily be dipped to control ticks carrying heartwater and redwater disease. Furthermore, the boer goat is highly resistant to diseases such as bluetongue and pulpy kidney. Its grazing habits also make it less susceptible to infestation by internal parasites. This goat is a good walker with sturdy legs, and it moves freely in rough, mountainous terrain and through dense shrub. Full-grown rams reach a live weight of 250 lb and ewes weigh around 180 lb.

The boer goat has an exceptionally high rate of fecundity. Kidding rates in excess of 170% are the rule, while researchers at the Animal and Dairy Research Institute at Irene have managed to achieve kidding percentages of 340% by means of a twice-yearly kidding and optimum feeding conditions. Does have enough milk to handle more than twins. Consequently, the growth rate of boer goat kids is excellent. Because so many does give birth to twins and triplets, it is common practice to keep the kids in a corral or shed during the daytime when the does are grazing. The does come back to the corral every night to stay with the kids. This is the time when pairs are checked to make sure that all kids get a chance to nurse. As soon as kids are strong enough, they are allowed to graze with their mothers. Kids are normally weaned from 2 mo to 3 mo of age and are marketed off the range at 8 mo to 12 mo when they weigh about 80 lb.

REFERENCES

Die Boerbok. 1975. Supplement to Landbouweekblad, October 17, 1975.

Hugo, W. J. 1968. The small stock industry in South Africa. Government Printer, Pretoria.

Van Rensburg, C. 1978. Agriculture in South Africa. Chris van Rensberg Publications (Pty.) Ltd. Johannesburg.

49

ANGORA GOAT MANAGEMENT AND MOHAIR MARKETING IN TEXAS

Jack L. Groff

INTRODUCTION

Angora goat production has been an important industry in Texas since shortly after this type of goat was introduced in 1849. Texas normally supplies over 95% of the U.S. mohair produced, which currently represents approximately one-fourth of the total world production (figure 1). Ninety-seven percent of the mohair clip is sold to foreign mills. Other major producing countries are South Africa and Turkey. Lesotho, U.S.S.R., Australia, and Argentina also produce mohair in significant, and in some cases, increasing amounts. Numbers of Angora goats in Texas have fluctuated widely over the years because of highly variable mohair prices and the influence of this variability on the demand for replacement animals. The present population of approximately 1.25 million is the lowest number since 1920 (figure 2). Favorable mohair prices in recent years have rekindled an interest in this industry, but low reproductive efficiency has limited the ability of the industry to respond with increased numbers.

At present, most of the Angora goats in Texas are on the Edwards Plateau where the industry began. Before losses to predatory animals became so severe and before the mohair market break in 1964, large numbers were also found in other areas such as the Grand Prairie and Cross Timbers (central and northcentral Texas) where they were used to advantage in land clearing or brush control. At the present time large areas of the South Texas Plains have excellent forage resources for goats but are not being utilized for this purpose because of predators.

PRODUCTION MANAGEMENT FACTORS

Low reproductive performance is a major problem for the Angora goat industry throughout Texas. This may be a result of breeding programs that have placed high pressure on selecting for mohair production traits. It is possible, however, to select goats that excel in producing both mohair and offspring.

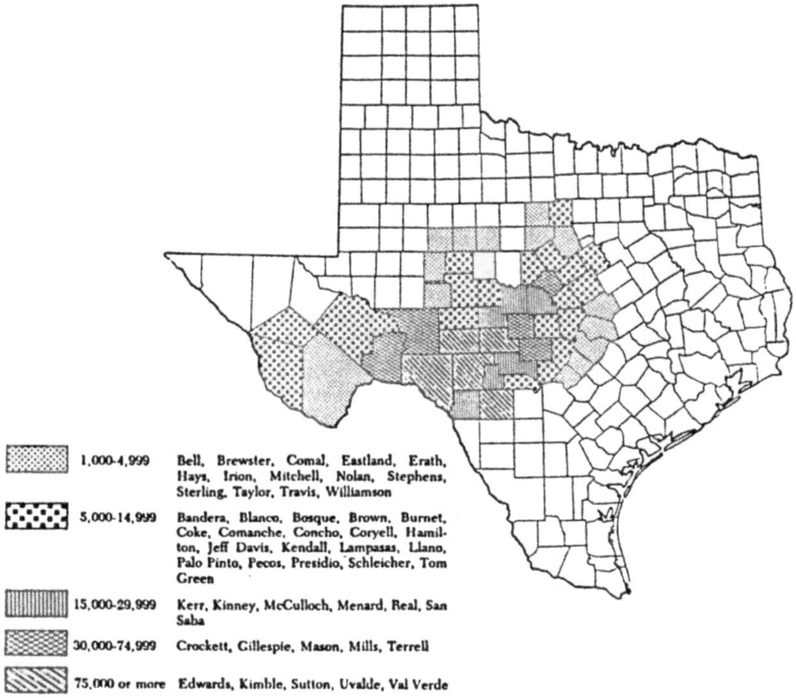

Figure 1. Major Angora goat-producing counties.

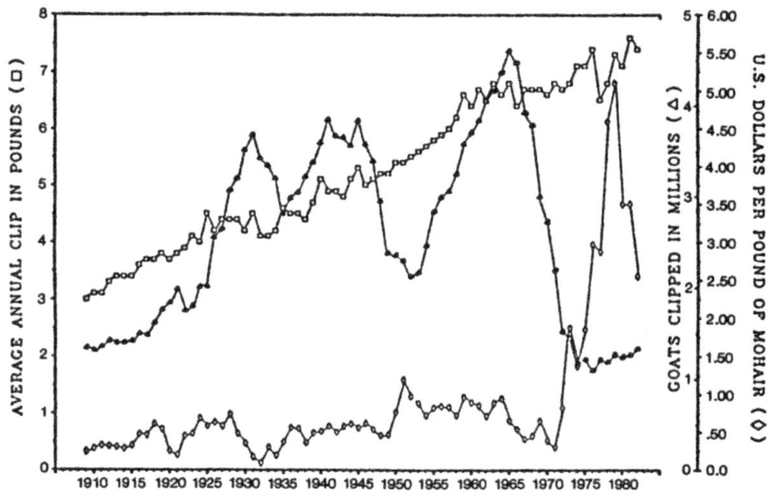

Figure 2. Trends in numbers, average clip, and prices received in Texas for years 1909 to 1982.

Improved management practices help overcome low kidding percentages. Most declining reproduction rates in an Angora flock are associated with: 1) failure of estrus and ovulation due to underdevelopment of the does (figure 3), 2) lack of vigor and strength in breeding males due to nutritional deficiencies, 3) abortion caused by poor nutrition or infection, and 4) death loss of kids (table 1).

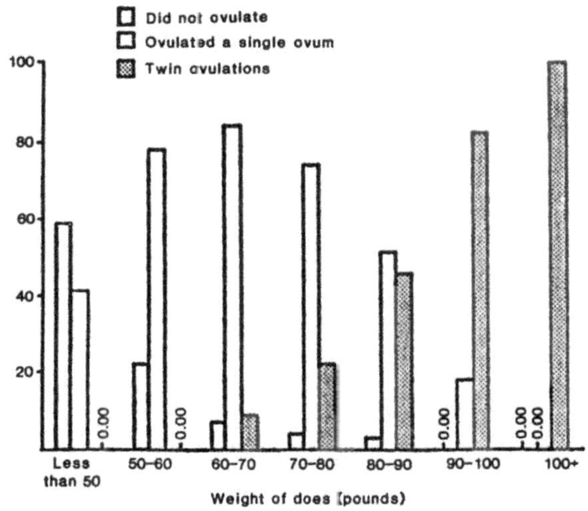

Figure 3. Breeding weight of does in pounds.

TABLE 1. INFLUENCE OF BREEDING WEIGHT ON NUMBER OF KIDS BORN AND RAISED IN TWO RESEARCH FLOCKS

Wt range (shorn body wt at breeding)	Sonora data		McGregor data	
	% kids dropped	% kids raised	% kids dropped	% kids raised
Below 60 lb	52.2	47.4	76.5	58.8
60 - 70	83.8	78.7	101.9	62.1
70 - 80	91.6	85.2	117.3	81.2
80 - 90	88.9	81.0	143.2	114.8
90 - 100	96.2	88.7	147.4	116.8
Over 100			115.4	113.8
Summary	74.7	69.1	128.1	94.6

Properly managed selective breeding programs, along with adequate nutrition and health care for improved physical development, increase the reproduction rate of any flock, with the exception of losses from extreme weather conditions or predators.

The average kid crop weaned in Texas is thought to be in the range of 50% and has probably improved somewhat in recent years. The knowledge is available for markedly improving the kid crop. However, this cannot generally be accomplished without some increase in cost through reduced stocking rate, increased labor input, increased supplemental feed, and improved predator control. The challenge to the individual producers is to implement these practices so that they provide an economic response, which is very dependent on mohair prices and, in consequence, their influence on demand for replacement kids.

Type of Production

Angora goat herds can be maintained for different kinds of production: 1) a doe and kid operation that consists of a herd of healthy animals of productive age -- 3 yr to 6 yr; 2) replacements raised to make herd improvements; and 3) a wether goat enterprise that replaces wether goats for mohair production when the mohair becomes coarse and loses its character. It is advisable to stock goats at a rate that will ensure maximum mohair production with a minimum of supplemental feed.

While the reproduction rate of the flock has the greatest impact on the potential net return for the Angora goat enterprise, the pounds of mohair produced annually per head are very important (table 2). Some flocks produce an average 12 lb of good quality fiber per head every 12 mo, although the average for Texas is about 8 lb for adult goats. Twelve pounds may be near the physical maximum when quality, length, and clean yield are considered.

TABLE 2. BREAK-EVEN MOHAIR PRICES TO RECOVER ANNUAL PRODUCTION COST PER DOE WITH VARYING PRODUCTION LEVELS

Annual production cost per doe, $	Kid crop, %	Income from kids @ $50/hd	Cost less kid value, $	Break-even mohair prices, $/lb		
				8 lb avg	9 lb avg	10 lb avg
46.58	25	12.50	34.08	4.26	3.79	3.40
46.58	40	20.00	26.58	3.32	2.95	2.66
46.58	50	25.00	21.58	2.70	2.40	2.16
46.58	60	30.00	16.58	2.07	1.84	1.66
46.58	75	37.50	9.08	1.14	1.01	.91
Estimates for your ranch						
_____	_____	_____	_____	_____	_____	_____

Breeding Practices

Improvement in both mohair production and reproductive rate can be achieved and maintained through selective breeding programs and good flock management. Mate the best does to the best bucks, the second-best does to the second-best bucks, etc. Save replacement does from the top two groups. An animal's adaptability to the area of production is probably the most important breeding factor.

Good bucks are essential to a good selective breeding program. Purchase bucks from one breeder whose goats possess the desired characteristics to produce a uniform flock. Select animals for quantity and quality of mohair but do not sacrifice size and vigor. Fleeces also should be uniform in quality and length over the entire body of the goat.

Use three to four bucks per 100 does, depending upon the size, brushiness, and roughness of the pastures. Avoid using one buck per pasture in commercial goat production. Condition bucks with supplemental feeding about 2 wk before breeding.

Flush does by supplying 1/4 lb to 1/3 lb of supplemental feed daily or move to a fresh, rested pasture about 2 wk before releasing bucks. Protein blocks may be used in flushing when range conditions are not too severe. When ranges are extremely dry, give does vitamins A, D, and E for 2 wk before breeding to possibly yield a positive economics response. Breed does in September and October for February and March kids.

Supplemental Feeding

Angora goats respond to supplemental feeding more than other livestock. They reflect this through heavier fleece weights. Increased supplemental feeding during dry periods and especially during the last 60 days of pregnancy is essential for high reproductive rates (figure 4).

The final birth weight will be dependent on such factors as sex and type of birth as well as on size and nutritional status of the dam. Under normal or optimum conditions, the rate of fetal growth during the 2 wk before birth appears to be approximately .1 lb daily. Thus, in extreme cases, birth weights could vary as much as 3 lb below the 6.5 lb plotted in figure 4. In reality, most kids weigh between 4.5 lb and 7.5 lb with the larger kids having better chances of survival. Dystocia (difficult birth) in goats is relatively rare.

Feed 1/4 lb to 1/2 lb of protein supplement, 1/2 lb to 1 lb of yellow corn, or 1/2 lb to 3/4 lb of 20% protein cubes per head daily depending upon the condition of the pastures and the does. Pregnant does require larger amounts of feed than dry animals. Abortion often can be prevented by increasing supplemental feeding.

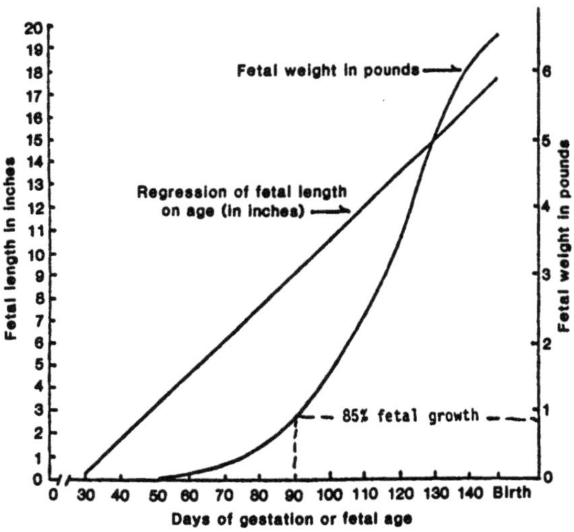

Figure 4. Rate of fetal development. (The age of the fetus or embryo can be calculated with reasonable accuracy by dividing the fetal length [crown-rump length] in inches by .149 and adding 30 to give the age in days.)

Self-feeding, using salt as an inhibitor, may be used in large, rough, or brushy pastures. Keep salt as low as possible and place the feeders 3/4 mi to 1 mi from water. Move feeders for better pasture utilization. A popular mixture is 3 parts of ground milo, 1 part cottonseed meal, and 1 part salt. Salt-controlled feeding is not recommended unless all other methods are impractical.

Feed kids during winter months to ensure growth and development (figure 5). This will improve the kid crop on 2-yr-old does. Cull underdeveloped kids that do not learn to eat. Protein blocks may be fed during kidding season. This method of feeding reduces the number of kids becoming lost from does.

Kidding

Three management systems have been evaluated on a limited scale in experimental flocks maintained on the Edwards Plateau area of Texas (figure 6). These are:

- traditional kidding on the range
- kidding in small traps or "small camp" kidding
- kidding in confinement

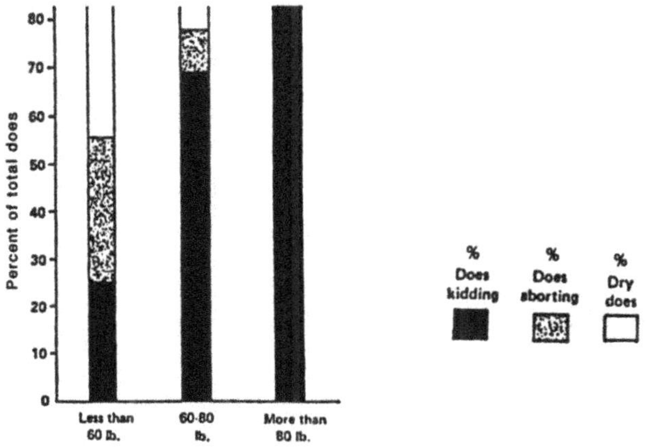

Figure 5. Reproductive performance of 218 does in the experimental flock for the 1969 season. Fifty-six percent of the does that mated in the below-60-lb weight class aborted as compared to 11.5% for those weighing 60 lb to 80 lb and .0% for those weighing above 80 lb.

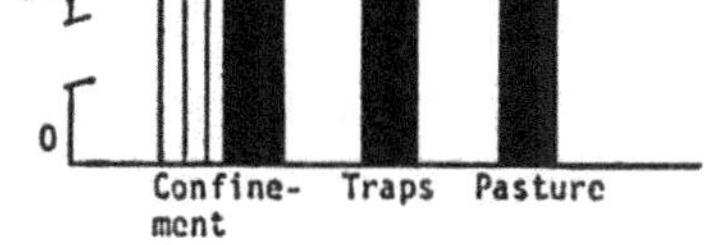

Figure 6. Percent kid crop raised in three kidding systems.

Kidding on the range. Kidding on the range has been defined as the conventional or traditional system. Use rested pastures for kidding and do not disturb does during the kidding season. However, extensive predator control efforts (considering several potential species) should be carried out prior to the start of the kidding season.

Preferably kidding will occur in pastures that give natural protection during inclement weather. Producers often report better kidding results in some pastures than in others.

There should be minimal disturbance in the pasture during kidding. Guard against feeding and(or) gathering stock in the pasture, the major sources of disturbance.

Traditionally good range management practices have called for placing feed, water, or salt at different locations in the pasture to encourage more uniform range utilization. During kidding, animals can be encouraged to congregate at one point, or to return to this point one or more times daily, by placing feed, water, or salt in one place. If the animals are to be fed, full or continuous feeding is suggested to encourage the does to remain near a common point.

In some cases it may be desirable to delay kidding until green feed is expected. This may reduce the tendency for does to travel extensively in search of forage.

Small traps or "small-camp" kidding. In limited experimental comparisons, small-camp kidding has had favorable economic results because a smaller number of does run in one bunch in a smaller pasture, which reduces the number of kids lost through mismothering when does bed down their kids and leave to search for food.

How big is a small trap and how many does can be put per trap? Definitive answers are simply not available. Research work was done with pastures of 30 acres or less and with 50 or fewer does per group. Most commercial producers would not have an adequate number of such small pastures, but they might be able to use the concept of kidding in smaller groups and smaller pastures.

Confinement kidding. Confinement kidding, which has increased in recent years, has the potential of maximizing the number of kids raised. However, a poor job of managing does in confinement can result in losses exceeding those with traditional kidding. The producers most likely to practice this are purebred or stud breeders and producers with smaller flocks that have barn or shed space available. The primary costs associated with confinement kidding are 1) increased feed requirements, 2) increased labor, and 3) buildings or facilities. If the producer has facilities available and is willing to do the work himself, cash costs will be only for feed. Feed costs can be substantial if the does are maintained in total confinement during the kidding season, but costs may be modest if the does are run in small traps or fields during the daytime and are confined at night and for a few days after kidding. The latter system is generally workable only for small numbers of does.

Producers with existing shed or barn facilities are encouraged to adapt them to confinement kidding. However, several producers have built special crates or jugs. Most of these are a 3x3 ft or a 4x4 ft construction of either wood or metal with watering and feeding facilities built in the cubical.

Normally crate numbers should equal approximately 10% of the breeding flock to ensure a workable situation. By using portable crates, barn space may be used for other purposes after the kidding season. Many producers practicing confinement kidding obtain kid crops of 100% or more.

Advantages of the confinement system are: 1) greater bonding between the doe and offspring during the first 2 days after parturition, 2) identification of kids that are not receiving milk and that need assistance, and 3) greater success for newborn kids to regulate body temperature during the first 24 hr to 36 hr after birth, thus reducing loss to cold stress.

Health Care and Marking

Vaccinate kids for soremouth and earmark for identification when most of the kids are large enough to travel. Plastic ear tags are a practical way to identify.

Castrate kids when they are about 9 mo to 10 mo old. This produces a larger, masculine wether goat that buyers prefer.

Shearing

Spring shearing time is January through March, depending on the area of production. Goats should be put in sheds during the first 6 wk to 8 wk of shearing to avoid weather losses. Fall shearing is from July through September.

Goats may be cape sheared. Caping is the practice of leaving a strip of unsheared mohair about 8 in. wide down the back of the neck and back of the goat for protection against temperature shock. The cape should be sheared a month or 5 wk after the regular shearing. If capes are not sheared then, they should be packed separately when they are sheared.

Some producers prefer to shear goats with special goat combs. These combs leave about 1/4 in. of stubble on the goat and provide about 2 wk earlier protection than regular combs.

Spraying

Spray goats after they leave the shearing pen and again in 12 to 18 days for the best control of external parasites. Change sprays occasionally to avoid resistance buildup. Spray so that goats will dry before dark. Use only recommended sprays or dips in strengths advocated by the Food and Drug Administration. Do not spray inside a shed or barn. Spray with the wind, not against it. Do not mix solutions with your hands.

Drenching

Watch animals closely for signs of internal parasitism and drench as necessary. Drench away from the shearing pen using one of the recognized anthelmintics. Change drenches occasionally so that parasites do not build up resistance to any specific chemical. Improved chemicals are being introduced for even better health-improvement measures. Persons applying chemicals should always follow container directions on all internal and external medications.

Marketing

Some producers sell kids after their first shearing, but most prefer to market yearlings after the second shearing. Producers should 1) sell through a reputable commission man or an auction that specializes in handling goats; 2) market mohair through one of the Texas wool and mohair warehouses; and 3) select a warehouse that provides services to meet requirements. If the bulk of your clip is finer than 24s (37 microns), it may pay to have your clip graded. Follow the recommendations of the warehouseperson in preparing and marketing mohair. Currently a new mohair marketing and grading service is being offered to producers that should make Texas mohair more competitive on the world market.

50

AN ECONOMIC ANALYSIS OF GOAT MILK PRODUCTION IN THE OZARK REGION OF ARKANSAS AND MISSOURI

Jim A. Yazman, M. D. Norman,
and M. Redfern

The Ozark Mountain region of northern Arkansas and southern Missouri is home to a large population of dairy goats. Fostered by a milk processing plant that produces canned evaporated goat milk at Yellville, Arkansas, the dairy goat milk production sector has developed over a period of 30 yr. Primarily as a part-time enterprise, the dairy goat industry allowed farm families and rural home-owners with jobs in the small cities of the region to make use of available family labor and surplus land of marginal quality. As the majority of the commercial farms are classified Grade C, investment in buildings and equipment has not been high, which has allowed for much movement into and out of the industry. More recently producers with larger investments in facilities, equipment, and dairy goats have become interested in expanding their market opportunities into Grade A pasteurized milk and processed milk products.

In 1980 a study of the goat milk industry of northern Arkansas and southern Missouri was initiated by the University of Arkansas Department of Agricultural Economics and Rural Sociology and Winrock International. The objectives of the study were 1) to characterize the goat milk production sector of the region and the current production costs for Grade C goat milk; 2) to review marketing opportunities for goat milk in the U.S.; and 3) to determine the feasibility of an expanded market. Based upon a survey of commercial and noncommercial producers, with the collaboration of goat milk production specialists, one output of the study was an intensive analysis using farm-budgeting methodology of the costs of producing Grade C goat milk and break-even milk prices.

SOURCES OF DATA FOR MODEL BUDGETS

Model budgets were produced using a modification of the Budget Generator computer program at the University of Arkansas Agricultural Experiment Station, Fayetteville. The sources of data for the program were dairy goat experts in the region and an extensive survey of 29 dairy goat milk producers. Of the 29 producers surveyed, 11 were commercial producers who sold milk to the Yellville plant and 18 were noncommercial who utilized the milk from their doe herd for home use and raising livestock.

For the purpose of the study, the standard Budget Generator program was modified to remove labor costs from variable costs of production. Labor was then listed as "labor opportunity cost" for family and operator labor. Labor opportunity cost was then subtracted from "net returns to land, labor, and management" to arrive at "adjusted net returns."

The nine budgets produced were for herds with 35, 50, and 80 milking does producing 1,200 lb, 1,800 lb, or 2,400 lb of milk per lactation. Of the budgets generated, survey results indicated that the combination of 50 does producing 1,800 lb per lactation was most characteristic of the producer population in the region. This report focuses on that budget.

COSTS AND RECEIPTS FOR A 50-MILKING-DOE HERD WITH AN 1,800 POUND LACTATION AVERAGE

Gross Receipts

Table 1 presents the budget generated for the herd milking 50 does and producing an average of 1,800 lb of milk per year. The products sold and considered in this budget are:

- Grade C milk
- Weaned doe kids of breeding quality
- Cull doe kids (3 days old)
- Weaned buck kids of breeding quality
- Cull buck kids (3 days old)
- Cull adult does

Additional sources of income, but too insignificant to be considered in the budget, are cull adult bucks, wethers (castrates), manure sales, buck stud fees, and premiums won at dairy goat shows. There are opportunities for increased income to dairy goat producers in the region through sale of goat meat products but to date the market channels have not been well developed.

Milk receipts are divided into "normal season Grade C" and "winter milk," -- the latter bringing producers a $2.00/ cwt (hundredweight) premium. As dairy goats are seasonal producers, December and January are months of little or no herd milk production under normal management conditions. The Yellville processing plant pays producers a premium on "winter" milk in these 2 mo to encourage producers to plan for out-of-season production by using artificial lighting, selecting for out-of-season kidding, and extending lactations through their nutritional programs. The total milk

TABLE 1. ESTIMATED COSTS AND RETURN FOR A DAIRY GOAT MILK OPERATION WITH 50 LACTATING DOES PRODUCING 1,800 POUNDS PER LACTATION

Description	Unit	Price or cost/unit	Quantity	Value or cost
Gross receipts				
Grade C milk	Cwt	$ 14.65	814.10	$ 11,926.55
Winter milk	Cwt	16.65	34.40	572.76
Breeding does	Head	150.00	15.00	2,250.00
Bucks (weaned)	Head	200.00	2.00	400.00
Bucks (3 days)	Head	5.00	33.00	165.00
Cull does	Head	25.00	8.00	200.00
Doe kids (3 days)	Head	15.00	10.00	150.00
Total gross receipts				$ 15,664.31
Variable costs				
16% dairy feed (lact. does)	Cwt	$ 9.20	9.00	$ 4,140.00
9% grain feed (dry does)	Cwt	7.90	.60	237.00
9% grain feed (bucks)	Cwt	7.90	2.92	92.27
16% dairy feed (breeding stock and replacements)	Cwt	9.20	.21	54.10
16% dairy feed (replacements)	Cwt	9.20	3.44	348.13
Alfalfa (lactating does)	TN	120.00	.50	3,000.00
Grass hay (dry does)	TN	60.00	.16	480.00
Grass hay (bucks)	TN	60.00	.57	136.80
Alfalfa (replacement)	TN	120.00	.09	118.80
Vet. and medicines	Head	9.00	1.00	585.00
Dairy supplies	Head	6.45	1.00	322.50
DHIA	Head	12.00	1.00	600.00
Electricity	Head	16.00	1.00	800.00
Hauling cost	Cwt	1.50	848.50	1,272.75
Machinery	$			515.81
Tractors	$			76.90
Equipment	$			523.10
Interest on oper. cap.	$	.15	98.88	15.42
Total variable costs				$ 13,318.52
Income above variable costs				2,345.79
Fixed costs				
Interest on livestock cap.	$	$ .13	6,839.99	$ 759.20
Interest on other equipment	$	.13	9,517.49	1,237.27
Depreciation on bucks	$			373.33
Depreciation on other equip.	$			3,048.00
Other FC (mach. & equip.)	$			500.97
Total fixed costs				$ 5,918.77
Total costs				19,237.30
Net return: land, labor, and management				$- 3,572.99
Labor opportunity cost				
Family labor	Hr	$ 2.00	1,816.00	$ 3,632.00
Operator labor	Hr	3.50	1,560.00	5,460.50
Total labor				9,092.50
Adjusted net return				$-12,665.49

sold in the normal production season (February-November) was adjusted for milk utilized in raising kids for replacements.

A "kidding rate" (number of kids born and surviving to 3 days of age per 100 does kidding) of 140% was assumed for the purposes of the budget in table 1. The 50-doe herd therefore produced 70 kids, 35 (50%) of which were female. With an assumed 20% replacement rate for adult does, 10 of the female (doe) kids were retained for the milking herd. Of the remaining 25 doe kids, 15 (60%) were sold as registered breeding stock at $150/hd and 10 were sold unregistered at 3 days of age for $15/hd. Of 35 males (bucks), 2 were sold for potential breeding males at $200/hd and the remaining 33 were disposed of at $5/hd at 3 days of age. Many of these buck kids are raised by families for meat or for controlling brush in pastures and woodlots.

A 4% death loss, or 2 does, was subtracted from the 10 adult does that left the milking herd each year. The remaining 8 cull does were sold at the end of their lactation in December for $25/hd.

Kid sales resulted in a total income of $2,965 and cull does added $200 for a total of $3,165. With total milk sales of $12,499.31, total income for the 50-doe milking operation was $15,664.31. Sale of milk accounted for 81% of total income and livestock sales for 19%. These figures closely agree with data from the producer survey.

Variable Costs

Total yearly variable costs for the 50-doe herd were $13,318.52, of which $8,606.30 (64.6%) was accounted for in feed and hay. Feed and hay usage by milking and dry does, bucks, and replacement kids was determined from nutritional requirements published by the National Research Council (NRC, 1981). Feeds used and their assumed costs are listed in table 2. All other variable costs were estimated from survey data or, as in the case of equipment costs, were generated by the budget program.

Fixed Costs

The machinery complement for the 50-doe herd was assumed to be a pickup truck, a small tractor (40 HP) with a front-end loader, and a three-point mounted, PTO-driven rotary brush cutter. Some producers in the survey did own a manure spreader but few owned hay equipment. Harvest of hay where land was available was mostly contracted by the surveyed producers. For the purpose of the study all hay was assumed purchased.

The machinery complement in a budget is usually calculated by defining replacement cost as that of new machinery. For dairy goat milk producers, used equipment seemed to be more appropriate.

TABLE 2. FEED SOURCES FOR 50-DOE MILKING HERD

	% TDN		Feeds			
			Grain mix		Hay type	
Class	Grain	Hay	%	$/cwt		$/ton
Adults does						
Milking	50	50	16	9.20	Alfalfa	20.00
Dry	20	80	9	7.90	Grass	60.00
Adult bucks						
Breeding	50	50	9		Grass	
Nonbreeding	10	90	9		Grass	
Doe kids						
0-3 mo	n.a.	n.a.	16		Alfalfa	
3-7 mo	60	40	16		Alfalfa	
7-12 mo	70	30	16		Alfalfa	

A two-unit bucket-type milking machine system was assumed to have been purchased for $1,500. A 345-gallon used bulk tank was purchased for $375. Both units of equipment were assumed to have a 5-yr amortization life.

Net Return to Land, Labor, and Management

Total costs (variable plus fixed costs) were $19,237.30 for the 50-doe herd with an 1,800 lb lactation average. With gross receipts of $15,664.31, net return for land, labor, and management was $-3,572.99.

Labor-Opportunity Cost

None of the producers in the survey hired labor, but all used "unpaid" family labor. Surveyed producers indicated a total of 65 hr of labor was used weekly to operate the average dairy goat farm. In table 1 the owner-operator was assumed to have off-farm employment but spent 30 hr/wk (1,560 hr/yr) on the goat milk operation. The remainder of the labor was provided by the spouse and children.

An "opportunity cost" is the expected "market value" of labor applied to the goat operation or what it could bring if applied to an alternative enterprise. In the budget for the 50-doe dairy, opportunity cost for family labor was calculated at $2.00/hr and $3.50/hr for the labor of the owner-operator. While these values may seem low, they represent the use of children in many tasks (feeding kids, moving the doe herd, etc.) and the low demand for the unskilled labor used in many dairy goat farm tasks.

Total labor-opportunity costs were $9,092.50.

Adjusted Net Return

When the $-3,572.99 net return to land, labor, and management was adjusted for labor-opportunity costs, "adjusted" net return was $-12,665.49.

Cash Flow

Monthly cash flow for the 50-doe herd (interest on operating capital of $98.88 was not included as a cash cost) is presented in table 3. The months of January, February, and March, when milk flow was assumed to be lowest, are periods of negative cash flow. Though milk production was also assumed low in December, the sale of 8 cull does and the $2.00/cwt premium paid on milk delivered to Yellville did result in a positive cash flow.

TABLE 3. MONTHLY GROSS RECEIPTS, TOTAL CASH COST, AND NET CASH RETURN FOR A 50-DOE HERD AND 1,800 LB LACTATION AVERAGE

Months	Gross receipts	Total cash cost	Net cash returns
January	$ 144.95	$ 463.45	$ -318.50
February	306.50	1,017.78	-711.28
March	742.46	1,065.33	-322.87
April	2,062.31	1,421.75	640.56
May	3,085.51	1,187.87	1,897.64
June	1,772.65	1,217.27	555.38
July	1,904.50	1,243.85	660.65
August	1,699.40	1,263.76	435.64
September	1,503.09	1,527.04	23.95
October	1,103.14	1,180.39	77.25
November	711.91	1,124.86	412.95
December	627.80	589.78	38.02
Total	$15,664.31	$13,303.09	$2,361.22

Comparison of Costs and Returns for 35-, 50-, and 80-doe Herds Producing 1,200 Lb, 1,800 Lb, and 2,400 Lb of Milk Per Lactation

Table 4 is a summary of the budgets of the nine combinations of herd size and per doe lactation production. Fixed and labor-opportunity costs were assumed to be a function of herd size and independent of per doe level of production. With the lowest producing does (1,200 lb per lactation), income above variable cost was negative for all budgets. Net return for land, labor, and management was positive only for the milking herds where does produced 2,400 lb of milk per lactation. This lends emphasis to the critical importance of doe productivity, especially where fixed and labor-opportunity costs do not decrease with lower levels of milk production.

All combinations of herd size and per doe levels of production resulted in negative adjusted net return.

TABLE 4. ESTIMATED ECONOMIC PERFORMANCE OF DAIRY GOAT MILK FARMS WITH DIFFERENT SIZE OF DOE HERD AND AVERAGE LACTATION YIELD

Lactating doe herd size	Average milk productivity (lb/doe)	Gross receipts	Total variable costs	Income above variable costs	Fixed costs	Net return: land, labor, and management	Opportunity costs of labor	Adjusted net return
35	1,200	$ 6,923.49	$ 8,761.52	-1,838.04	5,472.50	-7,310.54	7,066.00	-14,376.54
35	1,800	10,816.13	9,392.04	1,424.09	5,472.50	-4,048.41	7,066.00	-11,114.41
35	2,400	16,326.58	10,657.40	5,669.18	5,472.50	196.68	7,066.00	-6,869.32
50	1,200	9,992.68	12,250.29	-2,257.61	5,918.77	-8,176.38	9,092.50	-17,268.88
50	1,800	15,664.31	13,318.52	2,345.79	5,918.77	-3,572.99	9,092.50	-12,665.49
50	2,400	23,417.67	14,928.36	8,489.32	5,918.77	2,570.54	9,092.50	-6,521.96
80	1,200	15,663.78	19,243.20	-3,579.42	6,691.26	-10,270.68	13,144.00	-23,414.68
80	1,800	24,839.54	20,946.91	3,892.91	6,691.26	-2,798.35	13,144.00	-15,942.35
80	2,400	36,757.35	23,460.17	13,297.18	6,691.26	6,605.92	13,144.00	-6,538.08

Break-even Milk Prices

Break-even milk price is that price necessary to cover a given set of production costs assuming a certain level of production. In table 5, break-even prices needed to cover the various sets of production costs are calculated for each of the nine different combinations of herd size and per doe levels of production. Break-even prices range from $7.14 to $19.18 to cover variable costs at the 50-doe/2,400 lb lactation average combination. To cover all costs for the 35-doe/1,200 lb lactation average combination, break-even prices range from $18.19 to $52.07.

The producer-survey results indicated that the average commercial herd size was close to 50 does producing 1,777 lb of milk per lactation. The price paid for the Grade C milk marketed to the Yellville plant averaged $14.65/cwt F.O.B. farmgate. The analysis in table 5 indicates that one source of the high level of turnover in the commercial goat-milk-producer population is that price paid for milk is sufficient to cover variable costs but not to cover fixed and labor-opportunity costs. In the short run, if producers see their variable costs covered and can foresee the possibility of an increase in productivity so that fixed costs might also be covered, they will probably make the decision to continue to produce. However, in the long run, as opportunities for labor and capital in alternative enterprises are presented (especially for recreation and off-farm employment), the low to negative rates of return to family labor and management dictate an abandonment of the goat milk operation.

TABLE 5. BREAK-EVEN MILK PRICES ($/CWT) NECESSARY TO COVER DIFFERENT SETS OF PRODUCTION COSTS, BY DOE HERD SIZE AND AVERAGE LACTATION YIELD

	35 lactating does			50 lactating does			80 lactating does		
	Production level (lb)								
	1,200	1,800	2,400	1,200	1,800	2,400	1,200	1,800	2,400
Cover variable costs (VC), $	19.18	12.25	7.49	18.63	11.99	7.14	18.61	11.79	7.29
Cover variable costs and fixed costs (FC), $	33.02	21.45	14.40	29.09	18.84	12.37	26.01	16.71	10.99
Cover VC, FC, and operator labor, $	46.99	30.60	21.42	38.63	25.40	17.20	32.00	20.84	13.94
Cover VC, FC, and all labor, $	52.07	33.30	23.44	45.02	29.68	20.41	40.49	26.48	18.19

SUMMARY CONCLUSIONS

Economic analysis of Ozark region goat milk production enterprises using farm-budgeting methodology and data from a producer survey indicates the following:

1. At current market prices for Grade C milk, dairy goat producers in the Ozark region of northern Arkansas and southern Missouri must milk at least 50 (and probably 80 or more) does producing from 1,800 lb to 2,400 lb per lactation in order to have a positive return for land, labor, and management.
2. Cost and utilization of feed and hay account for 60% to 70% of the variable costs of producing Grade C goat milk in the Ozark region. Close attention paid to the nutritional management of the dairy goat herd should result in reduced feed costs and(or) increased returns from milk sales.
3. Even if net return for land, labor, and management is positive, receipts are not sufficient to cover family and owner-operator labor at minimum wage rates or lower.
4. Due to the seasonal nature of milk production with dairy goats, monthly cash flow, especially in the winter, can be negative. Dairy goat milk producers must be prepared to "ride out" periods of negative cash flow by subsidizing the operation with income from other sources.
5. A stable, progressive dairy goat milk production sector in the Ozark region with producers earning an adequate return on investment in land, equipment, and animals will require an expanded market for milk, meat, and breeding stock. Considering the costs of production of Grade C milk, a milk price exceeding $15.00/cwt will probably be necessary for Grade A milk production to be feasible.

51

SMALL ANIMALS FOR SMALL FARMS IN LATIN AMERICA AND THE CARIBBEAN

Donald L. Huss

INTRODUCTION

Between 9 and 10 million farms in Latin America and the Caribbean do not have the resources to raise large animals either for income or for food for home consumption. More than 31 million men, women, and children are involved and their generally poor diets are especially deficient in protein of animal origin.

The solution to this problem requires the development of animal production systems that can convert limited farm-grown forage and feed supplies into food and other consumable items. This could be achieved by raising a few head of the small animal species for home consumption to improve the diets of the people and, under some circumstances, to produce a surplus to sell for additional income.

ADVANTAGES OF SMALL ANIMAL SPECIES

Reproduction. The small animal species have many advantages over large animals for small-farm rearing (Huss, 1982). They reproduce at a younger age than cows so that the farmer waits a shorter time to have something to eat or to sell. The small ruminants can produce consumable products at 15 mo of age, and some small animal species do even better. Also small animal species generally have fewer fertility problems.

Feed. Large animals are usually selective consumers of forage whereas small animals can be raised on feeds, forages, and kitchen refuse that would otherwise be wasted. The space required for handling and feeding small animals is much less than that required for larger animals. A small plot of land planted to a high-yielding forage crop could provide an adequate year-long supply of feed for a few small animals, whereas a cow faced with the same supply would starve to death. Also, the small animals generally have more efficient feed conversion ratios, which makes them better suited for limited feed resources.

Production. The productivity of small animals can be similar to that of larger animals if compared on an animal unit basis. For example, the average annual milk production for F_1 Alpine x Criollo goats under semiintensive management in Venezuela was 425 lb (Garcia et al., 1979). This is equivalent to 2,550 lb for a cow based on the animal unit ratio that six goats are equal to one cow. The produce of small stock can be eaten by a family in one meal or one day, which eliminates the need for refrigeration. Finally, the relatively small size/low purchase price of small animals makes them more available to low-income households that have neither space nor capital for a large animal. In fact, they can be utilized by landless rural inhabitants in backyard production systems.

SMALL ANIMALS SUITABLE FOR SMALL FARMS

Goats

Goats, because of their wide range of adaptability, are suitable for nearly all of the different kinds of environments. The goat is often called the poor man's cow because two does bred at alternate intervals (which is possible in the tropics) can provide a family with a year-long supply of milk and some meat. They can be raised on a variety of forages and feeds that would not be consumed by other animals (figure 1). Normally, fertility is not a problem and twinning and three gestations in 2 yr are common.

Figure 1. This small farmer in the humid tropics is raising a small herd of goats on the by-products of the crops shown in the background and the adjoining shrubland (FAO photo by author).

In spite of these attributes, goats have received very little attention from policymakers, planners, and scientists. On the contrary, they have often been blamed categorically for the destruction of natural resources. It is now generally understood that resource destruction by goats is due not to the goat but to uncontrolled grazing permitted by man. In some quarters there is still opposition to goats. But the tide is turning as people realize that goats, under conditions where other kinds of livestock cannot survive (much less produce), are making a notable contribution to national meat and milk supplies.

Sheep

Sheep also have a wide ecological range, and they too can be raised on a variety of forages and feeds. They are valuable for meat, wool (some breeds), and hides. Other than the Churra breed in northern Chile (Herve, personal communication), there are no milking breeds in Latin America and the Caribbean. About 25% of this region's nearly 17 million head of wooled sheep are raised on small farms in Colombia, Bolivia, Ecuador, and Peru. They mainly subsist upon roadside vegetation or communal pastures where they are either tended by women and children or by tethering. Sheep are sheared throughout the year and the wool spun, woven, and made into cloth by the families. These sheep represent an important source of income for the farmers.

There are many different breeds of hair sheep adapted to tropical conditions (Mason, 1980). These are found on small farms in the Caribbean countries, Venezuela, northeastern Brazil, Colombia, and Mexico. An outstanding characteristic of these sheep is their prolificacy. Litter sizes of 1.7 to 2 are common with the Barbados Blackbelly breed (Fitzhugh and Bradford, 1983) and three gestations in 2 yr. are possible. The prolific hair sheep represent a valuable resource for meat production on small farms in tropical areas.

Pigs

While pigs do not produce milk for human consumption, they make up for this by their ability to produce meat. They are efficient converters of feed and can consume agricultural and industrial by-products, garbage, natural vegetation, and other such feeds that would otherwise be wasted. From the zootechnical point of view, there are few climatic limitations to pig production. Most pigs are raised under either extensive or semiintensive production systems. In the extensive system, criollo pigs scavenge freely or are tethered and receive no supplemental feed inputs. The semiintensive system involves criollo or criollo crosses with improved breeds and they receive minimum inputs of feed, housing, and management. Pig productivity under the semiintensive system is higher than the extensive, but both can be improved so that there are greater and more economical yields.

Guinea Pigs

The guinea pig is a rustic, tame, prolific, and fast-growing animal that requires very little space. The Andean countries have many different varieties with mature weights varying from 1 lb for the criollo up to 4.5 lb for the improved meat breeds. An estimated 18,150 t of meat were produced in Peru in 1970 (Zaldivar and Chauca, 1975).

Ten reproducing females, 2 males, and their nursing and growing young with an extraction rate of 15 animals every 3 mo can be raised in 2 cages of 6 sq ft each (Montoya and Demeure, 1981). The animals can be raised on green forage alone or on agricultural by-products and kitchen refuse. Some are raised in the homes on refuse alone.

Huss and Roca (1982) conducted a computer simulation to determine the possible benefits of a hypothetical guinea pig development center. It was assumed that the Center would maintain 500 reproducing females of a medium-sized breed. In the simulation, individuals were assigned 20 females and 2 males and were to maintain a herd of this size. Conservative calculations showed that the Center could interest 264 producers within 3.5 yr with each producing 202 lb of dressed meat per year. This little animal is capable of providing a lot of meat to hungry people.

Rabbits

Advantages of rabbits are: high reproductive rates, rapid growth, small body size, limited space requirements, and ability to use fibrous plant material and agricultural by-products for food. An individually caged doe (figure 2) can produce (depending on the breed) 70 lb to 95 lb of dressed meat per year. No other animal can be kept in a space of about 1 sq yd and annually produce 8 to 10 times its own body weight. While this level of production could not be expected under small-farm conditions, it does reflect the rabbit's potential.

Rabbits can be raised in backyard production systems using farm-grown or locally produced feedstuffs (figure 3). The required facilities are simple and can be constructed by the producers from local and scrap materials (FAO, 1981). This makes rabbits available to small farmers, landless rural inhabitants, and even some urban inhabitants.

Huss and Roca (1981) conducted a computer simulation to determine the benefits that could be derived from a rabbit development center. It was assumed that the Center would maintain 200 reproducing does of a medium-sized breed. Individuals were assigned 5 does and 1 buck and were to maintain a herd of this size.

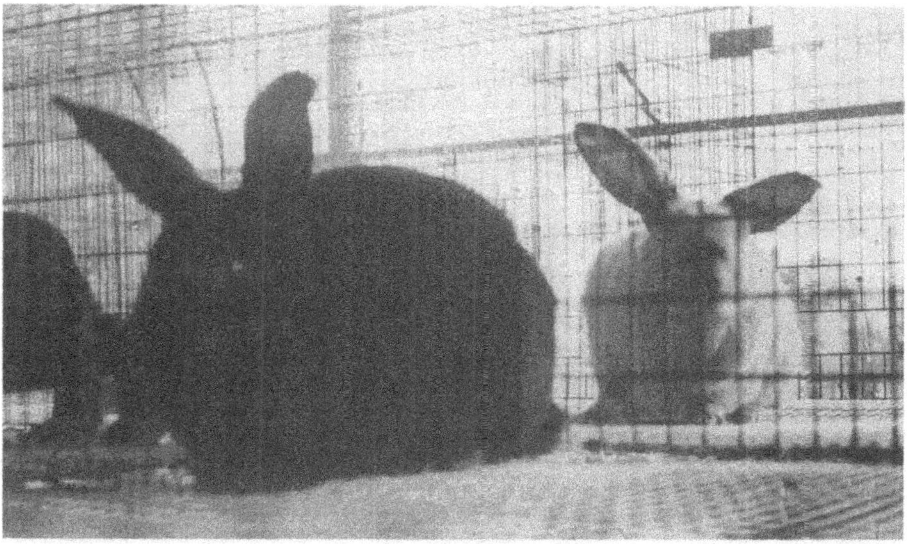

Figure 2. This caged doe can produce 70 lb to 95 lb of dressed meat per year. No other animal can annually produce 8 to 10 times its body weight in a space of 1 sq yd (FAO photo by author).

Figure 3. This farmer in Jamaica maintains a small rabbitry on crop by-products and on forage produced by the plot of grass behind his home.

Conservative calculations (based on reported average yields from some developing countries in Latin America, Africa, and the Near East) showed that the Center could interest 1,536 producers within 4 yr with each able to produce 123 lb of dressed meat annually. This little animal is capable of being an important provider of meat plus skins for the manufacture of garments.

OVERCOMING THE CONSTRAINTS

Few countries give a high priority status to animal development on small farms. This is probably because few realize that the small farmers now produce or could produce an adequate feed supply to raise some kind of a small animal species. Even if the farmers themselves realized this, they are not prepared to take advantage of it (figure 4).

Figure 4. The owner of this small farm probably does not realize that he is producing enough by-products to raise a few head of some kind of small animal species for home consumption (FAO photo by author).

Consequently, the FAO Regional Office for Latin America and the Caribbean has initiated an activity entitled, "Small Animals for Small Farms." In the identification of projects, the concept is promoted, information is disseminated, and technical assistance is provided whenever possible. An increasing interest in small-stock production is evidence of progress.

Providing farmers with animals is not sufficient; the technologies required for their successful rearing must also be introduced. Therefore, instigating training at both the technician and producer levels is essential. Training manuals and materials have been and are being prepared. Courses have been conducted and more are scheduled.

A constraint to the implementation of small-animal projects may be the reluctance of the individuals involved to eat a product that has not been a part of their traditional diet. This seems to be a serious constraint in the case of rabbit and guinea pig meat and, in some cases, goat milk and cheese. In other parts of the world, this constraint in the case of rabbit meat is being overcome through strong national campaigns. Home economics was a vital component of these campaigns. The rural schools could also be used for introducing students to the production and consumption of small animals. The technology could then be transferred to the home.

REFERENCES

FAO. 1981. Report of the FAO expert consultation on rural poultry and rabbit production. AGA-805. FAO of the UN, Rome, Italy.

Fitzhugh, H. A. and G. E. Bradford. 1983. Productivity of hair sheep and opportunities for improvement. In: H. A. Fitzhugh and G. E. Bradford (Ed.). Hair Sheep of Western Africa and the Americas. pp. 23-52. Westview Press, Boulder, CO.

Garcia, O., J. Bravo, J. Isakovich, and E. Garcia. 1979. Recomendaciones para la cria de ovinos y caprinos. CIARCO. Araure, Estado Portuguesa, Venezuela.

Huss, D. L. 1982. Small animals for small farms in Latin America. World Animal Rev., 43:24.

Huss, D. L. and G. Roca. 1981. A hypothetical rabbit development centre and its possible benefits. FAO of the UN, RLAC, Santiago, Chile.

Huss, D. L. and G. Roca. 1982. The guinea pig and a hypothetical development centre. FAO of the UN, RLAC, Santiago, Chile.

Mason, I. L. 1980. Prolific tropical sheep. FAO Anim. Prod. and Health Paper No. 17. FAO of the UN, Rome, Italy.

Montoya, J. F. and M. Demeure. 1981. Rentabilidad de la crianza de animales en la sierra a nivel familiar. Ministerio de Agricultura y Ganaderia. Proyecto ECU/78/004, PNUD/FAO, Ibarra, Ecuador.

Zaldivar, A. M. and L. Chauca F. 1975. Crianza de cuyes. Boletin Tecnico No. 81. Ministerio de Alimentacion, Lima, Peru.

INDEX OF AUTHORS

Donald L. Huss
Regional Animal Production
Center
FAO, Regional Office for Latin
America
Providencia 871. Casilla 10095
Santiago
CHILE
— Range Scsientist
Page 286, 304, 429

Craig Ludwig
Director of TPR & Research
The American Hereford Assoc.
715 Hereford Drive
Kansas City, MO 64101
— Animal Scientist
Page 93

George A. McAlister
President, Parthenon Corp.
8918 Tesoro Drive
Suite 320
San Antonio, TX 78217
— Rancher
Page 178

Howard H. Meyer
Associate Professor
Oregon State University
Corvallis, OR 97331-6702
— Sheep Scientist
Page 349, 375, 381

H. Allan Nation
Publisher
The Stockman
P.O. Box 9607
Jackson, MS 39206
— Cattleman/Journalist
Page 249

Hans E. Nel
Department of Animal Science
University of Wyoming
Laramie, WY 82071
— Sheep and Goat Specialist
Page 385, 406

Dieter Plasse
Facultad de Ciencias
Veterinarias (UCV)
Apartado 2196, Las Delicias
Maracay
VENEZUELA
-- Animal Geneticist
Page 32, 60, 137

Rodney L. Preston
Thornton Distinguished
Professor
Department of Animal Science
Texas Tech University
Box 4169
Lubbock, TX 79409-4169
-- Animal Nutritionist
Page 119, 202, 211

E. J. Richey
P.O. Box 114
Pawhuska, OK 74056
-- Veterinarian
Page 223, 228, 243

Allan R. Rider
Director of Product Testing
Sperry New Holland
New Holland, PA 17557
-- Agricultural Engineer
Page 322

Ervin W. Schleicher
Box 151
Whitney, NE 69367
-- Rancher/FAO Livestock
Officer (Retired)
Page 312

H. H. Stonaker
6529 East Highway 14
Fort Collins, CO 80524
-- Rancher/Animal Geneticist
Page 50, 294

R. L. Willham
Professor, Iowa State
University of Science and
Technology
Ames, IA 50011
— Animal Geneticist
Page 12

J. N. Wiltbank
Department of Animal Husbardry
Brigham Young University
Provo, UT 84606
— Physiologist
Page 148

Iain A. Wright
Hill Farming Research
Organization
Bush Estate, Penicuik
Midlothian EH26 OPY
SCOTLAND, U.K.
— Animal Scientist
Page 110, 195, 341

Roger D. Wyatt
International Minerals &
Chemical Corp.
P.O. Box 207
Terre Haute, IN 47808
— Animal Scientist
Page 253

Jim A. Yazman
Winrock International
Route 3
Morrilton, AR 72110
— Animal Scientist
Page 421

Other Winrock International Studies Published by Westview Press

Beef Cattle Science Handbook, Volume 19, edited by Frank H. Baker, and Volume 20 edited by Frank H. Baker and Mason E. Miller

Dairy Science Handbook, Volume 15, edited by Frank H. Baker, and Volume 16 edited by Frank H. Baker and Mason E. Miller

Sheep and Goat Handbook, Volume 3, edited by Frank H. Baker, and Volume 4 edited by Frank H. Baker and Mason E. Miller

Stud Managers' Handbook, Volume 18, edited by Frank H. Baker, and Volume 9 edited by Frank H. Baker and Mason E. Miller

Future Dimensions of World Food and Population, edited by Richard G. Woods

Hair Sheep of Western Africa and the Americas, edited by H. A. Fitzhugh and G. Eric Bradford

Other Books of Interest from Westview Press

Animal Health: Health, Disease and Welfare of Farm Livestock, David Sainsbury

Calf Husbandry, Health and Welfare, John Webster.

Carcase Evaluation in Livestock Breeding, Production, and Marketing, A. J. Kempster, A. Cuthbertson, and G. Harrington

Energy Impacts Upon Future Livestock Production, Gerald M. Ward

Livestock Behavior: A Practical Guide, Ron Kilgour and Clive Dalton

The Public Role in the Dairy Economy: Why and How Governments Intervene in the Milk Business, Alden C. Manchester

Other Books of Interest from Winrock International*

Sheep and Goats in Developing Countries: Their Present and Potential Role, A World Bank Technical Paper, Winrock International

Goat Health Handbook, Thomas R. Thedford

Sheep Health Handbook, Thomas R. Thedford

Management of Southern U.S. Farms for Livestock Grazing and Timber Production on Forested Farmlands and Associated Pasture and Rangelands, E. Byington, D. Child, N. Byrd, H. Dietz, S. Henderson, H. Pearson, and F. Horn

Potential of the World's Forages for Ruminant Animal Production, Second Edition, edited by R. Dennis Child and Evert K. Byington

Research on Crop-Animal Systems, edited by H. A. Fitzhugh, R. D. Hart, R. A. Moreno, P. O. Osuji, M. E. Ruiz, and L. Singh

The Role of Ruminants in Support of Man, H. A. Fitzhugh, H. J. Hodgson, O. J. Scoville, Thanh D. Nguyen, and T. C. Byerly

Ruminant Products: More Than Meat and Milk, R. E. McDowell

The World Livestock Product, Feedstuff, and Food Grain System, R. O. Wheeler, G. L. Cramer, K. B. Young, and E. Ospina

World Agriculture: Review and Prospects into the 1990's, Winrock International

Prediction of Genetic Values for Beef Cattle: Proceedings of a Workshop, Winrock International

A Study of the Beef Cattle Performance Databanks of the United States, Frank H. Baker, H. A. Fitzhugh, Will Getz, and Paula Gerstmann

*Available directly from Winrock International, Petit Jean Mountain, Morrilton, Arkansas 72110

www.ingramcontent.com/pod-product-compliance
Ingram Content Group UK Ltd.
Pitfield, Milton Keynes, MK11 3LW, UK
UKHW022208190926
475702UK00003B/253
9780367005894